Miami Virtue

Doing Arts Thinking: Arts Practice, Research and Education

VOLUME 9

The titles published in this series are listed at *brill.com/data*

Miami Virtue

Choragraphy of the Virtual City

By

Gregory L. Ulmer, Barbara Jo Revelle,
John Craig Freeman and William L. Tilson

TUTA SUB AEGIDE PALLAS · 1683 ·

BRILL

LEIDEN | BOSTON

This book was originally published by Small Cities Imprint as Vol. 3, No. 2 (2012).

Cover illustration: *Crossroads*, photograph by Barbara Jo Revelle

The Library of Congress Cataloging-in-Publication Data is available online at https://catalog.loc.gov

Typeface for the Latin, Greek, and Cyrillic scripts: "Brill". See and download: brill.com/brill-typeface.

ISSN 2542-9744
ISBN 978-90-04-53462-9 (paperback)
ISBN 978-90-04-53463-6 (hardback)
ISBN 978-90-04-53464-3 (e-book)

This book is printed on acid-free paper and produced in a sustainable manner.

To make a map of Thebes instead of performing Sophocles, to make a topography of obstacles instead of fighting against destiny.

GILLES DELEUZE AND FELIX GUATTARI (*Kafka*)

∵

Contents

PART 3
Miautre – Oracle

Foreword

In "Why I Teach Plato to Plumbers," Scott Samuelson reflects on his experiences teaching philosophy at a community college. At a time when disciplines like philosophy are thought to be in a state of crisis, he argues for the ubiquitous importance of arts and humanities disciplines for no less than the preservation of democracy itself. Having spent the last decade teaching writing in a wide range of learning contexts – from a community college and a small private institution to research I and II universities – I have witnessed firsthand the various ways that departments and their larger divisions are responding to the mounting tensions he describes, from an unfettered faith in techno-solutionism, to a growing focus on vocational training as the *telos* of education, to the elitism traditionally associated with arts and humanities disciplines that connotes a widening gyre between "the university" and "the public." Samuelson's solution is to stress the profound importance of student exposure to a wide range of disciplines like philosophy for everyone, including plumbers. I find the systemic problems that he identifies to be central to the work that I do as someone who researches, teaches, and practices environmental writing with students who are primarily interested in going into the so-called STEM fields (science, technology, engineering, mathematics). My joint appointment at the University of Rhode Island – 51% in Writing & Rhetoric, housed in the College of Liberal Arts & Sciences, and 49% in Natural Resources Science, housed in the College of Environmental and Life Sciences – provides the opportunity to reflect on these tensions almost every day. When explaining my joint appointment breakdown with colleagues, I sometimes refer to myself as "slightly-less-than-half-a-scientist" or "slightly-more-than-half-a-humanist" – my self-deprecating way of illustrating the difficulty of working across, between, and beyond disciplines as well as the imposter syndrome that often accompanies transdisciplinary inquiry. Samuelson places these educational problems in stark relief.

Yet, I remain somewhat dubious as to the solutions he offers, both for Higher Education specifically and more generally for what Gregory L. Ulmer refers to as the "society of the new metaphysics" brought about by the transition from literacy to electracy. Ulmer has spent the last forty years describing with prescience the ontological changes that characterize the move from print to digital media, which he argues has produced a cultural shift as profound as the transformation from oral to literate cultures. Rather than understanding this change in apparatus through literacy, which retrofits digital media with print-based logic, Ulmer forwards a provocative set of theories, methods, and practices to engage this new paradigm. He rejects the term "digital literacy" – which he compares to calling literacy "written orality" or science "nature magic" – and demonstrates that this apparatus change requires entirely new lines of flight (Jones and Greene, 2017). Ulmer provides us with not only a lexicon to describe this shift but also methods for identifying and addressing the vast problems we face as we become an electrate society. As such, he gives us the tools to surf the "new wave" of the avant-garde which is washing at the shoreline of the modern academy. In *Miami Virtue*, Ulmer and his collaborators forward a compelling example of how to apply place-based and digital methods to electrate fieldwork. This book situates the broader theories, methods, and practices of electracy within the specific virtual and material environments that compose the Miami River. The resulting project demonstrates the value of electrate research for interdisciplinary public advocacy and intervention. In my own research, I find that electracy indexes the most important sites of inquiry and reveals a wide array of interventions into the biggest problems our societies are facing today.

With these goals in mind, I founded the Digital Writing Environments, Languages, and Location (DWELL) Lab at URI. In the lab, we emulate Ulmer's motto (which he borrows from Basho), "not to follow in the footsteps of the masters, but to seek what they sought" (quoted in Ulmer, 2002: xiii). DWELL is experimenting with science and environmental communication practices informed by place-based theories, creative methods, and locative technologies. We endeavor to apply electrate theory to science and environmental communication to uncover how emerging technologies and creative methods can offer new ways to engage communities through the poetics of place. Recent student projects funded through the lab include an immersive creative art exhibit depicting the dangers of per- and polyfluoroalkyl substances (known as PFAS) and a podcast miniseries called *For Shore: Ocean Energy Explained* that introduces offshore wind to local residents in the United States. These student projects combine mobile and immersive technologies with creative methods to make technical scientific information accessible for non-specialist audiences. The lab is also participating in large-scale, ongoing collaborations with projects like *STEEP: Sources, Transport, Exposure & Effects of PFAS*, which is a partnership between URI, the Harvard T.H. Chan School of Public Health, the Department of Environmental Health, and the Silent Spring Institute that is researching and communicating about the dangers of PFAS to the public. These pervasive chemicals pose an immediate public and environmental health emergency, but they are also an example of the ongoing science communication crisis. As the conversations about the types of risks these chemicals pose to the public are still emerging, science communicators must work to translate the latest technical information into meaningful and engaging messaging. In response, DWELL is using place-based perspectives and mobile media to develop creative approaches to science communication.

In another ongoing project, a collaboration between a team of researchers at URI, the Manissean Tribal Council, the Tomaquag Museum, the Block Island Historical Society, and regional experts and culture bearers, DWELL is working to develop a critically curated mobile application that supports an augmented reality walking tour on Block Island representing Indigenous and African American perspectives of place and public memory. As a popular tourist destination, Block Island is a site of erasures. Indigenous and African American history on the island is often shown through the lens of settler colonialism, framed as something occurring in the distant past. Through a series of community co-creation workshops, we are developing content to test how location-based media can best support anti-colonial and anti-racist public humanities work designed by local communities to combat the erasure of Native American and African American peoples and identities from public space and cultural memory. While the question of who owns the rights to augmented space is increasingly being answered by large AR companies with heavy stakes in corporate interests, we are working together to acknowledge and support the sovereignty of the Manissean Tribe in both the physical spaces of Block Island and the virtual places of public memory and augmented space. As we deploy locative technologies in this project, we seek what the collaborators in this book sought. We are practicing the work of the theoria, as outlined in what follows, bearing witness to difficult truths and working to bring a more complex "image category" to bear on this iconic tourist destination.

Our work is situated within the context of the Florida Research Ensemble (FRE), an ongoing collaboration which began in the late 1980s between Ulmer and John Craig Freeman, Barbara Jo Revelle (and others) and from which this book is one result. As they describe it, their place-based project is the culmination of years of virtual consultancy, or "emerAgency," work that the members, or "egents," of the FRE completed along the Miami River zone. The resulting work has as much to teach us about activism as it does tourism as these practices might apply to addressing the most pressing environmental and social justice issues of our day. As you will find, the methods they outline in theorizing and applying choragraphy remain one of the most novel approaches to place-based advocacy using digital media. Further, they demonstrate the necessary shift from literate notions of justice, which they identify with ontology, toward electrate engagements with affect and the ethics of care. One compelling example of this work can be found in climate change discourse. For the past thirty years, scientists and science communicators have attempted to respond to the problem of carbon emissions and climate change by presenting the public with more and better data. Yet, Ulmer suggests that literate practices will prove ineffectual in solving problems in an electrate world. The prevalence of climate denialism, misinformation, and more pernicious forms of disinformation like greenwashing and deepfakes demonstrate the forethought of Ulmer's theories. In the face of massive problems like climate change, we come up against what systems thinkers term the "problem of scale." That is, climate change is at once both too vast and too mundane, and so the problem becomes "aporia" as it challenges our very notion of self and agency. As I have argued at length elsewhere, Ulmer's Mystory, choragraphy, and MEmorial methods are essential tools to confront the problem of scale in environmental communication (Jones, 2019). These electrate practices allow us to embody the distributed sense of self that digital technologies increasingly amplify.

This book suggests that electracy provides an important conduit through which researchers, educators, community members, and practitioners can begin to address some of the critical challenges of the modern university and to make progress toward the existential problems we face as a society. One keen example is the problem of "community outreach" or "extension" (terms which for me will always invoke a mental image of Michelangelo's *The Creation of Adam*), which science communication scholars are increasingly rejecting in favor of approaches that foreground inclusion, equity, and intersectionality in research, teaching, and practice (Canfield, et al., 2020). This is especially important for understanding how technologies are changing "the public," which has always been a vexed concept but is only increasingly so in networked environments. Ulmer builds from Paul Virilio to interrogate the "general accident" produced by the internet (Ulmer, 2011a). Virilio's neologism "dromosphere" describes the collapsing of spatiotemporal dimensions bought by digital technologies. Electracy compels us as researchers, teachers, and practitioners to directly engage with the conditions of the dromosphere as we confront the problems of our time. While I find any directive for educators to encourage students to value inquiry beyond the narrow confines of their particular academic and professional interests both laudable and appealing, I believe there is a larger effort needed beyond simply advocating for more expansive acronyms – which many universities have adopted – like STEAM (i.e. by adding arts) or STREAM (with the "R" standing in for reading, reasoning, wRiting, and/or rhetoric), or SHAPE (Social Sciences, Humanities, and the Arts for People and Economy). I might go so far as to suggest that, in the face of the recent sixth and most dire report to date from the IPCC, educators should move from STEAM to something more like SCREAM (Science, Communication, Rhetoric, Electracy, Arts, and Monumentality) as we attempt to voice a primal response to the various and intersecting emergencies we face in the epoch known as the Anthropocene.

In response to such calls for rearranging an alphabet soup of disciplines, Ulmer proffers what he calls the "H'MMM disciplines," reappropriating the

Campbell's soup slogan (as well as reappropriating Warhol's reappropriation), and emphasizing "Humanities + Movies Music Media" for electrate pedagogy. H'MMM suggests both the sounds of an inquiring mind and the joyful slogan "Mmm mmm good!" With its emphasis on H'MMM disciplines, electrate pedagogy works against what Ulmer refers to as "vocational tracks for Entertainment production" and focuses instead on education (Ulmer, 2012). He suggests that as educators we must move beyond treating tools, content, vocational training, and community engagement as separate elements of the modern university. Rather, as Sergio Figueiredo puts it, Ulmer brings together "the triad of knowing, doing, and making – *theopraxesis*" (58, emphasis original) in order to invent "the rhetoric of electracy (as distinct from the equipment)" (58–9). This is why Ulmer insists that electracy is a shift in apparatus, not simply in technology. For the sake of addressing the large-scale problems we face as a society, Ulmer highlights how electrate education "depends on a new collaboration and syncretism among all the disciplines now isolated from one another in the silo architectonics of literate learning" (59). In other words, rather than crafting an acronym which encompasses all of the disciplines or calling for emphasis on new or different sets of individual disciplines, electracy offers scholars and practitioners a conduit for participatory knowledge-making practices which bridges across institutions, disciplines, and communities and offers ways for each of us to make a more concerted effort.

That is why we as ecologists (and in the Anthropocene we earthlings are all ecologists) must study electracy. As Ulmer says,

> Consider what is at stake in our world today, registered in the basic fact that the Anthropocene coincides with electracy. We notice immediately that prayer and engineering do not suffice, and no one doubts they are trying. We desire effective change but are not able, we cannot despite our will and knowledge. Here is the modality specific to electrate metaphysics: to be able. Invention required. (Ulmer, 2017)

Rather than publishing more data documenting our impacts on the ecosystem, we must develop novel methods for locating ourselves within the affective dimensions of attraction and repulsion that govern those systems. Otherwise, we will be bystanders to the electrate phase shift from prescience to postscience. As Ulmer says elsewhere, "To comprehend the world as ecology, follow the body" (Ulmer, 2011b). As the networked technologies of the dromosphere transform our world and ourselves, we must all learn to attend to the systems we inhabit through electrate practices. We must come to better understand ecology itself as an electrate concept as it operates with the "ecosystem apparatus" (Pringle, 2019). When Arthur Tansley coined the term ecosystem, he drew heavily from thermodynamics to understand natural systems. Later, as ecology began to solidify into a "hard" science, its devotees like Eugene and Howard Odum drew upon cybernetics to innovate the study of ecosystems (Jones, 2021) and to collapse distinctions between technological and natural systems through holism (Woods, 2017; Pringle, 2021). To understand ecology and the Anthropocene, then, is to become electrate.

What I find enduring in this book is the rich and novel ways that it invites us to confront the limits of literate reason as we practice community-engaged and place-based digital field work. Networked technologies have made ecological thinking possible, even as they have begun to collapse our sense of space and time. The Anthropocene calls us together, even as it erodes our distinctions of inner and outer worlds. As we look upon the many overlapping social and environmental crises we face – from ocean acidification and sea-level rise to mass extinction, and from pollution to the COVID-19 pandemic (to name but a few) – we can more fully see the aporia: an unanswerable question, a wound, an impassible territory for literate reason. It was through the innovations of literacy that we produced these problems, and Ulmer demonstrates that we might uncover solutions through creativity. In 2005, M. Jimmie Killingsworth called for science communication to shift from a limited technical practice to one that is informed by place-based methods and ecopoetics. This book offers a compelling example of how to accomplish just that. In my research, teaching and community-engaged work with DWELL, I seek what Ulmer and the egents of the FRE sought in the EmerAgency outlined below. We are developing community-engaged science communication projects through practices informed by ecopoetics and place-based writing. The challenges we face are great, but if *Miami Virtue* is any indication, this aporia has also opened a space for invention and creativity. As this book demonstrates, it is at these sites of impasse and destruction that we may find the answers.

Madison Jones
University of Rhode Island
Kingston, Rhode Island

Works Cited

Canfield, K. N., Menezes S., Matsuda S. B., Moore A., Mosley A. A. N., Dewsbury B. M., Feliú-Mójer M. I., McDuffie K. W. B., Moore K., Reich C. A., Smith H. M., Taylor C. (2020). "Science Communication Demands a Critical Approach That Centers Inclusion, Equity, and Intersectionality" *Frontiers in Communication*, 5, 1–8, https://doi.org/10.3389/fcomm.2020.00002

Figueiredo, S. (2016). "Theopraxesis and the Future of H'MMM in the University: An Interview with Gregory L. Ulmer." *Journal for Cultural and Religious Theory*, 16:1, 58–73. Retrieved May 6, 2020, from https://jcrt.org/archives/16.1/InterviewUlmer.pdf

Jones, M. (2021). A Counterhistory of Rhetorical Ecologies, *Rhetoric Society Quarterly*, 51:4, 336–352. https://doi.org/10.1080/02773945.2021.1947517

_____. (2019) "(Re)placing the rhetoric of scale Ecoliteracy, networked writing, and MEmorial mapping." *Mediating Nature: The Role of Technology in Ecological Literacy*. Sidney I. Dobrin and Sean Morey (Eds.), Routledge.

Jones, Madison, & Greene, Jacob. (2017). Augmented vélorutionaries: Digital rhetoric, memorials, and public discourse. *Kairos: A Journal of Rhetoric, Technology, and Pedagogy* 22(1). Retrieved May 6, 2022, from http://kairos.technorhetoric.net/22.1/topoi/jones-greene/index.html

Killingsworth, M. J. (2005). "From Environmental Rhetoric to Ecocomposition and Ecopoetics: Finding a Place for Professional Communication." *Technical Communication Quarterly*, 14(4), 359–373. https://www.tandfonline.com/doi/abs/10.1207/s15427625tcq1404_1

Pringle, T. (2019). "The Ecosystem Is an Apparatus: From Machinic Ecology to the Politics of Resistance," in Thomas Pringle, Gertrud Koch, Bernard Stiegler (Eds.), *Machine*. Lüneburg: meson press, pp. 49–103, https://doi.org/10.25969/mediarep/12238

_____. (2021). "The Tech Ecosystem and the Colony." Heliotrope, Retrieved May 6, 2022, from https://www.heliotropejournal.net/helio/the-tech-ecosystem

Samuelson, S. (2014). "Why I Teach Plato to Plumbers." *The Atlantic*, Retrieved May 6, 2022, from https://www.theatlantic.com/education/archive/2014/04/plato-to-plumbers/361373/

Ulmer, G. L. (2002). *Internet Invention: From Literacy to Electracy*. Pearson.

_____. (2011a). "Avatar Emergency." *DHQ: Digital Humanities Quarterly*. 5(3), Retrieved May 6, 2022, from http://www.digitalhumanities.org/dhq/vol/5/3/000100/000100.html

_____. (2011b). "Florida Measure (Chora)" *Journal of Florida Studies*, 1(1), Retrieved May 6, 2022, from http://www.journaloffloridastudies.org/files/JFSms1-1-3FINALcb.pdf

_____. (2012). "Teaching in the Margins: Gregory L. Ulmer." *Full Stop: Reviews. Interviews. Marginalia.* Retrieved May 6, 2022, from https://www.full-stop.net/2012/10/04/features/the-editors/teaching-in-the-margins-gregory-l-ulmer/

_____. "Foreword: The Apparatus of Attention" in Jason Helms, *Rhizcomics: Rhetoric, Technology, and New Media Composition*, Ann Arbor, MI, Sweetland Digital Rhetoric Collaborative. February 2017. Open Access, https://www.digitalrhetoriccollaborative.org/books/rhizcomics_drc/

Woods, D. (2017). *What is Ecotechnology? Biopolitics and Trophic Writing in American Cultures of Science*. PhD Dissertation, Rice University.

Preface

The Florida Research Ensemble

The Florida Research Ensemble (FRE) was formed at the University of Florida in the late 1980s by colleagues from a number of different colleges brought together by a common interest in electronic technologies and new media. The intellectual question guiding our work was the possibility of using media for delivering knowledge to institutions outside the academy. Part of the object of inquiry from the beginning was the process of collaborative creativity itself. In the 1980s the focus was on public access television, activity that included making a tape with Paper Tiger Television (a mystorical critique of the PBS telecourse on poetry, *Voices and Visions*).

The present project began with an exhibition entitled "Imaging Florida," curated by John Craig Freeman in collaboration with the University Gallery (University of Florida, 1996) that brought together the current membership of the FRE. The exhibition (which traveled to several other venues) explored the possibilities of Florida tourism as an "interface metaphor," taking advantage of the didactic component of many tourist attractions as a point of entry for a new consultancy on image culture. A motive for working with tourism has to do with its obvious importance to the State of Florida, whose economy is dependent upon satisfying the eighty million plus visitors who pay a six percent sales tax on most purchases (Florida has no state income tax). The very scale of tourism, by some estimates the largest industry in the world, lends significance to almost anything one might do with it. Moreover, the global reach of tourism makes it a good candidate for an international interface design functional within the global context of the internet.

This book is a report and proposal addressed to potential members of a virtual, on-line, distributed consultancy called the EmerAgency (members referred to as "egents"). The report is a resource for those interested in participating in the invention of a learning practice relevant to the Internet as a civic sphere. The premise is that citizens participate as consultants in the formation of public policy with respect to issues that affect their communities, and that this civic practice is at the same time an experience of self-knowledge of a group subject. The institution of consulting already exists, and is one of the primary means for the distribution of expertise from concentrations of knowledge in universities and corporations to sites of need in the lives of citizens. As presently constituted, consulting restricts the amount and kind of information brought to bear on public problems. The conventions of cognitive jurisdiction regulate which disciplines are relevant to which problems. Moreover, these conventions also determine which problems are "public" and therefore in need of "solutions," and which may be ignored.

What counts as a problem and solution is governed by the criteria of instrumental reason, rationalism, utilitarian pragmatism, and other manifestations of the strict opposition between pure and applied knowledge. The critique of the epistemological, political, and ethical foundations of these conventions is well advanced in poststructuralist theory. What is needed now is the formulation of practices that test claims made for the functionality of alternative (heterological, pathic, machinic, aesthetic) epistemologies. EmerAgency consulting is "inventional" rather than conventional (Caputo, 1997: 42). It is "heuretic," using theory to generate texts, forms, ideas, rather than to interpret existing works (hermeneutics).

A larger purpose of this project is the invention of a practice for electracy, for an electrate society, appropriating consulting as a kind of pedagogy for institutions and collective subjects. One reason for publishing such a report is that the invention of electracy is not a matter for an individual or group, but for a field of knowledge, and even for a civilization. The point is to announce this experiment and invite participation. A further goal is to make a case for the relevance of Arts and Letters methods to public policy formation. The point of entry is the practice of calling witnesses to testify with respect to a given problem. It is common to call not only upon experts, but also upon those affected by the problem and any potential solution. This book recounts the rationale for basing the new practice on "testimony" (testimonial) and documents the experiment testing the form or method in a specific case (the problem tour of the Miami River). While it is assumed that this tour of a site is a necessary step in the process, it is only one part of a larger program leading to the design of an Internet site that is an image category (choragraphy) abstracted from the actual place. Miami Virtue proposes the terms of this transfer.

This book is a guide to an implementation of the EmerAgency vision – the formation of an image category (chora), and a pedagogy for teaching with the Internet. Such categories are the equivalent in electracy for the conceptual categories of literacy, of which the first to be invented – "Justice" – is attributed to Plato. The goal of electracy is to develop an image supplement for literacy. Literacy tells us what justice "is" (ontology), and electracy images its mood (its affect), addressing motivation (care). Literacy already provides the knowledge relevant to problems; electracy supplements that knowledge with feeling, by drawing upon the arts. The ultimate purpose of the testimonial is not to make art, however, but to articulate a general education mass electracy: to use arts methods and the networking resources of the Internet as means to allow egents to participate as witnesses in public policy decisions. The immediate benefit of this practice is a fluency in the image discourse of electracy (becoming electrate). The ultimate benefit is to organize a constituency that speaks on behalf of well-being, the general good.

PHOTOGRAPH: BARBARA JO REVELLE

With the spectacle egents witness a scene of problems, with the understanding that the problem "is" a scene, a situation, that is scriptible. A philosophical context is the argument made by Kant and developed by Hannah Arendt, that what made the French Revolution a world historical event was not the actors themselves but the general public of witnesses whose approbation determined the significance of the event. Miami Virtue is a proposal for a practice for an Internet fifth estate, intended to exploit for democratic effect the augmentation of witnessing made possible by a society of the spectacle. The impact on community decision-making process (politics) is not so much in the form of instrumental solutions, but through the impact of focusing collective attention. This impact is dialectical, or takes place in the "middle voice": the observer is also affected by the act of observation. In general, the clear and distinct separation and compartmentalization of our institutions (work, school, home, play) is blurred within the testimonial. Indeed, its function as a logic is to support an holistic mode of reason made feasible by digital technology.

The Project

To reach the present state of development of this proposal has taken over a decade. While some of the work related to the concept was pursued individually, most of it was collaborative, based on weekly meetings of the group, cooperative design of courses, joint presentations at conferences, joint authorship of grant applications and journal articles. It is possible of course to identify the areas of leadership relative to the strengths of individual members. William Tilson, Professor of Architecture, has used the Miami River as an object of study in his studios and seminars for more than a decade. John Craig Freeman, a digital artist, produced installations and recordings representing important historical sites in Florida, including the Miami River zone, using VR nodes combined with aerial photography surveys, as a digital "memory palace" for organizing the consultation. His VRML panoramas have been included in numerous exhibitions. Gregory Ulmer, a media theorist, developed in his books an electronic or hypermedia rhetoric, including the related practices of heuretics, mystory and choragraphy, that provided the methodology of the inquiry. Barbara Jo Revelle had extensive experience with photographic explorations of cultural environments, as well as with the design and production of monumental public mosaics, before taking up the Miami inquiry. Miami Virtue is a synthesis integrating these specializations in a whole that is greater than the sum of the parts.

It is important to emphasize the collaborative nature of the project since the book is authored in the voice of the theorist. This arrangement foregrounds not the completed work – which necessarily takes various digital forms – but the process by which the work was accomplished. The goal of this publication is to consolidate the lessons of the past decade into a form that could be shared, in order to develop partnerships for the next stage of development (targeting the Internet practice). This book may be used in a variety of ways. The framing assumptions, procedures, and readings have been used pedagogically in courses (studio and theory, undergraduate and graduate) in Architecture, Fine Arts, and Liberal Arts. It also models a collaborative mode of research, admitting that no one person is likely to have all the talents and experience needed for full electracy (imaging, design, programming, theory). The research goal is the invention of a metaphysics for imaging, with the understanding of "metaphysics" as category formation. What counts as "real" depends upon the classification system of a civilization.

Here is a summary description of the project published as an "artists' statement" in Leonardo 36: 3 (2003).

Image Emergency: A Pyschogeography of Miami

MIAMI VIRTUE is a theory of a logic of images, intended as an Internet practice. Working with the premise that the Internet makes possible a new relationship among the institutions of society, the Florida Research Ensemble (FRE) established the EmerAgency – an online, distributed, conceptual consulting agency – whose purpose is to introduce a new voice into the collective pedagogy of consulting. A practice called "testimonial" is proposed as the basis for a virtual civic sphere (a fifth estate), as a response to Paul Virilio's warning that the Internet and digital technologies in general are destroying the civic sphere and threaten to bring about a "general accident." The invention of a practice for the Internet involves features generalizable to electracy (literacy of digital imaging) as such. The analogy with the history of writing shows that the Greek inventors of the practices of alphabetic technology invented a new kind of category – the concept – resulting in a different way of classifying reality (metaphysics). The FRE experiment proposes to do for the photographic image what the Greeks did for the written word – define and test its categorial function. The practical stakes are high, since most studies on policy debates state that what constitutes a "problem" and "solution" is determined in part by the very categories of thought.

The case used as the prototype for the consultancy is the "image crisis" declared by the government in Florida during the early 1990s in response to a series of tourist murders. The assumption is that the experiment may be applied to any public policy issue, in any democratic society. An analysis of news and entertainment documents resulted in a focus first on the city of Miami, and then on the Miami River, a 5.5 mile channel at the heart of the city that is a "zone" – site of nearly every policy issue in the state. "Testimonial" is a montage of documents (journals, video interviews, photographs) by Barbara Jo Revelle made during her five-week stay at an Inn on the Miami River. Testimonial is composed using a hybrid mix of the "drift" made famous by the French Situationists, the poetic form of the "ideogram" (the modernist poetics initiated by Ezra Pound), and Roland Barthes's obtuse or third meaning (a detail in an image triggers an involuntary memory).

The method reflects a "secret history" of "situation" that brings into relation Plato's *Timaeus* (chora), the Chinese *I Ching* (hexagram), Ezra Pound's *Cantos* (vortex), and Guy Debord's maps ("plaque tournante"). The poetic ideogram, extended to photography, is used to produce an image category out of the documentation of the river region. This amplified ideogram is the basis for the electrate equivalent of the literate concept. The photograph "Crossroads" (showing a Haitian trading vessel impounded by the Coast Guard) is to this image category what the topic "justice" is to a conceptual category (a specific concept). The testimonial generated this image by juxtaposing Revelle's documents with a public policy issue – "Operation Safety Net" and the Caribbean Code, intended to exclude substandard boats from the port. The result is an attunement, a cognitive-affective map in which the Haitian trader, impounded by the Coast Guard, juxtaposed with Revelle's personal situation, evokes the mood ("Stimmung") of the river zone. The EmerAgency motto is: *Problems B Us.*

The name of the Miami chora – Miautre – refers to Jacques Derrida's discussion of "aporia" (an intractable dilemma). Miami is treated not as a problem but as an aporia. In his *Politics of Friendship*, Derrida concludes his critique of the history of political theory based on the model of "friendship," by proposing a politics of hospitality (not friend and enemy, but host and guest). Derrida's aporia of hospitality is proposed as a replacement for the dilemma that structured the Cold War (Prisoner's Dilemma). In this context, "Mi-Ami" translates as "my friend." Derrida proposes the "other" ("autre") as the guest in a new politics; hence, the testimonial works to switch the Miami River from "my friend" to "my other" (Miautre). The French terms are justified by a focus on the Haitian presence on the river (Creole). The Haitians who come to Miami are not enemies or aliens but guests. "Crossroads" becomes categorial when it is extended to cover any experience of "impoundment" (the attunement of the zone).

Outline

Beginning with an introduction that establishes our analogy –category formation in literacy – the book describes the construction of an electrate category through an unfolding of one photograph, "Crossroads," designated as the emblem of a chora. This photograph is to its chora what the term "Justice" is to its concept. This scene is sorted into three registers, borrowing the structure of the memory palace used in manuscript culture to memorize information for oratory: (1) "Miami" establishes the ground, the geographical place and the procedures used by the FRE to "consult" here. (2) "Myami" places "Crossroads" in a narrative that organizes the different levels of individual and collective experience into an allegorical pedagogy called "mystory." Mystory produces a pattern that organizes the diversity of the region into a groove, a coherent atmosphere. (3) "Miautre" shows how the operations of "Myami" in "Miami" may become intelligible and democratically accessible through an interface metaphor of divination. Virtual Miami is a time-image, supporting a collaborative composition of a global wisdom whose purpose is to help a community make decisions about its political and ethical existence: about its well-being, its capacity to thrive. Here the city reflects on its character, its virtue. Although sited within the culture and society of the United States, choral imaging, like conceptual reason, is an open-access process.

PHOTOGRAPH: BARBARA JO REVELLE

Acknowledgments

The debts incurred in any project are numerous, and multiplied in this case by the collective nature of the work. Students at the University of Florida have contributed to Imaging Florida, the EmerAgency, and seminars exploring the testimonial. A number of colleagues have helped test the approach including Ron Kenley and Miodrag Mitrosinovic. Will Pappenheimer joined the FRE in 2001 and has been instrumental in helping with the transition to the video and online phases of the project. Special mention should be made of the work of the late Gordon Bleach, who contributed importantly during the early phases (Imaging Florida). Gordon produced a series of postcards (*ghost card/ coast guard*) superimposing the clichés of Florida tourism (such as photos of palm trees) with the divination techniques of palmistry. Some of these photographs have been exhibited in national and international venues. The Office of Research, Technology, and Graduate Education at the University of Florida supported the project with a grant from the Opportunity Fund.

PHOTOGRAPH: BARBARA JO REVELLE

Acknowledgments

INTRODUCTION

The EmerAgency

> Despite the great anthropological isue of nomadism and of sedentariness that explains the birth of the city-state as a major political form in history, there is little understanding of the vectorial nature of the transhuman species that we are, of our chorography.
>
> PAUL VIRILIO (*Open Sky*)

The Internet Accident

This book proposes an institutional practice – the testimonial – that allows citizens to use the Internet as a civic sphere. The testimonial and its institutionalization in the EmerAgency as a new kind of consultancy is a response to concerns formulated most forcefully by Paul Virilio, who describes the Internet as the potential source of a general accident.

> Today, the new technologies convey a certain type of accident, one that is no longer local and precisely situated, like the sinking of the Titanic or the derailment of a train, but general, an accident that immediately affects the entire world. This is quite obvious when we are told that the Internet has a worldwide vocation. Yet the accident of the Internet, or that of other technologies of the same nature, also represents the emergence of a total – not to say integral – accident. And that situation has no reference. We don't know yet, perhaps with the exception of the stock market crash, what an integral accident could be, an accident that would involve everyone at the same time. (Virilio, 1999: 12)

Virilio's argument is that teletechnologies through their instantaneous interactivity have produced a "single time" – Real Time – whose milieu is speed. This unprecedented immediacy and ubiquity make democracy impossible, he argues. Public space in Real Time becomes an image in some medium – photography, cinema, television. These images replace the "trajectories" of the city, the face-to-face interaction of the public sphere and the encounter of subject with object in the agora, the forum. The question he raises is whether a virtual city is possible – whether it is possible to urbanize real time. "If the answer is no, then a general accident is inevitable, the accident of history, the accident of accidents

 | DOI:10.1163/9789004534643_001

that Epicurus spoke of regarding history. If we are not capable of urbanizing the real time of exchange, in other words the live city-world, the city-world in real time, through the globalization of telecommunications, then both history and politics will be called into question. This is an extraordinary drama" (40).

Virilio does not claim that the general accident is inevitable. Rather, he poses this possibility as a general emergency. The danger that he sketches begins with the formation of the "mental map" that each person develops. The threat to freedom, he says, begins with the reduction and impoverishment of this mental picture. As an urbanist he locates the source of the crisis as the displacement of the real city by the tele-city, the loss of a lived public space in favor of a virtual gathering on the Internet (45). The challenge is that the lived city as a space and place is no longer visible, its borders or shape lost to perception, a loss that affects border itself and hence identity as such. Intelligibility depends on borders, edges, limits: "the city is the law, and the law is the border" (76). The new technologies of electromagnetism and the video screen no longer concern sun rays, light, and geometric perspective, but wave optics. "Geometrical optics made the city centers and the suburbs. Centralized cities such as Palmanova are the result of perspective vision. The ideal city of Piero della Francesca is the vision of a city that implements perspective vision. Now, the invention of a perspective of real time and electromagnetic wave optics demands an understanding of what the city of that vision of the world will be –virtual city" (85).

> Alongside air pollution, water pollution and the like, there exists an unnoticed phenomenon of a pollution of the world's dimensions that I propose to call *dromospheric* – from *dromos:* a race, running. Contamination has in fact spread further than the elements, natural substances, air, water, fauna and flora it attacks – as far as the space-time of our planet. Gradually reduced to nothing by the various tools of transport and instantaneous communication, the geophysical environment is undergoing an alarming diminishing of its "depth of field" and this is degrading man's relationship with his environment. The optical density of the landscape is rapidly evaporating, producing confusion between the apparent horizon, which is the backdrop of all action, and the deep horizon of our collective imagination; and so one last horizon of visibility comes into view, the transapparent horizon, a product of the optical (optoelectronic and acoustic) magnification of man's natural domain. (Virilio, 1997: 22)

Kevin Robins agrees with Virilio's analysis of the crisis of our image culture. He cites Donnatella Mazzoleni's assessment of the loss of correspondence between the individual body and the body politic.

> Metropolises are no longer 'places,' because their dimensions exceed far the dimensions of the perceptive apparatus of their inhabitants. The widest sensory aperture, that of light, is shattered. It was the visual field, in some respect, which defined the city dimensionally: in the metropolis there is no longer pan-orama (the vision of all), because its body overflows beyond the horizon. In the metropolitan aesthetic the eye fails in its role as an instrument of total control at a distance." The city no longer reflects our identity back to us. Mazzoleni describes this in terms of the experience of immersion, loss of self, regression to a condition of pre-separation: "the most elementary distinction of space – the distinction between 'inside' and 'outside,' which is the distinction between 'I' and 'the world' – grows weaker. (Robins, 1996: 133)

This weakening of the borders of identity is the subject formation that Scott Bukatman called "terminal identity."

> Body and image become one: a dissolution of real and representation, certainly, but also of the boundaries between internal and external, as the interiorized hallucination becomes the public spectacle of the Videodrome program. Here *Videodrome* echoes *The Simulacra*, in which a character's psychosis results in a physical transformation.... In the postspectacle society delineated by Baudrillard, all such boundaries will dissolve, will become irrelevant through the imperatives of the model of communication (simultaneous transmission and reception): "In any case, we will have to suffer this new state of things, this forced extroversion of all interiority, this forced injection of all exteriority that the categorical imperative of communication literally signifies [Baudrillard, *The Ecstacy of Communication*].... The subject has "no halo of private protection, not even his own body, to protect him anymore." (Bukatman, 1993: 92)

What looks like a generalized schizophrenia from the point of view of literacy, is understood in electrate terms as symptoms of a change in subjectivation intrinsic to the new apparatus. In electracy, for better or worse, the borders of "identity" – of the group subject (between individual and collective) – are undergoing revision. The name of the art of category formation is "metaphysics."

Virilio's conviction is that we must not be seduced by the tele- or cine-city, but must recover the living city, the polis. "If there is a solution possible today, it lies in reorganizing the place of communal life" (Virilio, 1999: 52). The first step he claims is to counter the image by recreating speech, recovering language and writing (87). Meanwhile, that which makes this condition intelligible is the accident, which is an "inverted miracle" in that it has the effect of revelation ("when you invent the ship you invent the shipwreck") (89). The most important aspect of Virilio's argument from the point of view of electracy is his claim that the new technologies of the spectacle (that is, the technologies associated with the apparatus of electracy) produced the conditions of a "knowledge accident" in which each invention includes its own catastrophe. The accident of accidents at the level of knowledge involves the very category formation of literacy itself invented by the Classical Greeks.

PHOTOGRAPH: BARBARA JO REVELLE

The literate metaphysics of Aristotle articulated within the Real a certain reality in terms of substance with properties sorted into essences and accidents by means of the practices of definition. In the era of information the logical "accident" has now merged with the material accident of "what happens." Aristotle did everything he could to eliminate the pun from inference, since it created turbulence in the system of closed concepts. Thus it was not possible in his logic to "learn about the accident from the accident." This merger of logical and material "accidents" in contemporary science indicates that literate knowledge has reached and even exceeded the limits of the grasp of conceptual thinking. "If, according to Aristotle, 'the accident reveals the substance,' the invention of the substance is also the invention of the 'accident.' Seen this way, the shipwreck is indeed the 'futuristic' invention of the ship, the air crash the invention of the supersonic plane, and the Chernobyl meltdown, the invention of the nuclear power station" (Virilio, 2003: 6).

> In his *Physics,* Aristotle remarks at the outset that it is not Time as such that corrupts and destroys, but the *accidens – that which happens.* It is, then, *passage into Time* or, to put it another way, speed of emergence, which effects the ruin of all things, each 'substance' being, in the end, the *victim of the accident of temporal circulation.*
>
> One can easily imagine, then, the damage wrought by the Time accident – the accident that occurs with the instantaneity of the *temporal compression* of data accompanying globalization – and the unimaginable risks of the *synchronization* of knowledge.
>
> So, the "imperative of responsibility" of which Hans Jonas speaks, should be based, in the first instance, on the need for a new understanding of *accidental production* –that unconscious industry the "materialist" man of science refuses to face up to, even though the "military-industrial complex" showered us with incidences of the sudden militarization of the sciences throughout the last century, the prime examples being the fateful invention of weapons of mass destruction and of a thermonuclear bomb capable of extinguishing all life on the planet. In fact, the *visible*

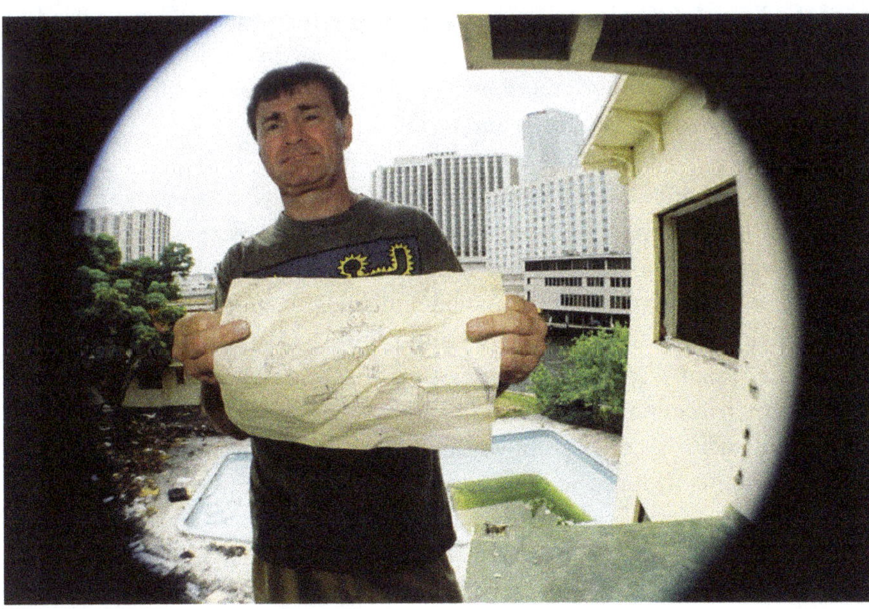

PHOTOGRAPH: BARBARA JO REVELLE

> *velocity* of substance – the velocity of a means of transport or the speed of calculation or information – is only ever the emergent part of the iceberg of the – *invisible* – velocity of the *accident*. And this applies both in the realms of road traffic and the circulation of values. (26)

The EmerAgency experiment responds to Virilio's warning not by returning to the lived city, but by designing a practice to address the loss of borders experienced in the virtual city. As a consultancy its point of intervention is precisely at the level of the circulation of values. The claim is that the new category system (the new metaphysics) emerging within electracy is capable of making an object of study of the metaphysics of literacy, and hence of "getting up to speed" with the Time accident. The unconscious or machinic dimension of the literate knowledge accident becomes accessible to reason again within the holistic category of the electrate chora, including the cinematic *time-image* (Deleuze) capable of writing the dromosphere that has exceeded the capacity of literate reason.

Information Infrastructure

The EmerAgency is a virtual consultancy whose purpose is to develop an Internet practice that responds to the challenge posed by Virilio. The basic premise of the EmerAgency is that "problems" no less than the policies devised to address them are "social constructions." Problem and policy formation, that is, are special cases of Western metaphysics in general, meaning literacy and the ontology of categories, concepts, and topics.

> Problems come into discourse and therefore into existence as reinforcement of ideologies, not simply because they are there or because they are important for well-being. They signify who are virtuous and useful and who are dangerous or inadequate, which actions will be rewarded and which penalized. They constitute people as subjects with particular kinds of aspirations, self-concepts and fears and they create beliefs about the relative importance of events and objects. They are critical in determining who exercises authority and who accepts it. They construct areas of immunity from concern because those area are not seen as problems. Like leaders and enemies, they define the contours of the social world, not in the same way for everyone, but in the light of the diverse situations from which people respond to the political spectacle. (Edelman, 1988: 12–13)

In other words, within literacy problems are treated as "topics." Ian Hacking noted, for example, that "women refugees" is posited as a category or idea, with certain consequences (and we could substitute the terms of our Miami case – the Haitian trader). "Ideas do not exist in a vacuum. They inhabit a social setting. Let us call that the matrix within which an idea, a concept or kind, is formed. 'Matrix' is no more perfect for my purpose than the word 'idea.' It derives from the word for 'womb,' but it has acquired a lot of other senses – in advanced algebra for example. The matrix in which the idea of the woman refugee is formed is a complex of institutions, advocates, newspaper articles, lawyers, court decisions, immigration proceedings. Not to mention the material infrastructure, barriers, passports, uniforms, counters at airports, detention centers, courthouses, holiday camps for refugee children" (Hacking, 10). "Matrix" is interesting in our context since it resonates with one of Plato's metaphors for chora (not to mention with a series of blockbuster films). The implication is that one needs to grasp not one "problem," but the matrix as a whole within which such an "idea" makes sense. In choragraphy "chora" replaces, or rather subsumes, "topic" to name how information is organized into usable sets on the Internet.

What public policies should regulate the borders of all things "American"? Problem entities become images, Joseph Gusfield argued in a discussion of the "idea" of "drunk drivers," created as "public facts." "They constitute a public reality in two senses. First, these are facts that are shared by many people in the society who have no, or very little, personal knowledge of individual cases making up the aggregated facts. As public facts, they are not peculiar to any particular class, cultural group, or educational status. Second, they compose a public reality in being an aggregated fact rather than the events involving particular people. They are not about anyone but about a society – the United States [for example]. Because such facts are not records of individual events but are rather aggregations of data, amassed and presented, the 'discovery' of public facts is a process of social organizations. Someone must engage in monitoring, recording, aggregating, analyzing, and transmitting the separate and individual events into the public reality of 'auto accidents and deaths.' At every stage in this process human choices of selection and interpretation operate" (Gusfield, 37). Testimonial includes a monitoring of this process within its role as witness. Through the EmerAgency, egents may help "compose" public facts.

It is a matter of the relations among ideology and values, politics and morality. "Multiple voices and silences are represented in any scheme that attempts to sort out the world. No one classification organizes reality for everyone" (Bowker and Star, 41). Bowker and Star warn that this inevitable interdependence of policy and metaphysics becomes dangerous in a condition of "convergence" – when the system disappears as such. "This blindness occurs by changing the world such that the system's description of reality becomes true.... It will be impossible to think or act otherwise" (49). A community will not notice when it has reached convergence until there is a breakdown in the system (34), when its solutions fail to pass a reality check. The events of 9/11 may be considered one such "reality check." The single greatest determinant of policy formation is "the dominant value paradigm." "For any problem at the regime or macro-level of discussion and analysis there are remarkably few alternatives actually under debate. In the United States a powerful and enduring political culture (as distinct from an arguably much more variegated popular culture) helps to whittle down the range of 'legitimate' alternatives to a pitiful few long before any quasi-pluralist 'conflict' over problem definition ever ensues" (Bosso, 184).

In Virilio's terms, the problem is that one of the fundamental brakes on change – a respect for the surprise of an unknown quantity – has been lost. "Since the beginning of the last century, breakdowns or failures have progressively become not exactly an unexpected surprise, but rather a subject for theoretical speculation, a thesis-subject, allowing the very term 'accident' to lose the philosophical definition that had been its own since Aristotle. Suddenly, the accident is no longer unexpected; it is becoming an –*a priori* scandalous – rumor, in which the presumption of wrongdoing tends to win out over that of involuntary action and yet, conversely, the quasi-certainty of willful harm is dissimulated beneath a concern not to provoke panic" (Virilio, 2003: 41). The challenge for the EmerAgency is to develop an approach to the civic sphere that opens a new perspective on literate society by means of a category formation native to the image apparatus of electracy. Given the interdependence of category (metaphysics) and policy noted above, the task of this study is to take up the question of the public sphere by proposing a new kind of category, specific to image technologies organized into the apparatus of electracy, and operating logically with a speed that matches the accelerated conditions of a wired planet.

From Chorography...

The problem articulated by Virilio concerning the visibility, representability, intelligibility of the city, is a matter of "chorography" (in its original sense within the field of geography). The invention premise is that chora becomes categorial by extending the natural process through which citizens form cognitive maps by internalizing an image of their city (their experienced environment). Chorography was distinct both from cosmography and geography because of its mix of the intelligible and the sensible in the mapping of space:

"Chorography's complex synthesis of scientific measurement and geometrical survey on the one hand and painterly and artistic skill on the other offered the opportunities and pleasures of mapping locally circumscribed land and life as a microcosm of a greater order" (Cosgrove, 1999: 18). As originally conceived by Ptolemy, chorography was to focus on "creating a recognizable image of the visible features of a single part of the inhabited world," thus rendering such locales as ports, rivers, villages. Geography was considered superior to chorography because it went beyond the sensual limits of the pictorial qualities of an individual site to grasp the abstract whole through the sciences of quantitative measurement (Nuti, in Cosgrove, 1999: 90). From the time of Alberti, however, there was an attempt to combine the two aspects of reality – being and appearance (associated respectively with architecture and painting) – into one practice, in order to form the natural image of the city (92–3).

The popularity of town portraits combining abstract structure with visual appearance had to do with their status as microcosm in chorography. As Lucia Nuti shows, the relation of an observer to an object has a history, and evolves with the visual culture of a place and period. She traces the different cultures of the South and North, of Italy and Flanders, with the former preferring an all-embracing view for the urban place from an elevated vantage point, and the latter favoring a profile rendering from a low, horizontal point of view. Each had its own solutions to dealing with the chorographic dilemma (the balance of being

PHOTOGRAPH: BARBARA JO REVELLE

and appearance, totality and vision). The 16th and 17th centuries dealt with the impossibility of putting a total vision into a single image by producing multiple views, with each one exploring a different experience in times and space of the city, all juxtaposed on one large sheet (104). The notion of "vision" in chorography was not restricted to sight, but included sense knowledge in general and beyond that historical and cultural interpretation, still striving for a total grasp of the place (108). The visual cultures of the South and North of Europe were equally transformed by the apparatus shift that took place in the 19th century.

> The continuing desire for an all-embracing urban view was satisfied in one of the nineteenth century's most spectacular graphic inventions, the panorama. The creation of a large-scale panorama required the achievement of a free horizon from a high viewpoint, such as the top of a tower or the upper story of a building. The delineator traced a series of views in a sequence, scanning through 360 degrees, and then joining them together. Conceptually, the panorama owes something to both the perspective plan and the profile, but it also diverges from both. The vantage point is a single elevated one and has to be purposefully chosen, as in the plan. The vision unrolls in a time-lapse at a constant eye level, as in the profile, but a significant change occurs in the way the object is focused. The observer becomes integral to the object and inspection is made form inside, rather than from a distance. (103)

PHOTOGRAPH: BARBARA JO REVELLE

As we know from Jonathan Crary's analysis of the shift in visual culture that took place in the nineteenth century (Crary, 1990), the panorama leads to a chorography of the cine- and tele-city that Virilio decries as precursors of the virtual image-city and hence of the general accident of the Internet. John Craig Freeman's VRML panoramas of the Miami River, displayed in the media of CD-Rom, DVD, and the Internet, take their point of departure from this tradition of "chorology" (Entrikin) to provide continuity for the process of the reinvention of chorography in electracy. The systematic extension of these techniques through Google Maps evidences the renewed relevance of this tradition. Chorology proper, however, including its contemporary application by Google, is inadequate to a dromosphere in which geography itself has come to an end (dimension collapse) (Virilio, 2003: 109). Time-Space (chronotope) today is "warped," saturated with human desire and drive.

...To Choragraphy

The challenge posed by the Internet accident is to bring the image of the city into digital intelligibility. This city-image will be to electracy what "topic" was to literacy – a "place" (with the aid of digital technology) to store and retrieve and invent what may be thought and communicated in contemporary civilization. The immediate inspiration for *choragraphy* (a neologism, spelled with an "a") was the collaboration of Jacques Derrida with Peter Eisenman and his team on the design of a "folly" for the Parc de la Villette in Paris (Ulmer, 1994). Derrida challenged Eisenman to give architectural expression to the notion of space or place introduced by Plato in *Timaeus*, the dialogue that is perhaps better known in architectural circles for ascribing the making of the cosmos to an architect-like Demiurge. Although this *folly* was never built, a record of the design process was published (Kipnis and Leeser, 1997). The difficulty of trying to build chora is that it is not representable directly as percept or concept. Metaphorically it is a *receptacle* making possible the interaction of the orders of Being and Becoming (Form and Matter). In Greek metaphysics the receptacle is a third kind, neither concept nor percept but generation whose function is to sort the chaos of Being-Becoming into categories (Earth, Air, Fire, Water). It is made intelligible indirectly (obtusely), by artistic ("bastard" Plato said) means of story and metaphor. The reliance on arts and letters creative devices to communicate chora extends to choragraphy as well. Chora may be articulated first as provocation.

Situations

What are the larger implications of the Miami experiment? The immediate purpose of our Miami testimonial is to produce a categorial image (a chora) that in turn may function online or digitally to constitute the civic sphere of a virtual city. The hypothesis is that the obtuse (Barthes's third meaning) is the point of emergence of the category formation or classification (we might as well say the "metaphysics," since categories determine "that which is") that is native to the image apparatus of electracy. Eric Havelock's account of the invention of the concept in literacy provides an important analogy for understanding what is happening in our own moment. The invention of chora as the category formation of electracy retraces the genealogy of the invention of the topic in literacy. The epistemic consequences and implications of the technological ability to record speech by means of the alphabet developed over several centuries, culminating finally in the invention of philosophy, dialectic, method, and their formal teaching in Plato's Academy – itself constituting the invention of a new institution – school. The first concept, the prototype of the new category, was "justice" (dike).

> Hesiod affords an initial example of a process which was to gather momentum later, when he chose the term *dike* (usually translated "justice") as the formal subject of a "discourse." The term occurs incidentally and not infrequently in orally preserved speech (as in Homer) but never as the topic of a formal discussion. The narrative laws of oral memorization would discourage such a choice.
>
> Having made his choice Hesiod cannot conjure the required discourse out of thin air. We could easily manage it today, because we inherit two thousand years of literate habit. He, on the contrary, must resort to the oral word as already known – the only preserved word that is known. He must build his own semi-connected discourse, either some pieces in which the term dike happened for whatever reason to occur, or others in which incidents occurred that he felt were appropriate to connect with the word. His decision is compositional (rather than ideological), or perhaps we should say re-compositional.
>
> If he must do this, he will be forced to continue to utilize the narrative forms which control what he is borrowing from. He still will not be able to tell us what justice is, but only what it does or suffers. He has taken one decisive step toward the formation of a new mentality by inventing the topic to take the place of the person. But he cannot take the second step of giving his topic a syntax of descriptive definition. It will still behave rather than be. So in seventy-three lines of hexameter we are treated to a panorama of dynamic situations in which justice singular or plural features as a subject performing or an object being performed on: a voice speaking, a runner in a race, a woman abused, a present you

> give someone, a thing you walk away from, a pronouncement of Zeus, a crooked weapon to inflict injury, a ward protected by guardians, a virgin goddess who sings her proclamations, a prisoner confined, a piece of property, a present given by Zeus to men, a cripple. The heterogeneity of the images may seem to be a symptom of confusion, but in fact it is the product of new invention, unavoidable at the first stage of a pioneering journey of language and of the mind. (Havelock, 1967: 101–102)

The gathering of disparate materials under the new principle of categorization at first seemed chaotic. When the oral narratives were recorded or written down in the alphabet, it became possible to discern a pattern, a repetition of theme and variation that went unnoticed in spoken discourse.

> The psychological push needed to bring this about must have been the use of vision as supplement to hearing. An architectural rearrangement has been performed on language as previously used. The various "justices" which perform one after the other in Hesiod's account echo each other acoustically to some extent, but they are also all look-alikes. The reading eye has been able to perceive them as located in an oralistic flow that has now been written down in the alphabet, which can be looked at, read, and "backward-scanned." Hesiod could have so composed only if he was able to "read" oral texts of Homer (and perhaps others). The first beginnings of the alphabetic revolution have occurred, in the creation of a topic as a subject of a "discourse" made possible by the conversion of acoustically preserved memorized speech into materially preserved visible artifacts that are capable of rearrangement. (103)

Choragraphy is a method for doing for digital imaging what the Greeks did for alphabetic writing. A chora gathers records of a specific social region, anchored in images, to form a category of electrate reason. If the first concept was "Justice," the first chora is "Miautre" (virtual Miami).

The Choral City

Since Miautre is an experimental test of an electrate rhetoric dubbed "choragraphy," the first issue is to clarify the nature of "chora" and to place it in a certain tradition with contemporary relevance. Plato introduced "chora" in *Timaeus*, one of his most influential dialogues, best known now perhaps for being the first telling of the story of Atlantis. The narrative diegesis has Socrates engaged in a series of conversations with several figures, including the Pythagorean astronomer, Timaeus, over several days. The first day was covered in the *Republic*, dealing with the kind of citizens required for an ideal city, and the final day is treated in the unfinished *Critias*, whose intention was to tell how this city might be realized in practice. In between, *Timaeus* explains the cosmology that produced the natural order to which the social order of the ideal city corresponds. A choragraphy of Miami, then, evokes this context of reflections on the nature and possibilities of the city as such, relevant to our concern with a virtual city that counters the society of the spectacle.

Timaeus is a major source for the doctrine of correspondences in the Western tradition – the idea, captured in the phrase "as above, so below," that the microcosm repeats the macrocosm. The Pythagorean principle of the "music of the spheres," expressing the mathematical structure of the cosmos, is manifest in *Timaeus*. A sticking point for Plato in working out the dialectical description of this theory concerned how the pure Forms or Ideas in the mind of the Deity (Being) interacted with the material substances of the world (Becoming). He reasoned that there is a third fundamental nature – a Receptacle – that mediates and enables the instantiation of the Forms in things. This receptacle he called space (chora).

> My verdict is that being and space and generation, these three, existed in their three ways before the heaven, and that the nurse of generation, moistened by water and inflamed by fire, and receiving the forms of earth and air, and experiencing all the affections which accompany

PHOTOGRAPH: BARBARA JO REVELLE

> these, presented a strange variety of appearances. And being full of powers which were neither similar nor equally balanced, was never in any part in a state of equipoise, but swaying unevenly hither and thither, was shaken by them, and by its motion again shook them, and the elements when moved were separated and carried continually, some one way, some another. As, when grain is shaken and winnowed by fans and other instruments used in the threshing of corn, the close and heavy particles are borne away and settle in one direction and the loose and light particles in another. In this manner, the four kinds of elements were then shaken by the receiving vessel, which, moving like a winnowing machine, scattered far away from one another the elements most unlike, and forced the most similar elements into close contact. (*Timaeus*, 52d–53a)

In Plato's cosmology chaos is a plenitude that is prior to the action of order introduced by the Deity. Since chora (space) is a third nature, neither intelligible nor sensible in itself (neither a concept nor a percept), it must be communicated by a kind of dream logic of metaphor, myth, and story. Chora supplements concept and percept as a means of contact and interaction with reality (Walter, 122). Our hypothesis is that choral thought is affective, emotional, correlated with memory and imagination. Chora is live (working with Real Time), performed, a generator of order. There is nothing "new" about choral knowing in itself. What is new is the hypothesis that electracy favors and

PHOTOGRAPH: BARBARA JO REVELLE

enhances it, and hence that it may be introduced into the schools and through them into society as part of the "skill set" of an educated citizen.

Some of the other metaphors Plato used to characterize the relations among the three kinds, in addition to the winnowing device noted above, included "family" – the eternal Forms as father, the created copies as offspring, and chora if not "mother" then at least "nurse." The function of chora in the cosmos is compared with that of the liver in the body. This latter analogy points to a key strategy of the choral interface: divination as an attitude toward information. In the analogy between the city and the individual person the hierarchy of value equates the mind or reason with the ruling ideas, and the heart with the guardian senses. The viscera are equated with the workers, and within these lower organs the liver is singled out as the seat of divination.

> And knowing that this lower principle in man would not comprehend reason, and even if attaining to some degree of perception would never naturally care for rational notions, but it would be especially led by phantoms and visions night and day – planning to make this very weakness serve a purpose, God combined with it the liver and placed it in the house of the lower nature, contriving that it should be solid and smooth, and right and sweet, and should also have a bitter quality in order that the power of thought, which proceeds from the mind, might be reflected as in a mirror which receives likenesses of objects and gives back images of them to the sight, and so might strike terror into the desires.... For the authors of our being, remembering the command of their father when he bade them create the human race as good as they could, that they might correct our inferior parts and make them to attain a measure of truth, placed in the liver the seat of divination. And herein is a proof that God has given the art of divination not to the wisdom but to the foolishness of man. (*Timaeus*, 71a–71e)

The mirror analogy alludes to another metaphor for chora as a receptacle that displays entities without participating in them. The choral function is mediation, and has been associated by various commentators with Plato's analogy of the cave. As part of bootstrapping literate practices out of orality, Plato retained some features of religious ritual while rejecting or demoting others. Thus the setting of *Phaedrus*, the first discourse on method in the Western tradition, is just outside Athens, near the sanctuary of Agrai on the banks of the river Ilissos. "Agrai was the scene of the Lesser Mysteries, which served as a preparation for the Great Mysteries of Eleusis" (Kerenyi, 45). The Lesser Mysteries, the "myesis," celebrated love, and were distinguished from the hidden mysteries given only to initiates in a revelation, the "epoptika." "Myesis" is another "my" term evoked in "mystory."

> *Myesis* can be rendered by the Latin word *initia*, "beginnings," or its derivative *initiatio*, or initiation, signifying introduction into the secret. For *myesis* comes from the verb *myeo* which denotes the action. The simpler verb *myo*, from which the noun derives, implies the element of secrecy. It means nothing other than "to close," as the eyes do *after seeing*. The self-evident first object of this verb is the subject itself: he closes himself after the manner of a flower. But a second object is also possible, which must be very close to the subject, his very own possession. Such an object is the secret. In German this close connection between the secret and the holder of the secret is expressed by the derivation of the noun *Geheimnis* (secret) and the adjective *heimlich* (secret) from *Heim* (home), the most private precinct. In Greek a number of composite words can be traced back to the hypothetical verbal adjective *myston*, which can signify only that which is shut up within itself: *mysto-dotes* is one who gives out such a secret. (46)

The concealed-revealed, canny-uncanny experience of the mysteries is an important dimension of "recognition" in testimonial. The Eleusinian Mysteries, celebrated every Autumn at the sanctuary of Eleusis near Athens, were based on the myth of Demeter's search for her ravished daughter Persephone. The myth associated the harvest with love or fertility in general. Plato's contemporaries would have recognized as an allusion to Eleusis the use of the winnowing fan or basket as a metaphor for chora. Scenes from the ceremony preserved in urn carvings show a priestess holding a *liknon* or winnowing basket over the head of the initiate: "a plaited winnowing fan, an instrument with which the grain was ordinarily cleansed and in which the accessories of the Dionysian rites were kept and carried about: the phallus or the mask. Infants, divine as well as human, were placed in such baskets" (57).

To communicate in an image the relationship of Being, Becoming, and Space, Plato alluded to this ritual and familial scene of the nurse carrying the infant in the *liknon*. As it happens, the *liknon* was also used in divination, not only in Greece but in other civilizations as well. "Nedembu diviners 'shake the basket' to 'cut a trail' to and from the dark springs of malice, hatred and the desire to hurt and control.... The basket diviner uses a winnowing basket and a set of 20–30 small objects that he makes or finds. These tell the story of misfortune, loss, and death, and the revengeful motives responsible for them. They articulate the disturbances and conflicts that surround the society's values. The story they tell explains suffering as a breakdown in human relations" (Karcher, 79). The diviner "sees" the situation of the querent by shaking the basket repeatedly and noting the relationship among the objects that come to the top each time. The harvesting symbolism in this context represented the winnowing of information to find the solution to the querent's problem (Roberts, 71).

The general significance of chora emerging out of this metaphor is that of an ordering matrix that sorts confused, heterogeneous materials into categories or kinds. Plato's chora sorts chaos into the elements or principles of Earth, Air, Fire, and Water, which are also geometric structures governed by mathematical ratios. Miautre chora must perform this sorting operation for the Miami River zone. In principle, a chora is a place that accesses the "secret" workings of order in a specific locale. Specifically (and this is the live or action dimension of the practice), chora in the digital prosthesis enables an egent to experience the workings of the macrocosm within oneself. In principle, this recognition opens a circuit that permits a revision of the system. The larger context for the project is this question of doing for electracy what the Classical Greeks did for literacy – inventing a practice of category formation based on the possibilities of digital imaging. A practical guide for the choral map, then, is the analogy of winnowing, filtering, panning for the signifiers of mood, the "accidental" properties of things that produce aesthetic effects associated with images. If the literate category began as an ideal pure Form, the electrate category begins as a living mutating city.

Topos

Plato's *Timaeus* in the context of the history of writing (grammatology) is a step in the heuretic process of inventing literacy. This history serves as an analogy for our own moment, relevant to our own task of inventing an institutional practice for electracy. Plato is credited as the inventor of philosophy since he brought to completion a long process in which alphabetic writing was used to extract from the oral epics a new mode of knowledge, a new manner of category formation – the concept. In short, he invented the institutional practices of the literate apparatus: school and method. A goal of Miautre is to call attention to the relativity of this apparatus in order to make explicit the possibility of inventing a different institutional practice, and a different way of ordering what *is*, within the emerging apparatus of electracy. In these terms electracy does not eliminate or "close" metaphysics, but produces a new metaphysics, understood as a new way of categorizing the world.

Aristotle continued Plato's literate project in his own school, the Lyceum, where he and his fellow peripatetics invented the "thing."

> In two of his early works – in the *Categories* especially, but also in the *Topics* – Aristotle presents a revolutionary metaphysical picture. This picture has had a peculiar fate. Its revolutionary theses are so far from being recognized as such that they have often been taken to be statements of common sense, or expressions of an everyday, pretheoretical ontology. The most stirring and far reaching of those theses is the claim that included among what there is, among the entities, there are *things*. Aristotle, famously, goes on to maintain that these things are ontologically fundamental. All the other entities are by being appropriately connected to the things, for examples, either as their features (their qualities, sizes, relations-to-each-other, locations, and so on), or as their genera and species, that is, the kinds under which the things fall. These claims and their interpretation have received considerable discussion. Yet the fundamental one has gone virtually unnoticed. To formulate it most starkly: before the *Categories* and *Topics*, there were no things. (Mann, 3–4)

When he extracted things from the world Aristotle was inventing a discourse, a way of talking (theory). "Ontology was born when someone realized that any view of this sort implies a distinction between individual things, on the one hand, and their properties, on the other. Ontology was born when someone realized that there are not only different kinds of individual thing

PHOTOGRAPH: BARBARA JO REVELLE

but also different kinds of entity. Plato's theory of forms deals with just these sorts of questions, and we think of it, therefore, as one of the first ontological inquiries" (Grossman, 3). Aristotle continued this project by establishing that which is constitutive of an entity, its essence (what is real about it and that makes it what it is). The task was to determine what something essentially is, as distinct from what it merely happens to be: to distinguish its form from its matter. Aristotle criticized the Presocratics for treating entities as if they were mere heaps of stuff. "Both a heap and a thing are one – for each is unified such as to be one heap, or one thing – the kind of unity characteristic of things is much stronger than that of heaps.... Things are genuine unities, whose cause of being one is intrinsic to them, while heaps are merely accidental unities, whose cause of being one is wholly extrinsic to them" (33–34).

Aristotle famously brought Plato's Forms down to earth, and in the process abandoned choral space. His refinements on Plato included a preference for "topos" over "chora" to theorize "space." Topical space is determined by its outer form and volume located in a certain place, which in turn must be in "an encompassing but delimited portion of choric space" (Casey, 1998: 35). Edward Casey traced Aristotle's rejection of Plato's chora – the amplitude of the receptacle – in favor of a more stringent container, as he passed from cosmology to physics. What constituted "place" was its capacity as a vessel independent of any body in it or passing through it. Aristotle's example of this essence or nature of topos adds symbolic resonance to the circumstances of Miautre.

> Place thus construed is "the inner surface of the innermost unmoved container of a body." As such, it contains-and-surrounds the body by furnishing to it an environment that, if not always stable (the immediate "spot" of a boat in the river is only a momentary locale, not a lasting locus), is nevertheless a defining locatory presence. Thanks to this presence, place is actively circumambient rather than merely receptive (55). In the case of the floating vessel, is this limit the immediately surrounding water regarded as an ideal perimeter (yet as flowing water, it is constantly changing with the result that the place of a stationary boat will be continually changing), or is it the river's bed-and-banks or even the river itself as a whole? This seemingly trivial but in fact momentous question was to engage over two thousand years of debate in Western philosophy. (70)

The status of the impounded trader on the Miami River resonates with this example from Greek metaphysics. The question concerned the distinction between the container as limit or as boundary.

PHOTOGRAPH: BARBARA JO REVELLE

The reason for going into this Classical context of "chora" is to establish by analogy the nature of what we need to invent for electracy – an educational practice that does for digital media what Aristotle's topics did for alphabetic writing. What has to be invented is not "only" a new kind of category, the related device of concept formation, but an institutional practice for learning and applying categories and concepts. The task is posed as a return to chora, prior to the formalization of predicate logic, in order to open a new direction that exploits the possibilities of digital imaging. Chora is not "against" topic, but supplements it in a collaborative integration of literacy into electracy. Miautre is "exemplary" in this sense: its purpose is to demonstrate an image category and its associated devices, so that more such categories may be created. Aristotle extended his insights into place (topos) from physics to logic, with the invention of the dialectical topic. The topic, based on Plato's dialogues, is a mnemonic system for classifying and deploying arguments.

> A topic is a "head" under which are grouped arguments, or lines of arguments; in a *topos* (place, *locus*, region) the speaker has a stock of arguments to which he may turn for a particular need. If he knows the *topoi* (regions, places, lines of argument) he will know where to find what he wants for a special case. The general topics, or *common*places, are regions containing arguments that are common to all branches of knowledge; these are the topics of *more and less*, of *magnifying and minifying*, of *past*

> *and future*, and of *possible and impossible*–the four commonplaces in the strict sense.... The topics or places may be indifferently thought of as in the science that is concerned, or in the mind of the speaker. (Cooper, 154–55)

The work entitled *Topics* is a handbook that teaches students how to argue in a structured debate (Slomkowski, 3). The competition begins with a problem, even an aporia, something about which there is puzzlement and disagreement. A fundamental aspect of the practice is "definition," the procedure of producing "predicables" for the topoi – what may or may not be said of a thing, concerning its proper attributes. The definition, that is, establishes essence. There are four kinds of predicates admitted: definition, property, genus, or accident. An accident is that which may or may not belong to a thing (a given horse may be awake or asleep, brown or white, large or small, without altering its essence, its true nature). "To define a substance means to establish, among various accidental attributes, the essential ones, particularly that one which causes the substance to be as it is" (Eco, 1984: 57). The difficulty confronted by those who undertook this work in the Western tradition, Eco noted, is to distinguish essence from accident: "who really knows which differentiae are strictly essential or specific?" (67). Acquinas provided a scholastic solution to this dilemma: "Essential differences cannot be known directly by us; we know (we infer) them by semiotic means, through the effects (accidents) they produce, and these accidents are the sign of their unknowable cause" (67).

This Medieval oxymoron of the "essential accident" evokes the choral reasoning by which Miautre responds to the challenge of the spectacle. Choragraphy proposes (borrowing Basho's Haiku formula) *to know the general accident by means of the general accident.* The phrase is not tautological but homonymic. The question concerns whether it is possible to understand the problem of a potential universal disaster by means of a category formed by (ontologically) irrelevant qualities of a place. It is a matter of a different way of moving between the particular and the general. In *Topics* Aristotle explicitly excluded homonyms from his practice, since puns undermined the carefully crafted distinctions among things constructed by topics and concepts. The consultant is challenged to apply to the problem site the operations of a fourth kind of inference – "conduction" – distinct from deduction, induction and abduction. "Conduction" carries inference outside the unknown into invention, as demonstrated in the following study.

Experiment

The photograph entitled "Crossroads" is an emblem testifying to the revelation experienced by the FRE through the Miami River. It evokes Miami as an image category named "Miautre," a virtual city. The goal of the consultation process was to learn how to bring the city into an internet practice that would make its mental reality accessible to democratic revision: "Miami" as a "transapparent," collectively-authored, dynamic poiesis, a poetics bringing to bear arts and letters learning on public policy. This book details what has been learned so far: how to take the choral measure of the place. The book is composed as a systematic unpacking (unfolding) of one photograph – "Crossroads" – promoted from its anonymous condition of one still among the hundreds taken by Revelle, a passing moment within the hours of video tapes recorded during the drift through the River zone, to become a metonym with categorial power. The book takes a "documentary" approach: it is argued in a literate way, but draws on the bricolage of image categories by including substantial citations not only from Revelle's field work but from theoretical and scholarly sources as well, one purpose of the book being to serve as a resource for the design of other choras in other cities. The categorial power of an image is dependent in part on the digital technology of an internet database, and in part on the rhetorical practice of choragraphy. Miami Virtue stands on its own as a theory and pedagogy, but the proof of what it claims requires electrate, not literate, means.

PART 1

Miami – Image

The past can be seized only as an image which flashes up at the instant when it can be recognized and is never seen again.

WALTER BENJAMIN (*Theses on the Philosophy of History*)

CHAPTER 1

Occasion

A Photograph

This book is a guide to making a categorial image. It is a genealogy of the development of an electrate mode of inquiry – the testimonial. It begins with one image which happens to be a photograph, standing for imaging as such. What makes images electrate is the reasoning practice in which they operate to support learning, understanding, knowledge, decision, and the digital environment that augments their power to organize experience and information into a new system of classification. Why a book? Oral peoples invented literacy; literate people are inventing electracy. We are undertaking a bootstrapping operation. At the gateway of the archive is one photograph, "Crossroads," whose scene will be enhanced (unfolded, enlarged) through the rest of this guide.

Image Emergency

The photograph was taken by Barbara Jo Revelle sometime during her five-week stay at the Miami River Inn, May–June, 1998. She was in Miami working with the Florida Research Ensemble (FRE) on a consultation for the EmerAgency. The point of departure for a consultation is the selection of a problem of concern to the community. The measure of this concern is determined by the attention the issue receives in the news media. The initial problem selected by the FRE was a public relations crisis: the damage done to the *image of Florida* by the murder of a number of tourists. Egents using this guide for their own consultation will select their own instance of an image emergency.

The crisis in Florida's public image was a good test case for the consultation, since it provided a situation that matched the expertise of the FRE with the qualities of the problem. There is a strict rule of cognitive jurisdiction in conventional consulting regarding which discipline knowledge is relevant to which kinds of problems. Arts and Letters disciplines are not considered relevant to most of the problems that beset the lifeworld, especially the kinds of problems that most concern citizens having to do with crime, the economy, the environment, and the like. Despite the fact that trouble in the world is an entangled ecology, literate expertise is specialized, imposing a fatal limitation on problem solving. The holistic nature of choral categories promises a metaphysics more adequate to problem ecologies. To get started, however, the FRE

 | DOI:10.1163/9789004534643_002

used its expertise in imaging to establish its credibility in a case concerning precisely an "image." The EmerAgency is concerned with the civic sphere created through the circulation of information (the imagined community). The purpose of the new consultancy is to develop a practice for the Internet that allows netizens to become agents of a "fifth estate" (egents), giving citizens a voice in the public policy process equivalent at least to that of journalism, whose agenda-setting role in a society of the spectacle has become a matter of concern.

A Sample of the Reports of the Murders that Produced the Image Crisis

TOURIST KILLING HAUNTS MIAMI
Uwe-Wilhelm Rakebrand was cautiously following the rules when he collided with the law of the jungle. His murder has become a symbol of exasperation for a metropolis that can never quite escape its own violent image. The German tourist's shooting death in the car he and his pregnant wife had just rented triggered a volley of suggested remedies for Miami's violent virus. Mayor Xavier Suarez proposed highway checkpoints to look for guns, similar to those used to catch drunken drivers; U.S. Sen. Connie Mack, R-Fla, urged tougher sentences for career criminals; a Gulf Coast-based citizens' group called Stop Turning Out Prisoners said it will place more billboard warnings. And there were calls for gun control from various quarters. But Rakebrand's death probably had little to do with what he did; a lot to do with where he was. Rakebrand was headed toward the lights of Miami Beach's glittering night life when he was caught in Miami's darker reality: it is a place not only of dance clubs, fashion shots and glistening beaches, but also of drug runners and inner-city desperadoes who mix guns and gore with the city's sun, surf and cosmopolitan glamour. Miami Police Chief Calvin Ross had a grim view of future prospects. "Until we come to grips with what that problem is, and keep these individuals off the street, or do something on the very front end with juveniles to ensure that they don't enter into a life of crime, we're still going to continue down the same path." Reruns of *Miami Vice*, a TV series that depicts reasons for both allure and alarm, are popular in Germany. Just the week before, newsstands displayed the international edition of Time magazine's cover story summary on Miami: Glitzy, violent, and muy caliente (very hot), it's taking the world by storm. "I don't know that there's a rational explanation for such an irrational and

heinous act," said Gary Stogner of the state's division of tourism. Besides lamenting a tragedy, officials are also casting worried eyes on the new wave of negative publicity for a state, and metropolitan area, built around image. (Dan Sewell, *The Gainesville Sun*, September 12, 1993)

What Public Agency Bears Responsibility for the Problem?

Although the EmerAgency operates "without portfolio," its work is nonetheless addressed to an existing agency with responsibility for the problem, whether that agency is part of government, business, a non-profit or volunteer organization. Any practical consequences of the report, however, are likely to depend on the response of public opinion. In the FRE case, the initial responsible party was the state government of Florida. Over a fourteen month period during the early 1990s nine tourists were murdered in Florida. Official reminders that this number was out of 41 million visitors, or that crimes against tourists made up just 3.2 percent of all crimes committed in Florida, did little to address what one official called a media-induced perceived tourist-safety crisis. The Commerce Secretary suspended tourist advertising temporarily, while countering negative news coverage. In the tourism business, he said, image is the business and has a direct bearing on the health of the industry. "We have a huge perception problem, but nonetheless a real problem." A number of steps were taken to treat the symptoms of the problem, such as posting armed guards at highway rest stops, and changing the license plates on rental cars to make them less conspicuous.

What Does It Mean to Be Safe? Here Is a State of Mind to Notice, This Feeling of Existential Safety, or Lack of It

The part of the policy formulation that first caught the attention of the FRE was the government's decision to use advertising agencies to treat the damage to the state's image. In 1991 the legislature created the Florida Tourism Commission (FTC) charged with devising a strategy for promoting tourism. One of the first acts of the FTC was to hire the New York consulting firm of Penn and Schoen which, for a fee of $250,000, agreed to assess what role the state should play in tourism promotion. The Florida Research Ensemble accepted the same mandate "without portfolio." Florida already attracted more than seventy million visitors each year. Nonetheless, the local tourism boards that were set up in the wake of the creation of the FTC posed as the primary question not the quality of the "Florida experience" but how and when to advertise and how to get the attention we need. A related question concerned what facilities and resources may be magnets for visitors. The State Commerce Department commissioned a campaign in which an ad agency monitored the weather in northern cities. After a blizzard in a given city, the agency placed ads in the newspaper. In the winter of 1992–1993 the ad depicted a piece of nicely browned toast, accompanied by the text: "just a reminder that it's nice and toasty in Florida."

In 1996 the legislature dissolved the Department of Commerce and its Division of Tourism, and the Governor created Visit Florida, a public-private marketing agency, that took over the $17 million budget raised by a two dollar rental car tax. For $800,000 the Fahlgren Advertising Agency created a TV spot designed to trigger memories for viewers in Chicago, New York, and Atlanta of the carefree Sunshine State vacations of their childhood. Amid colorful sun, sand, and water scenes, the commercials showed black-and-white shots that sought to stir memories of singing in the car, learning to fish, and seeing the ocean for the first time. "We ask them to reacquaint themselves with that memorable childhood experience," Austin Mott, president of the Florida Tourism Industry Marketing Corporation was quoted as saying in one newspaper account, "and incorporate it into their current lives." The ads ran on network shows and CNN, while similar print ads were placed in magazines such as Family Circle, Mature Outlook, Bon Appetit, and Conde Nast Traveler. Television time and print space for the initial round of ads cost $1.5 million. In hindsight the policy to address the perception apart from the actuality of the place seems misguided. In 1999 tourism again was in decline, due to the destruction of the Florida "ambiance" both culturally and environmentally.

News Coverage of the Problem and Related Issues Sets the Agenda

Once the problem is selected, it is treated as a motif in order to note the broader pattern it makes, or the context it evokes. News items are useful for determining which elements are within the public sphere, and for providing the point of orientation for the consultation by establishing the commodified voice and stance of the "fourth estate." One such item in particular formed a pattern when juxtaposed with the Rakebrand murder, and moved the FRE to focus its research more precisely. "MIAMI – Two boats believed carrying as many as 43 Haitians in an apparent bid to smuggle them into the United States sank Saturday off the South Florida Coast, and rescuers found only three survivors. The Coast Guard was alerted at 2 a.m. by a freighter after crew members heard screams from the water about 30 miles east of West Palm Beach." Scarcely a week goes by without such a story appearing in a Florida newspaper. There are Haitians appearing on our shores in the care of smugglers, jumping from ships, being rescued at sea, wading ashore, washing up drowned.

The symmetry of the cases – the deaths of a German tourist and a Haitian alien – gave initial form to the plans for the consultation. There is a Miami such that Uwe-Wilhelm Rakebrand, accompanied by his pregnant wife, set out from Germany to experience it. "Meanwhile" (the temporality of nation-building supported by the novel) Joel Estiverne (the only victim named in the follow-up stories about the nightscreams incident) boarded a smuggler's boat and was lost at sea. Samedi Florvil, with the Haitian Refugee Center in Miami, commented that the illegal immigrants often "pay $2,000 or $3,000 to get on a boat. It is to spend money to buy misery." The first point of interest, in addition to the question of the nature of this attractive power, was that Rakebrand's death precipitated an image crisis while Estiverne's death was ignored.

Uwe-Wilhelm Rakebrand. Joel Estiverne. They Were Not Safe, and Only Spent Money to Buy Misery. Screams in the Night. The Image Crisis Is Worse Than It Appears

The Cuban boat people have played a larger role in the history of Miami than have the Haitians. The "wet-foot, dry-foot" policy makes for colorful drama on the evening news. There was the case of the net-cam set up to monitor the site of the excavation for the condominium that unearthed the Tequesta burial grounds with its series of puzzling holes. More immediately relevant was the TV coverage of the Coast Guard attempts to enforce State and Federal immigration laws, that required them to intercept the six Cubans who leaped from their small boat and tried to swim to the nearby Florida shore. If picked up in the water the Cubans automatically would be repatriated. If they reached shore, they would be granted a hearing to determine their eligibility for political asylum. The live helicopter coverage of the drama brought out a large contingent of Cuban-Americans (and others) to the site of the incident in progress, resulting in a suspension of the laws in this instance (if not a change in policy). The Haitian community cannot muster the same level of support for its boat people, and this condition made their circumstances exemplary. The ultimate case revealing the politics of public policy is that of Elian Gonzales, which will go unmentioned here.

The Power of Attraction Presumes a Power of Revulsion. Indifference Is Also a Power. What about Ignorance?

The FRE research began to focus on the Haitian attraction to Miami as a neglected dimension of the image crisis. Our method assumes a polarity of attraction-repulsion. It is difficult to understand an attraction without taking into account the negative pole. The following story again sharpened the focus of the inquiry.

HAITIANS JUMP SHIP IN MIAMI

MIAMI – More than 150 undocumented Haitians jumped from a freighter into the Miami River Tuesday and were chased into and around riverfront businesses by U.S. immigration agents and police. "We had a lot of policemen running in and out with their guns drawn," said Este Garcia, owner of Garcia's Seafood Grill. "They were in the restaurant, jumping into the water. It was a madhouse. A couple of them hid in trash cans." At least 86 people were captured and many were ordered to sit against the wall of a warehouse near the river, which is frequented by small freighters from the Caribbean and the Bahamas. Forty teen-ager's were among the Haitians who jumped from the green-and-white freighter named Rose-Marie Express as it pulled into dock, police said. Some of the passengers ducked into waiting cars and escaped while others hid under the docks. "At first I thought it was gangs because I just saw kids running and Immigration with guns. I think it went a little overboard. They were pushing them and roughing them up," said Garcia's waitress Lisette Rivera. A U.S. Customs agent raced to the dock after receiving a call that a ship might be off-loading drugs, said U.S. Customs assistant special agent in charge Mark Bastan. Garcia said some immigrants were stumbling around as they tried to escape. "They were disoriented, they've been out on the water for days." The freighter set out from Cap Haitien on Haiti's northwest coast about four days ago, according to a volunteer from the Haitian Refugee Center. The U.S. Coast Guard in July began banning wooden freighters such as the Rose-Marie Express from entering the Miami River because they are not considered safe. (Perez, 1998)

Several Haitians Who Departed from the Unsafe Rose-Marie Express *upon Its Arrival at the Miami River Hid in Trash Cans*

The final selection for the site visit, the field work – the Miami River – was determined the same way that many screenplays begin. "Stories usually emerge from fragmentary ideas: a colorful character, a poignant or offbeat relationship, a childhood incident, a physical object, or a problem in your own family. Dozens of successful motion pictures originated in the pages of the daily newspaper. Remember that newspaper items seldom contain all of a screenplay's elements. You will have to find some ingredient in the news item that excites you, that stimulates your imagination, and then build your story on that. It's a game that professionals call 'what if'" (Armer, 1988: 17). The *Rose-Marie Express* (green and white, freshly painted, wooden) made it safely into the harbor of the Miami River all the way from Cap Haitian. The safe arrival of this boat pinpointed the location of the FRE inquiry into the *Florida Image Emergency* on the Miami River.

PHOTOGRAPH: BARBARA JO REVELLE

The Florida Research Ensemble (Introductions)

William Tilson (Architecture) on "Denizens' Map"

The denizen map was imagined as a "blow-up" of the street map insert of downtown Miami in the Michelin Green Guide, whose accompanying text hides everything in the word "atmosphere."

> A vibrant 1.5 sq mi surrounded on three sides by the warm waters of Biscayne Bay and the Miami River. Miami's downtown exudes the bustling atmosphere of a Latin city. Here retailers purveying electronic goods, jewelry, clothing and fragrances in Spanish and Portuguese draw visitors from Cuba, Puerto Rico and Latin America. The site of the city's first wooden commercial structures now features a dense array of government and office buildings in a host of architectural styles including Neoclassical, Art Deco, Mediterranean Revival and the stark contemporary design of the skyscrapers that illuminate Miami's night skyline. (Michelin, 1996: 204)

Ironically, the atmosphere of the river is practically unknowable to the general public except through representations supplied by institutions of security

PHOTOGRAPH: BARBARA JO REVELLE

and surveillance, the tourism and entertainment industry, the chamber of commerce and the news media. The denizen map stands in marked contrast to the river one can imagine from the Green Guide. The purpose of the map is to expand this "narrow band" information landscape through direct contact with the river zone and its denizens.

The Miami River is called a "working river" based on the predominance of water-related businesses and activities found along its banks. While this is an accurate description to a degree, the phrase is usually employed as a euphemism aimed at camouflaging or qualifying its nefarious reputation as the underworld of Miami. The river denizens, Haitian boat owners stranded in their vessels by the "Caribbean Code" maritime regulations, unemployed seamen living in abandoned buildings, "outsider" artists and smugglers, haunt the interstices of this world.

Contemporary maps, data graphs and aerial photographs provide some useful information about these conditions, but in the end they simply extend normative planning strategies that view the river zone through a picturesque lens. Current proposals for improving the public image of the river involve "before and after" drawings that fix social and environmental problems, i.e., Illegal aliens, contraband, abandoned buildings, the homeless, pollution, by recomposing them out of the scene. Rather than inventory the presence or absence of traditional typologies of the river zone such as housing, shops, places of work, recreation and public monuments and illustrate them in picturesque scenes, the densely layered construction of the denizen map accumulates evidence of the interaction between the real and the imaginary and attempts to present them poetically.

In spatial terms, the denizen map re-presents the dynamic and co-extensive relationship between the smooth space of the sea and the striated space of the city. The conditions of the river zone demonstrate an urban place where this sponge-like space exists as constantly shifting degrees of interpenetrating social and physical realities. The river is an exemplar of Deleuze and Guattari's position that "the principles of the mixture are not at all symmetrical" (Deleuze and Guattari, 474) – everywhere there is leakage and contamination of one kind of space into the other. Like the Haitian vessels listing at port along the river leaking bilge and taking on water (the photographic marginalia at the top and bottom of the map shows the inside-outside appearance of this world), the unchecked pollution of the city seeps into the river. Likewise, the diagonal cut of the river changes the complexion of Miami by creating contaminations in the efficient order of the grid.

The map addresses the graphic problem of imaging the interplay of smooth and striated space. For example, the black background of the map is etched with a tangled mat of contrasting lines of flight intersecting, competing for attention and canceling each other out but remaining on the threshold of legibility. The visual material for the resulting constellations is sampled from airport landing diagrams, vodun *loa* symbols, coastal navigation charts, manatee movement patterns, drug interdiction graphs, and archeological sites, which coexist in the map as they do in the zone. Images of denizens we met while walking in the zone are stuck in this matt like they are *in situ*. Their iconographic format recalls both tourist and war maps that locate events and landmarks in this fashion and Internet buttons that invite further exploration. The construction of the denizen map is consciously aligned with these genres in order to attract the viewer into the strange world of the river through familiar images. These associations are also chosen because the image is a prototype for an interactive cognitive map within the Miami-Miautre website (a public extension of the FRE's "problems be us" philosophy).

The process of mapping the river as a field of vectors and itineraries – as much haptic as visual – is essentially a choragraphic activity. The identity of place constructed through the ninety-odd Photoshop® layers of the denizen map does not depict the Miami River as a paradise that can be reclaimed by cleaning up the image. What emerges is a sieve-like network of place-making possibilities floating and drifting within the striated order of the city grid. A neologism playing on the Ptolemaic form of mapping called chorography, *choragraphy* exposes our disciplinary methods and ideological positions as much as it documents the physical reality of the river. If chorography (with an "o") "relocates us on the surface of the earth" (Cosgrove 107), choragraphy attunes our interior life with the concrete reality of place.

The practice of choragraphy, of which the denizen map is an example, is a playful process that maps the relationship between a state of mind and a state of place. Like montage, in which new meanings arise out of the collision of unrelated images, choragraphy represents places only as they are provisionally occupied and invested by particular subjectivities. *Chora* is place constructed through experience and action as opposed to the empty neutral space of *topos* into which ideas are poured. The denizen map records the energy of the *chora* rendering it visible as a cultural construction: a process, which according to Julia Kristeva, must precede any serious discourse on place. Using imaging to transduce Miami into Miautre produces chora–a site that manifests the ordering practices of a society.

John Craig Freeman (Digital Art), on "Imaging the Miami River"

We applied what we knew about the transition from oral to literate civilization to develop new uses of interactive electronic media, which would adequately respond to the transition from literacy to electracy. This transition includes a shift from a politics of friendship (Derrida) – Mi-ami as Creole "my friend" – to a politics of hospitality (collective relations not of friend-enemy but host-guest). The Haitian shippers' approach to understanding the world stood in stark contrast to the rationalist Western approach which was so dramatically embodied on the Miami River by the Coast Guard. We came to believe that to understand the Miami River it was necessary to approach it with an open mind that could incorporate both paradigms of reason.

A summer storm had just cleared as I drove into downtown Miami on I95 through Overtown. The water of Biscayne Bay was a clear turquoise green. The water of the Miami River was not so clear. I once saw manatee on this part of the river but not this day.

Just last week refugees scrambled for freedom in South Florida, jam-packed on a freighter. They sailed right up the Miami River. Freighters like that are part of the commercial shipping industry on the Miami River that creates thousands of jobs every year (Local TV News).

We collected a set of tax appraisal aerial photographs, shot by the United States Geological Survey, from Dade County. There were five tiles, which when laid out diagonally mapped the river from its source at the Miami Canal, 5.2 miles Southeast to its mouth at Biscayne Bay. Barbara Jo had already been living at the Miami River Inn for several weeks before I came down to do the VR fieldwork. The rain was no reprieve from the relentless June heat and humidity.

But along with its economic growth, the River is plagued with some problems.

Most divination systems have a framing device to help define what is to be included and what will be left out and then a chance component. The aerial images not only provided a framing device; they also provided a logical way to organize the space and the material we produced. We came to refer to each of the areas defined by the five aerial photographs as Divination Zones. The chance component became relevant once we were at the site. Subjects of investigation were allowed to emerge from anywhere throughout the zones but not beyond.

Barbara Jo had arranged to meet Simon Lubin, a 38-year-old captain of an impounded Haitian shipping vessel. He had been stuck on the river for about a year after the Coast Guard had cited him for multiple Caribbean Code safety violations. The Coast Guard would not let him go back to Haiti until he made the necessary repairs to his boat. He could not make the necessary repairs to his boat if he could not return to Haiti to sell his cargo. All the while he was accumulating an astronomical docking fee debt. He had wanted to write a letter of his plight to the Miami Herald, but his English skills were not sufficient. Barbara Jo had agreed to write the letter for him. This case provided the focus for our consultation on a public policy issue.

I think it's beautiful. I love sitting here by the river. But, what I'm looking at beyond the water, is disgusting to me (Concerned Resident).

I began building interactive computer interfaces, which would guide the user from global perspective to documentary virtual reality scenes on the ground. In a practical way, this allowed us to tell the complex stories of the River with the non-linear structure that the theory of electracy predicted. It also worked in a metaphoric way to connect the abstraction of theory with vernacular experiences of people on the ground. We drove across the swing bridge, under which the old bull alligator was rumored to live. All of this was once Everglades.

As disgusting as it is on the surface, underneath the historic Miami River, a dumping ground, made up of batteries, tiers, pipes, furniture and mounds of unrecognizable garbage.

I began to treat the camera as a kind of divining rod. The virtual reality scenes were produced by setting up a tripod, and shooting 360° panoramic nodes. Each node was connected by walking the camera in straight lines from node to node. These panoramas provided the ground for transducing chorography into choragraphy. The practice of moving through the zone guided by

PHOTOGRAPH: BARBARA JO REVELLE

an intuitive pattern began to resemble another ancient practice related to the Chinese divination, Feng Shui. I hate to even use the term, because it has been so bastardized by American New Age culture. The virtue of this context, at the same time, is that it underwrites the practicality of divination as an interface for our virtual civic sphere.

We boarded Simon Lubin's boat over a patchwork plywood dock. The boat and the dock were filled with cargo – old mattresses, bikes, plastic buckets and used clothes. The vessel was painted white and blue on the outside and bright red in the cabin with a painting of La Sirene, the Haitian Voodoo mermaid goddess of the sea, on its side.

Along with the smuggling of illegal immigrants and drugs onto the Miami River, Coast Guard officials say Haitian vessels have illegally discharged oil, bilge water, even human waste, into one of the few natural rivers on the Southeastern side of the United States.

We saw that in addition to superstitious misrepresentations and pseudo-science, divination involved a practical, hypermedia epistemology. Although there was an interior design component to the practice of Feng Shui in ancient China, most of it had to do with a study of the landscape and its relationship to the cosmos. It held that there was an underlying geometry to the landscape made up of power centers or crossroad vortices connected by straight lay lines. We wanted to extrapolate from the structure of the *I Ching*, in which each hexagram

PHOTOGRAPH: BARBARA JO REVELLE

represents an archetypal situation–from the ability of a querent to apply this formation to any possible life experience through an intuition of the mood of this situation–to choragraphy as reasoning with categorial images. The Miami River as chora is a hexagram in a potential online global divination system: not the book of changes but the internet of changes, based on patterns discovered through close observation of the contemporary natural and social worlds.

Barbara Jo Revelle (Creative Photography), on "The Pool"

Ron is Barbara's partner. He visited her numerous times during her stay on the river. This conversation took place in June, 2002.

> Ron: So this is the photograph, of all the hundreds of images you took, which you say now is the quintessential one? The one that most fully characterizes the Miami River experience for you? Five weeks just wandering around, talking to homeless people, drug dealers, the coast guard, the Haitian shippers who worked on the river? This torn-down-apartment one is not very flashy.
>
> Barbara: It's the one I'm the most attracted to now, four years later, because my burning question has changed. I took it that day we were at the site of the apartments undergoing demolition at the mouth of the river, so they could expand the Sheraton Hotel. I think its the drained

swimming pool that really gets me – with the river behind it. That's what I *recognize*. And across the river the big new hotels.

Ron: A lot has happened in four years, I mean in your (in our) life.

Barbara: Yeah, like my father died – in a shocking way. I have to come to terms with it without benefit of the belief in an afterlife. It might be better to use other words: hieroglyphics? constellations? Dowsing comes to mind. Since my father lived without much religious consolation either, I think the whole idea of death was, for him, an obscene mystery. You know how big my dad was on controlling things. He was always the great athlete, the Olympic high jumper, the champion swimmer, fancy diver, extreme skier, big game hunter, all that hero stuff. So I was looking through all those photos – Cuban workers, Haitian workers, boats, mattresses and bicycles piled on boats, but as soon as I saw that one with the drained swimming pool it gave me a plan. I just saw myself going up to my parent's home, finding this old piece of 8 mm footage my mom shot sometime in the 50's, and the feeling that when I saw it, I would understand something. I call it the "treading water film."

Ron: All this is from Ulmer's "mystory" system, right?

Barbara: Yes. We agreed to treat the river as an image, to discover its obtuse or third meaning (from Roland Barthes). You have to remember that, in my mystory, the Miami River equals the lake I grew up on – Pleasant Lake. Doing the mystory, I remembered how my father forbid me to cross to the other side of Pleasant Lake, the shanty town poor side of the lake which was, according to him, dangerous. He called it Skeeversville. But I did secretly cross to the other side of the lake, all the time.

Ron: You and your friends played in Linda McDonald's basement? In her drunken Dad's upholstery shop. Games with boys and blindfolds and getting naked, right? So that equals the dark dangerous forbidden river – Miami Vice and all that– where there are mostly black men and they carry guns and you can't understand things –they're speaking Creole and doing voodoo. And the Coast Guard's safety rules are like your father's control. And the swimming pool?

Barbara: Near our side of the lake we had a community swimming pool. The specific memory is of watching this home movie footage made by my mother, who was always an obsessive documentarian of our lives. My parents made their living as lifeguards at this public pool in the summers. I was just little, eight or nine. The actual six or seven minute clip is of my father at the pool performing in a nighttime diving show. He was the "clown diver." In the film he staggers to the end of the diving board. It's dark out and the shot is lit by green light. Daddy is waving a bottle of beer in an exaggerated, drunken way and wearing red long-johns and an old floppy granny hat pulled down to mask his identity. He pushes the straight man, Billy Gillard, a former Olympic diver, who has just done a breathtaking two and a half gainer in the preceding shot, off the diving board. Then Daddy mimes that he can do a triple flip better than Billy. He holds up three fingers. Pan to the crowd, laughing, booing, going wild. Then I remember a shot of him stumbling, falling almost off the board. He pulls himself back on, struggles up, takes a huge bounce, flails through the air all twisted, lands with a big splash, goes under, and does not come up. My brother and I knew that he's hiding in the dark end of the pool, treading water under an inverted bucket where air for him to breathe has been carefully trapped. Several minutes go by. Finally many lifeguards dive in to search for the drunk drowning victim. Its part of the show. There's a brief unsuccessful search. All the lifeguards come out of the pool, doing that arms out-stretched, palms-up "I don't know" gesture. Then Billy shrugs and says: "Oh well, he was just an old drunk anyway."

Ron: I thought it was 8 mm. How can there be sound?

Barbara: The film is silent, but I remember what Billy was saying. And in spite of the crowd's incredulity, the fancy diving show just goes on. My father, meanwhile, has shed his long-johns, hid them in the bucket, and come out of the pool with the group of lifeguards, undetected.

Ron: You and Jackie are thrilled to know the secret – that this "drunk' is not really drowned, but is your larger-than-life father? The guy who beats death?

Barbara: Something like that. The hero. Like he tried to beat death this last time by ignoring his diagnosis and going off to Africa on one more big game hunting adventure. Remember how we found all those eviction letters and family photos and children's toys in the debris? All the things the people left behind? Thinking of that loss, everything that tells you where you are in your own life story slips away. Just briefly you have thoughts with no sense of who is having them; you remember there is death but that this moment now is not your death, yet.

Ron: They can drain your swimming pool and kick you out of your apartment in order to make a bigger hotel, and you have to leave all these precious things behind, but you're still alive.

Barbara: Until you're not. Until you swim up under the bucket for air and – surprise! your careful treading-water plan has gone awry. There's been some mistake and there's *no air*.

Method (from Revelle's Journal)

Two days ago I filmed Winston Murdock (the Jamaican I met in El Capitan Restaurant) while standing on what is left of his cargo boat – burned mysteriously in what he said was "some kind of foul smelling play, you can know that for the truth." Winston mentioned a Haitian boat owner whose boat, Lady Bernadette, was seized and held on the river for three months by the Coast Guard, declared "derelict and unseaworthy," then finally sent back to Haiti empty. Its intended cargo was left in a riverside warehouse that was eventually broken into by thieves. Last night I dreamed about going into some kind of boarded-up warehouse on the river with crates and bundles of old clothes and trash, but also gigantic, hypnotic floats like in some Fellini film –the same floats with all the Snow White and Speedy Kilowatt kinds of characters I photographed across from the 17th St. Bridge shipyard last week. Inside the dream warehouse were Haitians speaking Creole and doing voodoo ceremonies, and two dachshunds so skinny and dehydrated that when I put down a bowl of water they sank their heads all the way under the water to drink like some kind of fish or manatee. People say there are manatee in the river. I keep thinking I see one – a manatee – but then it turns out to be a submerged paper bag. The dachshunds kept melting into other kinds of creatures, which makes sense because the heat was so bad yesterday that the headlines read: "Miami Melts."

Because of this dream, today I went looking for the warehouse that Winston told me about – the one broken into after the boat was sent back to Haiti. I met and filmed a glassy-eyed black woman who was moving like she was in a trance, looking through the piles of ripped-open boxes, broken-open bales of discarded clothes, old mattresses, bicycles, refrigerators, bags of onions, and other undifferentiated debris, all smelling of shit and rot. She said she comes here every day from Little Haiti, where she now lives, trying to find something she tried to send to her family in Haiti. After that, because of reading about the tent city in Jose Marti park in a book on the history of the river – how it had been the site of a "tent city" when Castro sent over a boat full of criminals and insane asylum inmates – I went to that park and just sat by the river until someone came up to talk to me. I met Frank Paone, a taxi driver born in Italy, who was walking his small, sick dog. He explained that a vet had predicted Lacey would die soon from Parvo. Frank makes "architectures" on his "commode" so I went with him to the small room he rents to film his constructions.

Gregory L. Ulmer (Media Theory), "Abstract"

The photograph entitled "Crossroads" is part of a picture theory developed by the Florida Research Ensemble (a group of colleagues teaching at the University of Florida) during the development of a prototype for applying arts practices to public problem solving as part of a deconstruction of the institution of consulting. The Miami River, a 5.5 mile channel through the heart of the city of Miami connecting the Everglades with Biscayne Bay, is administered by some 34 public agencies, addressing every conceivable policy issue. The image shows a vessel being loaded with cargo intended for Haiti, consisting mostly of used goods–mattresses, bicycles, ten-gallon plastic buckets. The ship has been impounded by the Coast Guard because it did not meet the safety standards imposed by the "Caribbean Code." This picture theory functions as a categorial image, exemplifying the logic of a new apparatus–electracy–that does for digital imaging what literacy did for manuscript and print technology.

Miami Virtue is an experiment involving the following elements:

1. The Internet.
   The Internet has been defined by Paul Virilio as the means of a potential "general accident"–a global catastrophe taking place everywhere simultaneously. The internet is an institutionalization of the technology that has produced our "society of the spectacle," in which the image is said to have destroyed the civic sphere existing within the cities of modern nation states.
2. The EmerAgency.
   A virtual, distributed, online consultancy, initiated at the University of Florida in the 1990s. The purpose of this conceptual agency is to coordinate the formation of the civic spectacle as a "fifth estate" that would allow citizens direct participation in public policy formation by means of the Internet.
3. Choragraphy.
   This experiment tests a rhetoric of hypermedia formulated by Gregory Ulmer in *Heuretics: The Logic of Invention* (1994). The rhetoric adapts for electracy the mnemotechniques central to pedagogy in the pre-print era of literacy. It is a hybrid of inventional memory palaces and chorological cartography in which plastic arts methods were used to map regional locales. The neologism ("choragraphy" with an "a") signals the genesis of the rhetoric out of the encounter of architecture with deconstruction in the collaboration of Jacques Derrida and Peter Eisenman on a folie for the Parc de la Vilette in Paris. In this prototype the Miami River is the basis for the design of an online "chora"–a holistic category made operational within digital imaging.

4. Testimonial.
The proposed virtual consulting practice. The creative photographer Barbara Jo Revelle lived at an Inn on the Miami River for five weeks in order to attune the site–to discover the mood or atmosphere of the River zone. "Attunement" alludes to the tradition of "Stimmung" from Plato's *Timaeus* (the dialogue that introduced "chora" as the space of generation) to Heidegger's phenomenology. Revelle's mapping of the zone drew upon the poetic encounter perfected by modernist poets in Paris from Baudelaire's tableaux through Breton's NADJA and Walter Benjamin's dialectical flaneur to the drifting of the Situationists. The new element is that the modernist epiphany is literalized in the photographic image. "Crossroads" evokes an inner experience of recognition in the consultant by means of an external detail of the scene. The consultant witnesses the catastrophe of the zone, in an image that locates a virtual border or boundary–logical, psychological, ideological.
5. Interface Metaphor.
The ultimate goal of the consultation is to design a website that supports a virtual witnessing of the Miami zone (Miautre), and that is portable to zones in other communities. This interface between public life and the Internet supportive of community witnessing of policy negotiations contributes to the formation of a critical civic sphere within the society of the spectacle.

CHAPTER 2

Place

Setting

The "Crossroads" photograph was taken on the banks of the Miami River. This region or river port is represented by a sampling of documents, selected to provide a sense of the place as it exists in the discourses of history, tourism, and policy formation. The intelligence of chora as image category functions through atmosphere, so even in this theoretical account it is important to accumulate the material details. It may happen one day that this image of the Miami River and its database will become a "mirror world." Mirror worlds "are software models of some chunk of reality, some piece of the real world going on outside your window. Oceans of information pour endlessly into the model (through a vast maze of software pipes and hoses: so much information that the model can mimic the reality's every move, moment by moment" (Gelernter, 3). "So Mirror Worlds function in part as fire walls opposing the onslaught of chaos. But they aren't mere fire breaks. They are beer halls and grand piazzas, natural gathering places for information hunters and insight searchers. Most important, they are microcosms – intricate worlds come alive in small packages. Whether in the shape of a Victorian winter garden, an electric train layout, a Joseph Cornell shadow-box or a mere three-inch plastic dome with snowflakes softly settling inside, microcosms are intriguing. They show you patterns and help you make discoveries that you'd never have come across otherwise. At their best, they are thought-tools of great power and evocativeness" (7).

The Miami River scenography is designed as a microcosm, perhaps as a mise-en-abyme of all possible crossroads, nor is it only a matter of, or even is there a necessity for, the technological computing power of Mirror Worlds for this design to become operational. A Mirror World without a practice of reading, reasoning, and writing, is as useless as a library to a pre-literate civilization.

Miautre is a microcosm of Miami, itself a microcosm of global America.

 | DOI:10.1163/9789004534643_003

History: The Miami River, Miami, Florida

Howard Kleinberg, Miami: The Way We Were

Miami's historic river has hidden origin.

"We started in our boats up the river and crossed the falls by towing canoes over it," wrote U.S. Coastal and Geodetic surveyor F. H. Gerdes to his Washington department head on Feb. 5, 1849. Gerdes was writing from what he referred to as the Upper falls of the Miami.

He enclosed maps of the river.

Before civilization's incursion into this area, Florida's Everglades began as close to Biscayne Bay as what now is NW 32nd Avenue in the vicinity of 18th Terrace. At the ridge line, where the Everglades met hammocks of oak trees, the waters from the river of grass dropped several feet in a fall and, through a rapids, fed the approximately 4-mile-long river known as the Miami.

Today the headwaters of the North fork of the Miami River are difficult to find – but they are there – and that which we commonly think of as the river west of NW 24th Avenue really is a man-made canal dredged to drain the Everglades beginning in 1909. The South Fork of the River is all but gone, but was referred to in an 1894 article by Charles Richard Dodge in Scribner's Magazine in this manner:

"The Miami river is one of the principal outlines from the glades on the east coast, and through a sluggish stream at its mouth, it tumbles over the coral rock near its source in splendid rapids against which a boat is dragged, not rowed, with difficulty. We entered the glades by the south fork of the Miami, as beautiful a stream as ever flowed through an unbroken wilderness, the trees in places almost arching the water, its banks clothed with strange vegetation and stranger flowers, the bottom presenting a kaleidoscopic picture of many-colored grasses and aquatic vegetation."

The river has a long recorded history. Juan Ponce de Leon may have drawn drinking water from it in 1513 when he visited Biscayne Bay and Pedro Menendez de Aviles, the conqueror of Florida, recorded in 1565 that a very large lagoon called Maymi (Lake Okeechobee) emptied itself into the country of the Tequesta which would have been in the area of the Miami River (124).

Stream still trickles where real river ran.

The draining of the Everglades in the first and second decades of this century all but eliminated what was the real Miami River west of NW

24th Avenue, as dredges dug a different course toward lake Okeechobee. All but lost to Miamians' eyes are the headwaters of the river, once flowing from falls in the Everglades – which were in the vicinity of NW 32nd Avenue – and churning as a rapids to bring excess waters into Biscayne Bay. But the headwaters do exist.

Covered by heavy underbrush and trees and reduced to a trickle fed either by storm drains or underground seepage, the little river still flows ever so slightly. Her banks are half her original width. Fill and debris mar her once natural beauty – described by surveyor F. H. Gerdes in 1849 as being "wild country, woods, hammocks, seemingly impenetrable, interwoven among themselves like spider webs, where in all probability, never human foot trod."

A small City of Miami park – Paradise Point – which is not even listed among city parks in the telephone directory, fills the distance between NW South River Drive and the original riverbed to the south. From the edge of the park, it is several hundred yards west to the base of the old river, just below the firefighter's hall.

The river is unnoticed, except by those who live or work there, the area on the south bank fenced off by an industrial paring lot.... Archeologist Carr and historian Arva Moore Parks, walking just east of the park, know they are on the ground where G. W. Ferguson operated a coonty starch mill in the rapids more than a century ago. Carr laments the condition of what remains of the headwaters of the mother river and professes the desire to see restoration and cleanup of Miami's real roots – the Miami River (128).

∴

Signing Miami's birth certificate.

Julia Tuttle's father, Ephraim Sturtevant, had come to Biscayne Bay from Ohio in 1870. Several years later, he brought here the woman who was to become the Mother of Miami.

Julia Tuttle arrived with her two children and immediately liked the area.

Soon, however, her husband, Frederick Tuttle, became ill in Cleveland and she returned to run his ironworks business. Frederick died in 1886 and in 1891, Julia moved her family here. Rather than live on the property of her late father she purchased a section of land on the north side of the Miami River.

Almost immediately she sought to open the almost virgin area to the outside world. Even before moving from Cleveland to Miami, she had sought, unsuccessfully, to interest Henry Plant's railroad in expanding to

Miami. She then turned to Henry Flagler, who had begun extending his railroad south of St. Augustine. Flagler initially ignored her but continued to build south, getting close to Palm Beach. The hard freeze of 1894–95 turned out to be a Godsend for Tuttle and what was to be Miami. With the crops in the rest of Florida destroyed by the freeze, Tuttle sent Flagler orange blossoms to prove the Biscayne Bay area was unaffected.

Flagler had to see this for himself and first came to Miami in June 1895, doing the last 70 miles by boat, then by mule-drawn cart. Julia Tuttle wound up making Henry Flagler an offer he couldn't refuse. In exchange for Flagler's bringing the railroad to Miami, building a first-rate hotel and platting and building the streets of the town, Tuttle turned over to him half her lands, much of it on the bayshore. Their agreement was drawn up in contract form on Oct. 24, 1895 and signed by both parties. That contract was Miami's birth certificate (32).

∴

"In the Happy Days of Yore on Beautiful Biscayne's Shore," by Col J. W. Evans.

And the moral of this story is that many who wake after the long journey from the icebound north and look on the tropical scenery in and near the progressive, high strung Magic city of Miami feel as he did, that to a great extent it is heaven. In his day it was nature. Now it is nature, energy and enterprise combined. If nature is not crowded too hard enough can be retained to ensure our citizens of health, wealth and happiness, and the Magic City will ever be, if not heaven, yet nigh unto the gate thereof.

If some one reads the foregoing they may want to know what we had here in those days. We had a mail once a week by sail vessel from Key West, a store on each side of the river, a Catholic church, built by Mr. Wagner, an intelligent citizenship of men who had been all over the world and had settled here because of climate and surroundings. We had a noble body of pioneer women. Among those who have crossed the river are Mrs. Hunt, Mrs. Sturtevant, Mrs. Tuttle, who visited here often….

The Indians visited us often. They were in numbers far more than the white people, were always welcome, and among them old Tiger Tail, Tusteneege and Emathla, towering up in such a stately way that the words savage grandeur might well be applied to them.

What did we live on? Turtle, terrapin, venison, bear meat, wild turkey, game birds, manatee, fish, wild hogs – and the meat of these can not be excelled. The Indians raised the choicest beans, pumpkins, sweet potatoes, bananas and chickens. Nearly all the settlers had a small fruit and vegetable patch near. And then, Providence and bad machinery and navigation often sent us a lovely wreck of wine, varied edibles and much to talk about. The population of Miami for a mile north and south was about twenty people" (21).

The Magic City, city of magic. Magic is a classifying system, a mode of knowledge.

Tourist Guides

Insight Guide: Miami, The Everglades & Florida Keys, 4th Edition (London, England: Insight Guides), 2000.

PHOTOGRAPH: BARBARA JO REVELLE

Touring the Area around the Miami River
Downtown Miami

It is a toddler in the family of America's big cities. Some of downtown Miami's older buildings are younger than many of the folks who have come here to retire. But since April 15, 1896, when Henry Flagler brought his East Coast Railway into Miami, literally putting the town on the map for subsequent waves of settlers, the little devil has turned into a spunky, scrappy city. With its notable, excitable Latin edge, downtown feels more like the northern tip of the southern hemisphere than the other way around.

From a distance, especially at night, downtown Miami looks like a throbbing megalopolis, its bank of skyscrapers floodlit and twinkling in colored neon, creating a magic-show skyline quite unlike any other in the United States. Yet for the visitor walking along Biscayne Boulevard, through the central grid of one-way streets, across the Miami River and into the gleaming Brickell Avenue quarter, Miami runs very much on a human scale. (155)

A walker's Delight

For the true heart of downtown, head west from [Bayfront Park] on Flagler Street. The city's geographical center is at the intersection of Flagler and Miami Avenue. The two streets divide Miami-Dade county into the four quadrants of the compass, and it's from here that all of Miami gets its bearings. This is the eye of Miami's urban storm.

PHOTOGRAPH: BARBARA JO REVELLE

Two blocks east of this crossroads is the best place to start a walking tour of downtown Miami's landmarks. Here, at the intersection of Flagler and NE Second Avenue sits a collection of the wonders of the city's brief past. The streamline Moderne architectural gem on the southeast corner, now a Sports Authority athletic goods store, has a curved facade, and a three-story atrium inside. Until 1995 it was a branch of Walgreens. (158)

Little Havana

Havana, Cuba, never looked like this hodge-podge of car dealerships, strip shopping centers, furniture stores, flower shops, run-down 1950s motels and well-kept Mediterranean-style houses. This is Havana, Miami, an immigrants' launching pad, a constantly changing monument to new beginnings and hope. Today, one can find *guayaberas* in Hialeah, dance *salsa* in key Biscayne, sip *ajiaco* in Kendall. The neighborhood's Cuban flavor has become diluted in recent years as Central American refugees have taken over entire blocks. Still, Little Havana remains the symbolic center of South Florida's thriving Cuban community, a place to savor a good meal or a good memory. Very roughly, Little Havana is bounded by the Miami River to the east, 37th Avenue to the west, State road 836 to the north and Coral Way to the south. (167)

River Scenes

To get to Little Havana from Downtown Miami you first have to cross the Miami River. Once a waterway to the Everglades plied by Tequesta Indians in cypress canoes, today's Miami River is a busy, gritty shipping center. Spilling on to its banks are boatyards, marinas, terminals and fisheries.

One doesn't need a boat to appreciate the river. Try visiting the public parks that line its banks. Most outstanding is Jose Marti Riverfront Park. On the eastern edge of Little Havana, a vision of pink stucco buildings and red-brick walks sprawled along the river's south bank. Built on what was once a Tequesta Indian settlement, the site became known in 1980 as Tent city, a temporary shelter for homeless Mariel refugees. In 1985, it became a park dedicated to Jose Marti, the Cuban writer and philosopher, and leader of Cuba's independence struggle against Spain. The park includes a pool, a fountain, a bust of Jose Marti donated by the Cuban government in 1952 and a statue of the late South Florida congressman Claude Pepper. (168)

Haitians in Miami

In other US cities, where Mexican-Americans and/or Puerto Ricans comprise the bulk of the Latin population, there is an emphasis on issues that are typical of a "minority" agenda: public housing, access to education and public services, employment, etc. The "minority" worldview predominates – the view that the dominant society discriminates against members of the minority and that there are entrenched barriers to the advancement of racial and ethnic groups. These are concerns and views shared with African-Americans and are an expected response to many decades and generations of discrimination.

Miami's Latins, however, are recent immigrants. They have not accumulated the experiences that would lead them to think like members of minority groups. They came here because they believe this place is a better place than where they came from. Their view of US society is a positive one: a land of opportunity and political freedom (60).

An immigrant orientation is one of the factors that differentiates foreign-born blacks in Miami from native African-Americans. The Haitians are the most numerous of the foreign black community, numbering in excess of 200,000. Little Haiti is perched on the southeastern territorial limits of Liberty City.

Although Haitians, as blacks, have been subjected to much prejudice and discrimination, especially in negative treatment by the US Immigration and Naturalization Service, they still share many patterns with immigrants in Miami that are distinct from African-Americans. The most evident is the importance given to self-employment and entrepreneurship. The business district of Little Haiti exemplifies this. During the 1970s the area became predominantly black and during the 1980s those blacks were increasingly Haitian. The neighborhood began to appear neglected. Previously crowded stores were empty and streets littered. Tumbledown houses with absentee owners showed the wear of years without care. In the last half of the 1980s, however, change unfolded. Little Haiti began to take shape as an enclave of Caribbean culture. The area's vivid tropical colors and community pride are now most evident in the folk murals seen throughout Little Haiti, especially along NE Second Avenue and NE 54th Street.

The Haitian community leaders are both businessmen and political organizers. The businessmen are energetic and optimistic. They envision Little Haiti becoming another Little Havana or San Francisco-style Chinatown. It is hoped that the only Little Haiti in the US will become a cultural and tourist attraction based on the Haitians' drive and enthusiasm and their unique cultural attributes – world-renowned painting, wood crafts, music, French-inspired cuisine, architecture, and numerous skilled trades....

Miami's Haitians are one of the most persecuted and suffering of immigrant groups in the US. While Cubans have always been encouraged to come to Miami and have received substantial government assistance, Haitians have always been rejected. The government has labeled them as economic refugees, in contrast to the Cuban political refugees.

In reality, the conditions and motivations impelling both the Haitian and Cuban flows are similar: underdevelopment, desires to improve oneself and political repression. For reasons of US foreign policy and racism, Haitians have not been welcomed into Miami.

But Haitians have resisted US governmental efforts to send them back to their island, and with the assistance of organizations and lawyers they have achieved partial victories. Few have been deported, and a community has gradually emerged. For a while they were identified as one of the main sources of Aids. Subsequent research revealed that Haitians were not a primary at-risk group and that Aids was probably first imported to Haiti from the US by so-called "sex-tourists." (61–62)

The crossroads formed by the quadrant of Miami geography is aligned in Miautre with "Crossroads" as a category of choragraphy.

Public Policy

Miami River Quality Action Team: Annual Report 1996–97
The Miami River

The Miami River is 5.5 miles long and serves as a multi-use federal navigation channel. Its average channel width is 150 feet with as little as 90 feet width at the limit of navigation. Miami river depth is 15 feet at high tide, down two feet from its original dredged depth of 17 feet. vessels drawing over seven feet of draft are towed up river dead ship, and the largest river vessels when fully loaded move only at high tide. The river is crossed by 10 bascule and three fixed bridges, including interstate Highway 95 and Metro-Dade Metrorail line.

Unique among ports, the Miami River is banked by properties that lie within the boundaries of the City of Miami and Miami-Dade county at large. Thirty-two terminal, all privately owned, are primarily concentrated in Miami-Dade County's westernmost navigable river channel. More than 80 facilities with Miami Dade county Marine Operating Permits stretch the full length of the river.

Ports of Call Served by the Miami River [Detail]

Anguilla, Leeward Islands. Antigua, British Windward islands. Aruba, Netherland Antilles. Barbados. Barranquillia, Colombia. Belize. Bimini. Bonaire. Caicos. Cartagena, Columbia. Isla de Cozumel. Curacao. Dominica, Leeward Islands. Eleuthera, Bahamas. Georgetown, Bahamas. Georgetown,

PHOTOGRAPH: BARBARA JO REVELLE

Guyana. Grand Cayman. Grand Turk. Grenada, Windward Islands. Guadaloupe, French Leeward Islands. Isla de Margarita. La Guaira, Venezuela. Guanta, Venezuela. Isla Mujeres, Mexico. Kingston, Jamaica. La Romana, Dominican Republic. Long Island, Bahamas. Montego Bay, Jamaica. Montserrat, British Leeward islands. Nassau. Panama....

The Quality Action Team

Despite its economic success, the Miami river has historically been plagued with a variety of problems and practices that threaten its overall welfare. Because these problems have been diverse in nature, they have not traditionally or clearly fallen under the aegis of any one particular agency. The formation of the Quality Action Team as a public-private partnership was an effort to centralize a focus on the river's problems through voluntary planned improvements and improved practices that would further strengthen the local economy by assuring that the river remains a viable navigational center for international trade....

Problems identified by the QAT fall into four categories: environmental, marine safety, commercial viability, and the need for enforcement efforts. All areas document a need for greater fiscal and staffing resources.

PHOTOGRAPH: BARBARA JO REVELLE

the following problems were specifically addressed by the 1997 Quality Action Team:

Funding for a variety of enforcement efforts and agencies; agency staffing for emergencies, nights and weekends; the need for a single, central reporting telephone contact number to process and respond to all river problems; hazards to navigation posed by substandard vessels; improper rafting of vessels; bridge openings and obstructions; oil and hazardous spills; sewage discharges; illegal dumping, marine debris and solid waste; crime; illegal immigrants; abandoned and/or derelict vessels; fire/explosive hazards and 'hotwork' practices; education of the general public; need for posted waterside addresses."

A typical policy issue.

Marine Safety

PROBLEM: Substandard vessels operating on the River. The Miami River is vulnerable to the potential safety and environmental consequences posed by substandard foreign flag shipping. In general, a ship is regarded to be substandard if the hull, crew, machinery or equipment used for lifesaving, firefighting, and pollution prevention is substantially below the standards required by U.S. laws or international conventions.

SOLUTION: Targeting of high-risk vessels – Operation Safety Net. In April 1994 in response to a Congressional mandate, the Coast guard implemented an initiative to eliminate substandard commercial ships from U.S. waters. The Seventh Coast guard district, which includes the Miami River, instituted a three-year interim boarding program for foreign flag vessels under 500 gross tons. Many of these smaller freight vessels routinely operate between the Caribbean, Central American, and the Southeast United States. Named "Operation Safety Net," this interim program allowed for continued operation of these vessels so long as they met minimum interim standards and were working toward meeting acceptable international standards or U.S. regulations by the end of three years.

During this same general time frame, a Regional Port State Control System was established in the Caribbean. A comprehensive set of regulations, the Caribbean Cargo Ship Safety Code, was developed as a standard for inspection for vessels under 500 gross tons. This code would provide the first true international standard for freight vessels under 500 gross tons that are flagged in Caribbean countries signatory to the new code. In the Spring of 1997, the Coast guard Captain of the Port held several town meetings with ship owners, operators, agents and marine surveyors in order to address inspection requirements and the forthcoming enforcement deadlines. The Captain of the Port also attended a meeting of Haitian shipping interests and civic groups which was convened by the then director of the Haitian Port Authority. As a large number of vessels under 500 gross tons operating on the Miami River are Haitian, numerous parties attended to present their views and concerns. These meetings resulted in a meaningful dialogue and understanding between the Coast guard and the regulated community.

Beginning January 1, 1998, all vessels under 500 gross tons trading with U.S. ports in the Coast guard's Seventh District must have a flag state Certificate, attesting to compliance with the new vessel standards of the Caribbean Cargo Ship Safety Code. Vessels failing to meet this requirement are denied entry into U.S. waters.

Operation Safety Net has had a dramatic impact on eliminating substandard ships from operating on the Miami river. Many wooden-hulled freighters and former fishing vessels could not meet acceptable standards and are no longer operating in U.S. waters. On September 1, 1996, 51 foreign freight vessels were on Coast guard and customs hold pending resolution of discrepancies. As of January 1, 1998, that figure had been reduced to only 17. Preliminary vessel arrival information also suggests a trend of larger vessels over 500 gross tons engaging in the Haiti/U.S. trade, perhaps in response to the elimination of smaller substandard ships from ability to trade in U.S. waters.

Statistics summary for July 1, 1997-December 31, 1997. Of the 41 total ships given orders to depart, 22 were wooden-hulled; 36 were Haiti flag; four Bahamas flag; one Belize flag.

Safety, this is the issue. We do not feel safe, we Americans. Even less so, after 9/11. Homeland security. Here is an aphorism (a puncept, using a form borrowed from Basho): One learns about standards from the standards.

Revelle Interviews Winston

I'm inside Murdock's Diesel Machine Shop, which is like a big open warehouse, near the 7th Avenue bridge. I came here to look for Winston, the Jamaican man with the burned boat whom I met at the restaurant near the boat with all the old bicycles. I try to make a tape of the Haitian women going through bins of old clothes and tying them into bundles to ship to Haiti, in the building next door, but they are violently opposed and afraid – hold their hands over their faces – so I make a tape in Winston's shop. Arty shots of the density – diesel engine parts on the floor, a table with dominoes, a filthy bathroom with a very old white great Dane who appears to be dead curled around the base of the toilet. It must be 98 degrees in here. Reggae playing on the radio very loud; a man moving engines with a tractor.

WINSTON: I'm from Jamaica. Eighteen years here, seen everything on this river, know everything on this river. This river is getting out of hand. A small guy can't do anything. I have a diesel engine shop and I am very good at what I do. I fix engines, rebuild engines for the boats on this river. I buy old junk engines and rebuild them, sell them to the Haitians people cause they can't afford to buy new ones. This is how I make a living. I could take a thousand bags of flour about five hundred bags of rice on my boat. On a big boat they charge you twenty dollars to take a bag of rice but on a small boat like this, only eight dollars. You need a smaller boat to go to Haiti. You can't cut out the small people.

MIKE: I'm the guy that runs around and gets all the parts. My boss here, he's well established on the river.

GEORGE: I transport all the finished engines, get em outta here. The green shows they are done, ready to go. We paint 'em.

TYRONE: I'm the general manager. I do the paper work and I'm Winston's brother. Winston was in a terrible accident. He went to Haiti and while he was over there he had a terrible accident; broke his head, broke his neck, and there is no medical attention over there. Don't wanna be in Haiti when you almost gonna die, no way. So he come back and I come over to help out and here I am.

SPENCER: Winston done a lot of work for me over the past twenty years. I'm his best customer. I don't use nobody else. I don't have no breakdowns. I'm from Bimini, see this shirt – Bimini. I come for general cargo. I got an 80 foot cargo boat and I go back and forth from Bimini and Catkey and Haiti. But right now they all stop. It's what they call the Caribbean Code. They try to get you to follow all their damn rules. They give you all these things you gotta fix on your boat, the boat that just come in the river perfectly safe for crossing for many years. I been waiting for them fuckers to call me back for three days now – tell me what size cable, and how long, and the one says one thing and the next says another. They don't even know their damn selves. One day its cable but the next day its you gotta have chain. I never had no accident, no problem, no incident, no nothing. Over two hundred trips going back and forth. The ferry boat that goes to Fish Island sank two boats already right across from the Coast Guard. Me, I never run down nobody, cause nobody no problem, nothing. But I can't go out. They say I can't go out.

BARBARA: Why do they stop your boat?

S: The Caribbean Code – they got this big book of new rules and shit you supposed to have but the main thing is to get the little man out of business. What the Coast Guard fail to understand is that all those bicycles and refrigerators, cars and mattresses, they help to clean up South Florida. Every boat come to buy stuff, not to steal. They need it in their country. Customs come with a dog this trip, make me take off my tank covers, drain my oil, drain my water; they don't find nothing. Now I got water in my generator from those fuckers. That what I'm up against. Everybody up against that shit.

W: I seen it all, the whole plan for the river, up at Mrs. Brickel's. Brickel's Boat Yard. They gonna close her down, you know, up there on the 2nd Avenue Bridge. They shot MIAMI VICE on her property. She got this map of how they plan to make a shopping center, link another Bayside, with shops and fancy little restaurants. They got this whole big plan. Only boats that gonna be on the river be the tour boats, water taxis and like that, taking the tourist tootin all round. River-walks and tours about history and shit.

B: Is there anything you can do about it?

W: It's mostly Haitians. The Haitian is difficult. The problem – they don't face the problem, they go around the problem. Coast Guard tell not to bring their wood boats in here no they just sit there on their boat waiting for some message from their gods about what to do or something, or they get together get a bigger boat, steel boat, instead of show the Coast Guard where they're off base. But then the big boats get seized too if they have Haitian flags. The other day a boat come in here with a Haitian flag – big boat, U.S. Navy built boat and they seized it. Said it was unsafe. U.S. Navy built boat. I mean, so we got boats going up the river now that is too big, get stuck in there at low tide, tip right over on their sides and dump all the cargo, but that's the kind of solution the Haitians think up. Don't go at it head on. If the boat is from Cuba, now that's a different story, or say the Bahamas or someplace American's like. Then they can come in, go out, no problem.

S: I know this: somebody is getting some kickbacks. Big boats give big kickbacks. They just want the little man out. If ten Haitians come in here, they send them back. If twenty Cubans come in here, its, oh, hello, can we help you? with a big escort and immunity or some shit. What they call it?

W: OK you wanna know what's really going on; turn off the tape. Here is what's really going on. Between about 1979 and say 1988 Cubans running drugs, that's what. Now Haiti do it. Cuba got too hot. Everybody know Haiti do it now. Nobody looking at Haiti that's why. All the drugs go through Haiti now. They can't catch Haitians.

B: Why not?

W: The Americans look at the Haitians like they're dumb and stupid. But they are very very smart people. But the American people think they are stupid, that's how they class them. That's why they can't catch the Haitians, they too smart, they are so dumb they're too smart. Do you know what Voodoo is?

PHOTOGRAPH: BARBARA JO REVELLE

B: Yes, Haitian religion, derived from Africa.

W: Yeah well everybody there have the Voodoo so they can bring the shit in. No problem. The other day, there's a boat, a wood boat. They can't bring wood boat no more, a wood boat come right down the river past the customs, the Coast Guard, the bridge guys even put up the damn bridge for them. They dropped off onehundred-fifty Haitians right in the river.

B: Oh yeah, the Rose-Marie Express. Wasn't it at night?

W: No, in the middle of the day; they caught less than half of them. You can believe whatever you want but how come they come down the river like they was invisible and half the people disappear into the city if they don't have something working for them that's more what the newspaper tells it. Them blowing right by all the Coast Guard, they have satellites on you and all. How they do that is what I wonder. They got some power with them, I know that much. Then three, four months ago you have a Haitian boat come in here. It had over four hundred people on it. the Coast Guard got after it before it got to the mouth of the river, way out in deep water. You know what the Coast Guard did? They took their steel boat and they rammed into that little wood boat, with all them people crowded into it with babies and crying and old people. They called it shouldering.

S: Now that human life right there. They have no respect for human life. They say they wanna protect human life so they makeup all these rules how

PHOTOGRAPH: BARBARA JO REVELLE

your boat isn't safe so to protect you. But that's not it. They send the boats back half the time, so how they say they fair about if it seaworthy? Send them back, three four days, with all the people with no food if they want, no water.

W: I'm the one who built the engine for that boat, that one they rammed. That was a good boat. They just hate the Haitians, any God Damn Haitian boat they just hate it. Big, little. Why only Haitian boats they send back? They hate the black man that's why, and they afraid of him.

S: Other boats come in from Cuba, they don't make them take off their water tank, don't bring no dog. Other boats come in, they carry all kinds of shit, every place in the fucking world has drugs. Satellites follow me. They put the dog, they put the cocaine machine, they check the money. What the fuck is that, man? My Daddy run this trip over fifty years. Twenty years I been working for them, but now they got this Caribbean Code and I got to suffer. They can plant something on your boat too, they done that to the Haitians plenty times. Everybody say that the wooden boats are dangerous, but that just show they don't know about boats. If a wooden boat gets a hole you can put a towel, clothes or a mattress, you can plug it with a mattress. Fiberglass boat crack, has to be totally dry to fix it. Out to sea it just go down. You can't plug it less it's totally dry. It's just the Haitian boat they want, nothing about dangerous. Wood boat is just as safe as steel boat, fiberglass boat. These Haitian people build boats without a blue print or nothing, from scratch, and beautiful, smart, I would go with them in a storm over any of these big fancy boats. Haitians come in here without radio, compass, anything. Coast Guard say you have to have six lights, they got one, two, they never have no accident, no problem. They know what they're doing.

W: One thing, they only speak Creole. They're supposed to have one guy on the boat speaks English but he can't read English and they give him this long list of, fix this, fix that. And how they supposed to get the forty thousand dollars to fix the wench and buy some more boats and all that. The cargo just sits and rots by the river. Gets stolen. Boat goes back empty or gets seized or just gets stuck here trying to figure out how to get up the money, so there's no mystery how they get into drugs, right?

S: They just take out all the garbage the city dumps anyway, they take it all back home, they smart people, see? There used to be a big line of them taking all this junk back to Haiti but that's all over now. Nobody to take out the trash no more. Common sense, my old man use to tell me. They got common sense.

W: Or Voodoo, something gets them across the sea. But the American people, they don't get it, they think the Haitians is diseased or something, think they gonna get AIDS. The Haitian boats, I been on the boats when the Coast Guard come on, afraid to touch 'em. Think they're dirty, they'll get sick to just be on the boat. Their idea is to take the boats up on the banks and crush 'em. I seen it. they just take them up and crush 'em and haul it all away. Beautiful boats all painted up nice inside, bright colors: crush em.

From Revelle's Journal: Winston's Party

Winston literally can't go home again, because he belongs to an Africa which is part of the Caribbean imaginary. Burning Spear and Bob Marley. The legacy of Marcus Garvey. Pinnacle, Jamaica where the first Rastafarian settlements was established. Dispossessed people all over Kingston. What Africa has become for Winston is a symbolic journey necessary to all of us, or at least that is what I figure out after many rum and cokes at his party. Winston wanted to see if I would come to his party – said there would be more than one hundred people. It was a party to celebrate his divorce but I felt like it was to celebrate the end of the Miami River drift. Felt like it was the right thing. So I go to this party. At first Winston is not even there but I hang out with Winston's brother Tyrone's girl friend Dianne who spends a lot of time telling me about how her Cuban parents don't like that she is with a Jamaican black man but that she thinks Tyrone is so hot but what do I think? The party picks up. Lots of rum drinks, an outside bar, dancing. George feeds me several goat dishes. The music is not only reggae, but African I am pretty sure. Almost everyone is black and mostly Jamaican. Spencer, from Bimini, was there and said he had lost his boat since the last time he saw me. Said he had to sell it.

Africa deferred. Africa must be reckoned with but it cannot be recovered. It was only in the 1970's that Afro-Caribbean identity became historically available to the majority of Jamaicans. I read the word "creolisations." Am I the white European colonizer at this party who both fears and desires what the dark "other" has? Nah. Not that simple. America, Terra Incognita. Diaspora aesthetics – what does Winston know or care for that? Winston finally shows up at almost midnight and comes right over to me. Says he was up north doing a diesel job which took longer than he thought it would. Doesn't seem at all concerned that he missed three hours of his own party. Introduces me all around to people I have already met, speaking English to me and patois to them, and whatever he says to them must be a lot funnier than how he translates it to me because everybody is laughing nonstop.

I don't know about him but I am pretty drunk sitting on his boat. Not his burned one, but a small one he has tied up behind his house by the canal. He and his brother go fishing on it. I was invited to go once but didn't and I remember wishing I had. He talks about Jamaica and missing it and invites me to go on a trip there with him if I want to. He talks about Africa, which he says he wants to visit some day. Then he goes into his house for a long time and finally I get bored talking to Dianne who is treating me like her sister confessor and I go in to find him. In his living room is a huge TV and a couch and nothing else. He is not there. Someone says he is in the bedroom. After a while I say goodbye to Dianne and George and go out to my car to leave. Winston comes out right as I am getting in my car and kisses me good-bye on the mouth saying, "I am a simple man but I can see in your eyes how you feel." I think this is hopelessly corny and all I want to do is go back to my room and sleep but I get lost trying to drive there and it takes forever because there are no direct routes from one place to another along the river.

From "The Caribbean Safety Code"
Fundamental Problem

The fundamental problem underlying this situation is not old ships, but ship owners who do not have the financial resources to maintain their vessels in a seaworthy condition or to act promptly when the vessel has a casualty. Even well maintained vessels break down occasionally, however their owners take prompt, effective action to bring the ship out of danger and correct the problem. The irresponsible ship owners in the Miami area have come to rely on the Coast Guard to take responsibility for their shipping problems. our outstanding Coast Guard Search & Rescue program has actually facilitated these substandard vessels to operate.

The FRE fieldwork in the zone came just at the conclusion of Operation Safety Net (Summer, 1998). Return visits proved the effectiveness of the policy: No wooden trading vessels remained on the river. Nor did we encounter any Haitian traders during those visits. The chora archive is accumulating here, piling up like the second-hand goods collected in the warehouses of the Miami River. At what moment does this pile become a heap, that is, a categorial configuration?

CHAPTER 3

Situation

For an image to become categorial in choragraphy involves a mapping operation. This mapping is an image poetics developed by modernist poets and artists from Baudelaire through Guy Debord. The "poetic encounter" emerged most explicitly among poets but it is a practice that is inherently "photographic." The categorial power of an image derives from its ability to provoke, evoke, a mood. In order to image the mood of the Miami River Barbara Jo Revelle used as a relay for her movement through the zone the drift (*derive*) made famous by the French Situationist psychogeographers. For the scene of policy problems to matter to anyone, and to Revelle and the FRE first of all, it had to become a "situation." For five weeks in the summer of 1998 she drifted in and out of place.

Theoria

The Situationist drift resonated with a "relay" for our consultation – the ancient institution of theoria – that seemed ideal for taking advantage of one of the founding metaphors of the Internet. This metaphor, not surprisingly, is reflected in the language of a technology challenge grant: "While the numbers of computers and connections to the Information Superhighway have increased in the schools," states the Department of Education brochure, "the capacity of the teaching force to use this technology in instructional practice has not kept pace." The scene of travel brings together the elements of our experiment – the tourist crisis, the Internet as information highway, the "meta-hodos," the "way" of method as it has evolved in disciplinary theory guiding the consultants. Like any consultants, we would have to tour the site of our designated problem. The question was, in what way, in what spirit, should this inquiry take place? How would our tour differ from that of the empirical investigators? "Theoria" located our response in an intellectual tradition. "The Greeks designated certain individuals to act as legates on certain formal occasions in other city states or in matters of considerable political importance. These individuals bore the title of 'theoros,' and collectively constituted a 'theoria'. They were summoned on special occasions to attest the occurrence of some event, to witness its happenstance, and to then verbally certify its having taken place" (Godzich). Others could see and make claims, but these would have merely the status of "perceptions"; only the

 | DOI:10.1163/9789004534643_004

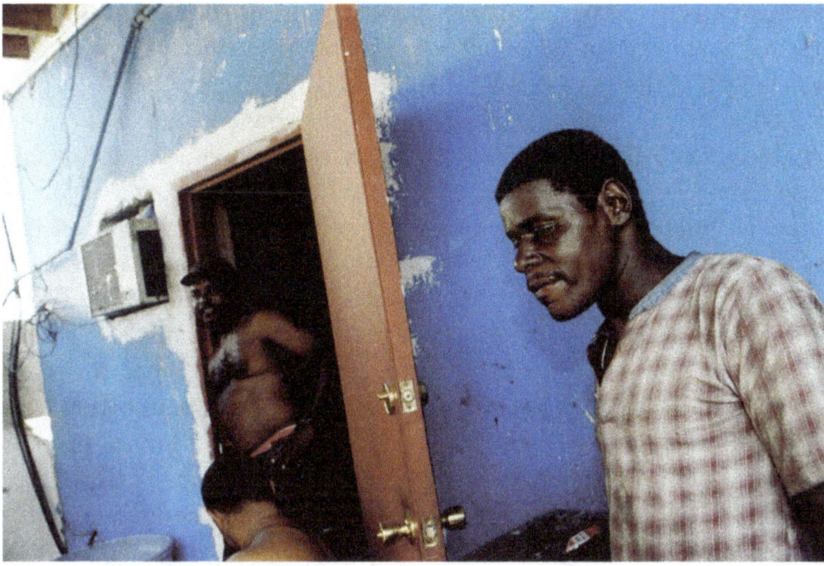

PHOTOGRAPH: BARBARA JO REVELLE

report of the theoria provided certainty, certifying the attested event such that it could be treated as fact. "What it certified as having been seen could become the object of public discourse." Another duty of a theoria was to consult any oracles located in the regions through which they passed.

PHOTOGRAPH: BARBARA JO REVELLE

Several aspects of this precedent recommended themselves: the function of "witnessing" some happenstance that was a matter of public concern, and of acting on behalf of the general public. But how was the "attestation" to be performed? Bertolt Brecht may have had something like the theoros in mind when he introduced his "Alienation" effect. To illustrate the intended effect of his epic theater he used the example of an eyewitness demonstrating to a collection of people how a traffic accident took place. The "demonstrator" acted the behavior of driver and victim in such a way that the bystanders were able to form an opinion about the accident. "A technique of taking the human social incidents to be portrayed and labeling them as something striking, something that calls for explanation, not to be taken for granted. The object of this effect is to allow the spectator to criticize constructively from a social point of view" (Brecht, 1964: 122). The demonstrator is not concerned with "illusion," but still performs the actions in question, such as how the victim stepped off the curb. "The demonstrator achieves the A-effect by paying exact attention to the victim's movements, executing them carefully, probably in slow motion; in this way he alienates the little sub-incident, making it worthy of notice."

Social science, however, has shown that eyewitness testimony is no guarantee of certainty. One study "shows how memory can be radically altered by the way an eyewitness is questioned, and how new memories can be implanted and old ones altered in subtle ways under interrogation" (Loftus, 1979). The theoria relay suggests another, less authoritative or less juridical position from which to testify. "Originally 'theoria' meant seeing the sights, seeing for yourself, and getting a world view," E.V. Walter comments. "The first theorists were tourists – the wise men who traveled to inspect the obvious world. Solon, the Greek sage whose political reforms around 590 B.C. renewed the city of Athens, is the first 'theorist' in Western history" (Walter, 1988). "Theoria did not mean the kind of vision that is restricted to the sense of sight. The term implied a complex but organic mode of active observation – a perceptual system that included asking questions, listening to stories and local myths, and feeling as well as hearing and seeing. It encouraged an open reception to every kind of emotional, cognitive, symbolic, imaginative, and sensory experience" (Walter, cited in Ulmer, 1992). The questioning did not neglect "second" sight as well, in that the theoria had recourse to divination.

From the institution of theoria we concluded that two practices that today seemed to be as far removed from one another as possible (academic theorist and leisure tourist) were at one time parts of one and the same experience. The Greeks believed that "leisure" created the conditions that allowed people to realize their essence through knowing (Blumenberg, 1983: 280). Commentators on intellectual curiosity in any case have argued that the same appetite for exploring the unknown addressed to the most worthy objects (observation of the heavens, in ancient times) also "drives masses of people to the spectacles" (261), or moves them to inspect "a mutilated corpse" (314). The continuum of curiosity linking scientific inspection and vulgar or morbid voyeurism marks out an area of educational intervention whose common ground is "media." A choral practice is meant to open up the dead interface metaphor of "information highway" by means of the relay of theoretical tourist (modernizing the stance of ancient wisdom).

What we wanted to learn from the poets (especially from the Situationist "poetry without poems") was how to grasp the whole of the zone in an image, the way an epiphany reveals in a flash the nature of a situation. In the culmination of this tradition of "flash reason," the Situationists "sought to understand that moment when people gain insight into the rationalized and alienated patterns of their everyday lives. Lefebvre's interpretation of this eruptive 'moment' as embodying 'fleeting but decisive sensations (of delight, surrender, disguise, surprise, horror or outrage) which were somehow revelatory of the totality of possibilities contained in daily existence' could stand just as well for the situationists' notion of 'situation'" (Sadler, 1998: 45). Most importantly, the Situationists believed that it was possible not just to recognize situations, but to construct them, once one had defined the appropriate "formulary."

Theoria occurs as an intervention within a diegetic "Myami." In this diegesis (the space and time of a fictional world) the conventions of the classical discourse of wisdom overlap with those of narrative in that both are organized around "problems." "Judgment can be represented in the form of a syllogism, called 'practical' because its conclusion is an action. The major premise contains a good the agent wants to achieve; the minor premise contains the description of a concrete situation facing the agent. The conclusion is the action the agent should take in the concrete situation viewed from the perspective of the good that is wanted" (Kekes, 1995: 26). Entertainment films are structured by such syllogisms, as illustrated by this comment in a handbook for writing screenplays: "Because audiences need to worry, writers devise problems with which to confront their protagonists. Study any well-written motion picture or television show, and you will find a succession of harrowing obstacles that prevent the protagonist from reaching his or her goal" (Armer, 1988: 9).

Both the wisdom and entertainment dramas, however, idealize the feeling of "control" (safety). "Contingency, conflict, and evil are adversities that stand in the way of our efforts to live good lives," a philosopher states, naming the ideals of moral wisdom as self-sufficiency, invulnerability. "Socrates recognized their existence, but he supposed that we can eliminate them by being sufficiently virtuous. He was mistaken about this. Contingency, conflict, and evil are permanent adversities" (Kekes, 1995: 51). Entertainment narratives with their fantasies of power and "happy" endings make a similar point, against which the testimonial defines itself. Our premise is that when Myami is designed as a chora within the prosthesis of the internet, this place (topos) and its defaults (cliches, commonplaces, stereotypes) become accessible to revision. The moral syllogism and the narrative conflict converge in our own life project. The Internet chora puts us inside the dialogical struggle for Myami.

A testimonial consultation approaches policy formation indirectly, given its concern with image, category, apparatus. "The most subtle forms of social control are those we least recognize as such. Precisely because the categories of understanding and meaning provide so powerful a constraint to what we experience and how we think about that experience, they prevent awareness of alternative ways of conceiving events and processes. Because they lead us to 'see' the accustomed forms as the only reality they minimize and obscure the possible conflicts and the volitionary decisions that have helped construct that 'reality'" (Gusfield, 1981: 29). As Gusfield explained in the case of drunk driving, a "problem" is a "container" joining certain events with certain ideas. One way to intervene in policy formation is to redefine not just what is in a given container (what cannot be imagined cannot be desired), but the container itself ("problem" not as topic but as chora). The testimonial is an opportunity to intervene in the policies affecting the Miami River not only at the level of image (the struggle for Myami) but at the level of category (the photographic Whatever that reopens situation as such). In addition to opening up the categories predetermining the policy issues of the Miami River, the larger goal is to reconsider the category of "problem" as such. Of course, as Thomas Kuhn warned, new paradigms do not solve the problems of the old paradigm; they just make them irrelevant.

The Drift

"Situation" evokes the philosophical weight given to this term in existentialism and the French Situationist experiments. A difficulty of such terms is not that they are complex, but that they are a bit "tired." Even if Situationism has not lost its appeal as a fashion, few want to be reminded of the Sartrean notions that supplied the name.

> The situation is the subject illuminating things by his very transcending, if you like; it is things reflecting to the subject his own image. It is the total facticity, the absolute contingency of the world, of my birth, of my

> place, of my past, of my environment, of the fact of my fellow man – and it is my freedom without limits as that which makes there be for me a facticity. It is this dusty, ascending road, this burning thirst which I have, the refusal of these people to give me anything to drink because I do not have any money or because I am not of their country or of their race; it is my abandonment in the midst of these hostile populations along with this fatigue in my body which will perhaps prevent me from reaching the goal which I had set for myself. But also it is precisely this goal, not insofar as I clearly and explicitly formulate it but insofar as it is there everywhere around me as that which unifies and explains all these facts, that which organizes them in a totality capable of description instead of making of them a disordered nightmare. (Sartre, 1965: 274)

The consultant (egent) at the Miami River is in an existential situation, with the "project" of the EmerAgency. The conventions for how to tour a city in a poetic mode were codified over the course of the century from *Paris Spleen* to the Situationist *Memoires*. An important relay for the theoria's problem tour is the Situationists' drift through a city in search of those "unsanitized" places that afforded a "unity of ambiance," an atmosphere coherent enough to attract or repel a visitor, of which Les Halles (the old market center of Paris) was the prototype. The Situationist experiments with psychogeography are a resource

PHOTOGRAPH: BARBARA JO REVELLE

for learning how to mediate a zone and its image in a chora. "*Memoires* sudden bursts of picture and text suggested the rewards of the drift: the flowerings of consciousness, the sudden 'comings together' of space and architecture, knowledge and social interaction. Chtcheglov left us with a cautious analogy: 'The drift (with its flow of acts, its gestures, its strolls, its encounters) was to the totality exactly what psychoanalysis (in its best sense) is to language'" (Sadler, 1998: 98). Chtcheglov formulated the drift as being to the city what free association is to a psychoanalytic session. Here is a relay for Revelle's passage through the Miami River zone.

The poets model how to grasp the whole of the zone in an image, the way an epiphany reveals in a flash the qualities of a situation. In the culmination of this tradition, the Situationists "sought to understand that moment when people gain insight into the rationalized and alienated patterns of their everyday lives. Lefebvre's interpretation of this eruptive 'moment' as embodying 'fleeting but decisive sensations (of delight, surrender, disguise, surprise, horror or outrage) which were somehow revelatory of the totality of possibilities contained in daily existence' could stand just as well for the situationists' notion of 'situation'" (Sadler, 1998: 45). One purpose of applying art methods directly to life was to augment and extend the epiphanic moment into a sustainable experience of intense living. Most importantly, the Situationists believed that it was possible not just to recognize situations, but to construct them, once one had defined the appropriate "formulary."

Turntable

The Situationists defined the crisis of the spectacle, and one of their chief responses to the commodification of everyday life was to develop a new behavior out of the poetic tradition of the *flaneur* – the drift – leading to the creation of situations in urban life. In protest against the rationalized urbanism of modernist planners (whose prototype they found in Le Corbusier), the drift (*derive*) cut across the established circulation patterns and zoning of the city in order to seek out places that possessed a specific ambiance, mood, atmosphere, aura. Such energy nodes they dubbed *plaques tournantes* (translated as "turntables" or "hubs") which they registered for Paris in perhaps their best-known works – the collage maps, "The Naked City" (the title borrowed from a film), and "Guide psychogeographique."

> The situationist maps described an urban navigational system that operated independently of Paris's dominant patterns of circulation. "The Naked City" (collage map) helpfully explained itself on the reverse: "The arrows represent the slopes that naturally link the different unities of ambiance;

PHOTOGRAPH: BARBARA JO REVELLE

that's to say the spontaneous tendencies for orientation of a subject who traverses that milieu without regard for practical considerations.... The earlier of the two situationist maps, the "Guide Psychogeographique," had taken the situationist interest in charting each unity's "exits and defenses" and "fixed points and vortices" to obsessive lengths. Its arrows restlessly danced between, in, and around specific streets.... "The Naked City" evoked beautifully the way in which some unities of ambiance acted as stations on the drift, junctions in the psychogeographic flow of Paris. The situationists coined a term for these junctions: *plaques tournantes*. The term punned on so many meanings that it is not possible to translate it straightforwardly. A *plaque tournante* can be the center of something: it can be a railway turntable, or it can be a place of exchange (in the same way that Marseilles is sometimes described as a *plaque tournante* for trafficking, or that Paris as a whole has been celebrated as a *plaque tournante* of culture). As a center for markets, drinking, prostitution, and drugs, Les Halles was clearly a *plaque tournante* in all these senses. Zola had described it as "the belly of Paris," an idea incorporated in Khatib's designation of it as "the transition zone of Paris," a place not only of commercial exchange but also of "social deterioration, acculturation, and a mixing of populations which is the favorable environment for cultural exchanges." (Sadler, 88–89)

The maps with their whirling arrows indicate that the turntable or hub is a Vortex (Ezra Pound). The maps are expressive, relational, not mimetic. "Debord and Jorn pasted down the chunks of map at some odd angles, as much as forty-five degrees from the mean. Reorienting the chunks in this way helped to represent the drifters' passages across the city as smooth flows. In topological reality, the districts down the left side of the maps lie along a straight north-south axis. But left like that, the reader of the map would have gained little sense of the supposed vortex of psychogeographical flow around the *plaque tournante* of the little *Carrefour de Buci*. Debord and Jorn instead kinked the chunks around Buci, showing it as a *carrefour* (crossroads) in a larger sense, a pivot for psychogeographic as well as traffic flows" (90).

The title of our primary image ("Crossroads") resonates with this *carrefour*. Sadler states that the various attempts by SI members to research the precise qualities that produced the atmosphere of a place – Debord in London, Michelle Bernstein in the *Square des Missions Etrangeres*, for example – had aesthetic value only. "With its detective-style iconography, Ralph Rumney's situationist photographic record of Venice's streets, made as he stalked American Beat author Alan Ansen, looked suitably systematic. But it failed to yield anything remotely like 'data,' its author struggling to explain the significance of his encounters with children and with old acquaintances, account for the romance of Venice, and identify 'sinister,' 'depressing,' and 'beautiful' zones" (78). Khatib's more substantial study of Les Halles shows the emphasis on tracing the ambient regions on the map: the method remained geographically explicit and psychologically implicit. "The zone of central interference, the turntable of the different ambient directions of Les Halles is, as we have pointed out, the Bourse du Commerce-Place des Deux-Ecus complex. This area is found at the western extremity of the block constituted by the juxtaposition of the large pavilions of the Halles Centrales. But since these edifices do not act as a link, but on the contrary as a break, the Rue Careme which traverses them longitudinally does not participate in this relation. The different directions which intersect at this turntable strongly affect the path any individual or group will, with apparent spontaneity, follow inside as well as outside Les Halles" (Khatib, 1996: 75).

Such assessments may be relevant to geography, but are misplaced in the case of choragraphy, since the latter relies on aesthetic methods to record the "warped" (psychological) dimension of place. In one exemplary drift Debord and Wolman set out across Paris, choosing their path intuitively, spontaneously, arriving finally and "unexpectedly" at the ruins of a rotunda designed by Ledoux, next to an elevated subway line. "Upon studying the terrain the Lettrists felt able to discern the existence of an important psychogeographic hub [*plaque tournante*] – its center occupied by the Ledoux rotunda – that could be defined as a Jaures-Stalingrad unity, opening out onto at least four significant psychogeographical bearings – the canal [Saint-]Martin, boulevard de la Chapelle, rue d'Aubervilliers, and the canal de l'Ourcq – and probably more. In conjunction with the concept of the hub, Wolman recalls the intersection in Cannes that he designated 'the center of the world' in 1952. One should no doubt link this to the clearly psychogeograpic appeal of the illustrations found in books for very young school children; here, for didactic reasons, one finds collected in a single image a harbor, a mountain, an isthmus, a forest, a river, a dike, a cape, a bridge, a ship, and an archipelago. Claude Lorrain's images of harbors are not unrelated to this procedure" (Debord, 138). And with these examples, Debord has placed the drift explicitly in the tradition of chor(o/a)graphy.

Part of the value of Situationist drifts is as a relay, transposed into cyberspace by considering a city zone as an "information chronotope." Any interface may be designed as a drift. Like constructed situations, the EmerAgency testimonial is not fixed, permanent, but "'ephemeral, without a future, passageways,' syntheses of those sublime moments when a combination of environment and people produces a transcendent and revolutionary consciousness" (105). The very forms of the society of the spectacle targeted for overcoming by the Situationists, such as narrative film and TV, also offered lessons in the construction of situations. "For the audience to find its place, it only needs the space of one sequence or scene (and this is particularly noticeable with child spectators, who quickly become interested in a film, fragment by fragment, without fully understanding the overall plot or psychological motives). All the audience requires to become inscribed within a scene is a structured network of relations, 'a situation.' Consequently, it is unimportant whether the spectators know the characters yet: within this rational structure mimicking any intersubjective relationship, the audience will readily organize a certain number of places, arranged in a certain order and fashion, all of which is the necessary and sufficient conditions of identification" (Aumont, 1992: 223). This capacity even of children to recognize situations, however it might be exploited within the modern conditions of spectacle, is also an opportunity for the educational (ethical and political) effects of testimonial.

The Internet with its digital tools supporting asynchronous, distributed, collective interaction, offers a technology compatible with "vortextual," choral category formation. A choral category may be constructed by a group since its coherence is a result, not an intention. "Every link in hypertext creates a category. That is, it reflects some judgment about two or more objects: they are the same, or alike, or functionally linked, or linked as part of an unfolding series.

The rummage sale of information on the World Wide Web is overwhelming, and we all agree that finding information is much less of a problem than assessing its quality – the nature of its categorical associations and by whom they are made" (Bowker and Star, 7). A feature of Miautre chora is to test the practical consequences of creating a choral Vortex online. A *vortext* (sic) does not define but discovers (invents). The first application of the logic is to help discover more germinal situations for a global Web of Changes.

Warped Space

The Situationist map is a point of departure for a practice of image reason. Aristotle invented the topic as a figurative combination of physical and logical space. Both kinds of space have mutated in the era of electracy. Image reasoning brings into category formation (metaphysics) the qualities of modern (electrate) space, which, in our terms, is not an abandonment of topos but the situating of topical space within the region of chora.

> This is the outcome of an epoch-making event so generally ignored that we have to be reminded of it at every moment. The fact is that around 1910 a certain space was shattered. It was the space of common sense, of knowledge (savoir), of social practice, of political power, a space thitherto enshrined in everyday discourse, just as in abstract thought, as the environment of and channel for communications; the space, too, of classical perspective and geometry, developed from the Renaissance onwards on the basis of the Greek tradition (Euclid, logic) and bodied forth in Western art and philosophy, as in the form of the city and town. Such were the shocks and onslaughts suffered by this space that today it retains but a feeble pedagogical reality, and then only with great difficulty, within a conservative educational system. Euclidean and perspectivist space have disappeared as systems of reference, along with other former 'commonplaces' such as the town, history, paternity, the tonal system in music, traditional morality, and so forth. (Lefbvre, 25)

A task of Miautre chora is to support the compiliation of a global common sense, to help citizens recognize and experience warped space. The emergence of electracy is dated from the rise of the industrial city in general, and the invention of new recording technologies, just as the epoch of literacy dates from the invention of the alphabet. Paradoxically, the machinery of mechanical visualization exploded the illusion of the alignment between the

PHOTOGRAPH: BARBARA JO REVELLE

standpoint of an observer and the truth (the cone of vision). The capacity of the camera to separate itself from the human physical and mental eye, combined with the theories of psychoanalysis, produced the notion of the "optical unconscious" (Walter Benjamin). The attitude towards truth as standpoint in the image apparatus of electracy is that clarity is an *effect of repression*, blindness (tragic Até). In the electrate apparatus the clear distinction between subject and object disappears.

> Space, in those various iterations, has been increasingly defined as a product of subjective projections and introjection, as opposed to a stable container of objects and bodies. From the beginning of the [twentieth] century, the apparently fixed laws of perspective have been transformed, transgressed, and ignored in the search to represent the space of modern identity. Thus the body in pieces, physiognomy distorted by inner pain, architectural space as claustrophobic, urban space as agoraphobic, all warpings of the normal to express the pathological became the leitmotivs of avant-garde art. The vocabularies of displacement and fracture, torquing and twisting, pressure and release, void and block, *informe* and hyperform that they developed are still active today, deployed in work that seeks to reveal, if not critique, the conditions of a less than settled everyday life. (Vidler, 2000: 1)

As Rosalind Krauss explained, applying the poststructural psychoanalysis of Jacques Lacan, the "extimate" inside-outside nature of the human subject installs an opaque obstacle "within the very heart of a diagrammatic clarity that is now a model both of vision's claims and of vision's failure.... The graph of an automatist visuality would show how the vaunted cognitive transparency of the 'visual as such' is not an act of consciousness but the effect of what is repressed: the effect, that is, of seriality, repetition, the automaton. Which is to say, it is a function of a caesura in vision, a gap" (Krauss, 1994: 74).

PHOTOGRAPH: BARBARA JO REVELLE

The classic standpoint of the literate observer, organized in terms of the subject-object binary, still exists within electracy, but now with the observer displaced from the central position to being just one point within a field network (a matrix) of which s/he is not the master.

> Lacan's discussion of vision in *The Four Fundamental Concepts of Psychoanalysis* is centered on a scopic drive structured by the distich, the rupture, the schiz. He calls the opening chapter "The Split between the Eye and the Gaze." But the whole of his account is called "The Gaze as *objet a*," or the gaze as part-object, the object that, because it marks the subject with the individuation of his existence, is most fundamentally the object of desire. Lacan wants to show the dialectic between deictic and distich, between pointing and screen, between this! and absence.

> The space of pointing, or of deixis, is the space that Lacan terms "geometral," namely the space of perspective, a space that as Diderot shows in his *Letter on the Blind* is actually a tactile space, a space mastered by the subject as though he were reaching out to grasp it, to palp it, running fingers over its front and sides, manipulating it.
>
> In contrast to the tactile "visuality" is the space of light, which Lacan calls "dazzling, pulsatile": an atmospheric surround that illuminates the viewer from both back and front, so that from the start there is no question of mastery. And in the context of this space of the luminous, the viewer is not the surveyor – standing at a point just outside the pyramid of vision – but, caught within the onrush of light, he is what blocks the light, what interrupts its flow. In this interruption the "viewer" invisible to himself enters the "picture" created by this light as a "stain" or blind spot, as the shadow cast by the light, the trace, its deictic mark. And from this place the subject, in all his exposure to view, can neither see himself nor see the source of the light. His position is one of dependence on an illumination that both marks him (the deictic) and escapes his grasp (the distich). The illumination Lacan calls the "gaze." It is the part-object operating within the instinctual field of vision, forever unlocatable, out of focus, in metamorphosis, pulsing. (87–88)

The only glimpse of this optical unconscious within the empirical view is that offered by optical illusions (137), such as the famous duck-rabbit, through which viewers become aware of their participation in the act of seeing (aspectuality). The literate production of truth that is clear and simple is an epistemic and ideological "optical illusion." The challenge is to learn how to reason from the standpoint of being a figure within a field of warped space. From within the prosthetic image (that is, one's choral scene online), the mind's eye of imagination sees through the gaze. In terms of electracy, the challenge is to devise an educational practice that corresponds to the time-image (the mode of reason within which may be figured the time-accident). Chora augments "feeling" the way topic augmented "analysis."

> If feelings are ages of the world, thought is the non-chronological time which corresponds to them. If feelings are sheets of past, all these sheets, the continuity which rolls them up and unrolls them like so many lobes, preventing them from halting and becoming fixed in a death-position. According to the novelist Belyi, "we are the unrolling of a cinematographic film subject to the minute action of occult forces: should the film stop, we will be fixed for ever in an artificial pose of terror." In cinema, Resnais says, something ought to happen "around the image, behind the image

> and even inside the image." This is what happens when the image becomes time-image. The world has become memory, brain, superimposition of ages or loves, but the brain itself has become consciousness, continuation of ages, creation or growth of ever new lobes, re-creation of matter as with styrene. The screen itself is the cerebral membrane where immediate and direct confrontations take place between the past and the future, the inside and the outside, at a distance impossible to determine, independent of any fixed point (which is perhaps what creates the strangeness of *Stavisky*). The image no longer has space and movements its primary characteristics but topology and time. (Deleuze, 1989: 125)

The time-image becomes the basis of practical reason (of a new prudence) in conditions of the dromosphere (of Real Time).

Formless

Mystory (the genre used to produce Miautre) articulates obtuse meaning, developing a new inference procedure using the sting of an image, to register the arbitrary and singular body that constitutes, in its obstinacy, the shadow blocking the light of empirical instrumental consulting on public policy issues. The proposed EmerAgency retains the tourist guidebook as an interface metaphor for learning electracy, but in a deconstructed version within the larger analogy used to generate an Internet civic practice. The counterintuitive claim is that choral categories are warped "singularities," not universals. The classifying gesture is movement not of abstraction but of absolute embodiment in an individual position. Each egent functions online as a living category. "Me, I exist – suspended in a realized void – suspended from my own dread – different from all other being and such that the various events that can reach all other being and not *me* cruelly throw this *me* out of a total existence.... The empirical knowledge of the structure this *me* has in common with the other *mes* has become an absurdity in this void where my dominion manifests itself, for the very essence of the *me* that I am consists in the fact that no other conceivable existence can replace it; the total improbability of my coming into the world poses, in an imperative mode, a total heterogeneity" (Bataille, 1985: 131).

This heterogeneous standpoint is at the core of the methodology of testimonial, taking up the paradox of group subjectivity in electracy. A consultancy that applies heterogeneous (un)knowing to problem-solving may be a self-refuting enterprise, since Bataille defines the heterogeneous as the remainder excluded from science. The invention by the Classical Greeks of the dialectical "topic" to replace the epic hero as the vehicle of thought and language clarifies what we are attempting when we propose chora as a supplement for topic in electracy.

A topic sorts and gathers information according to similarity and consistency of theme (it is conceptual), whereas chora gathers heterogeneous materials holistically by means of a feeling or atmosphere. Chora becomes writable in the digital recording equipment evolving in the modern period. This new categorization has been anticipated, dreamed, imagined for some time (an apparatus emerges within a matrix of which technology is only one part, interacting with institutionalization and subjectivation – identity formation).

A beginning point for chora as method is to deconstruct a set of topoi by blurring their conceptual borders. For example, the transformation of the Miami River into a chora involves assembling the topoi organizing the discourses of consulting, advertising, neo-noir narratives, and wisdom, and to gather them into a new set with its own emergent properties. As Deleuze says of the time-image, it gathers not things but cliches (Deleuze, 1989: 32). The topoi coexisting in the accumulated representations of a city become integrated through the work of formless (Bataille) into one atmosphere, a mood of a chora, appearing as a pattern (a vortex, a constellation) in a data set. The function of formless as a logic is to declassify, "an action that simultaneously (1) lowers or debases objects by stripping them of their pretension – in the case of words, their pretensions to meaning – and (2) declassifies, or attacks the very condition on which meaning depends, namely the structural opposition between definite terms.... The categorial blurring initiated by the continual alteration of identity within this work is precisely what Bataille means by formless. It is not just some kind of haze or vagueness in the field of definition, but the impossibility of definition itself due to a strategy of slippage within the very logic of categories, a logic that works according to self-identity – male, say, or female – stabilized by the opposition between self and other: male versus female, hard versus soft, inside versus outside, life versus death, vertical versus horizontal" (Krauss, 1999: 5, 7). Miautre does not stop with this deconstructive action of formless, but adds to it a constructive performance of the image category, making sensible the peculiar perfume of the blended topoi accumulated in the spectacle of Miami.

Beyond Psychogeography

With the new extremes of Virilio's dromosphere (contemporary space and time) in mind, choragraphy takes up where psychogeography left off. The Situationists used the drift or walk in the city and the poetic encounter of the epiphany (privileged moment) not to make texts but as a method for what some called the "science-fiction of urbanism." The Situationist maps proposed to include in urban studies the feeling or mood of the city. "So *Internationale situationniste* tried another juxtaposition, between an aerial photograph of Amsterdam ('an experimental zone for the drift') and a 1656 *precieux* 'map of the land of Feeling,' alerting readers to the possibility of mapping states of consciousness and feeling during the drift. The *precieux* had mapped their ideal of tenderness as a debate between love and friendship, the Situationists mapped the drift as a 'discourse on the passions of love,' the subtitle of the *Guide psychogeographique*, reminiscent of the way Andre Breton had compounded his excursions through Paris and through a love affair into his surreal novel *Nadja*" (Sadler, 85–86).

PHOTOGRAPH: BARBARA JO REVELLE

Despite the emphasis on geography in the surviving works and reports, in practice the drift attempted to overcome alienation through encounters with people, especially those denizens of the *demimonde*, bohemian, working-class, dilapidated regions which the SI members preferred. Debord devoted most of his "Two accounts of the Derive" to a tale of meetings in a bar. "On the evening of 25 December 1953, the Lettrists G.I., G.D., and G.L. enter an Algerian bar in the rue Xavier-Privas that they have long referred to as 'Au Malais de Thomas' and in which they had spent the entire previous night. They fall into conversation with an approximately forty-year-old West Indian man, unusual in his elegance amidst the regulars of this dive" (Debord, 136). At one of the meetings with this strange man and his entourage an allusion is made to "Voodoo." "Remarking that a ring that J. was wearing the night before is now on the finger of his wife, G. I. says rather quietly to G. L., alluding to their commentary from the previous evening that had not failed to evoke zombies and the identifying signals of secret sects, 'Voodoo has changed hands.' J.'s wife hears this phrase and smiles with complicity" (136). This mention of Vodou is a moment of passage between that early moment of SI history and the FRE mapping of the Miami River zone.

The consultant drifts in a zone, along the river. She is open to relationships with people and objects and all features of the setting. The warped dimension of place passes through the filter of her sensibility and history to come into appearance in photographs, video tapes, journal entries.

PHOTOGRAPH: BARBARA JO REVELLE

Both the SI and the EmerAgency employ mapping not as an end in itself but as a basis for intervention in city life. A difference between the Situationist goal and that of the EmerAgency is that the former led to direct action in the streets, from carnival to riots (culminating in the carnival riot of May, 1968). The EmerAgency consultation has a different approach to politics, in its long-range goal to develop the Internet as a medium for a fifth estate, creating a kind of "direct democracy" intervention in public policy formation (the virtual civic sphere).

Situationism was for actors; the Emeragency is for spectators. Or rather, egency dissolves this oppostion into "spector," since the Internet in the dromosphere becomes the necessary site of decision. What choragraphy shares with psychogeography is the mapping of ambiance through an arts relationship with objects. In choragraphy the city regains its coherence as an augmented image.

A further relay is the Situationist appropriation of modernist arts devices, which in electracy continue this evolution into the conductive inference procedures of flash reason. To appreciate the extent to which the Situationists drew upon the entire previous experience of vanguardism one need only note their reference to De Chirico's paintings. "Chirico remains one of the most remarkable architectural precursors. He was grappling with the problems of absences and presences in time and space.... In De Chirico's paintings (during his Arcade period) an empty space creates a *full-filled time*. It is easy to imagine the fantastic future possibilities of such architecture and its influence on the masses. Today we can have nothing but contempt for a century that relegates such *blueprints* to its so-called museums" (Chtcheglov, 4).

De Chirico in turn demonstrates the capacity of art to access one's fundamental orientation to the world, one's disposition ("genius"). This signature emblem is what enables an egent to function as category. In the context of his inquiry into "compulsive beauty" Hal Foster refers to a memory reported by de Chirico, who elaborated the scene in numerous early paintings. Foster notes that this scene ultimately is elaborated in all de Chirico's work. "One clear autumnal morning I was sitting on a bench in the middle of the Piazza Santa Croce in Florence. It was of course not the first time I had seen this square. I had just come out of a long and painful intestinal illness, and I was in a nearly morbid state of sensitivity. The whole world, down to the marble of the buildings and the fountains, seemed to me to be convalescent. In the middle of the square rises a statue of Dante draped in a long cloak, holding his works.... Then I had the strange impression that I was looking at all these things for the first time, and the composition of my picture came to my mind's eye. Now each time I look at this painting I again see that moment. Nevertheless, that moment is an enigma to me, for it is inexplicable. And I like also to call the work which spring from it an enigma" (de Chirico, in Hal Foster). This obsessive memory scene illustrates how an epiphany creates a necessary interface, an extimacy, between an external scene and a state of mind.

Magical Sobriety

The De Chirico "blueprint" relates psychogeography to the history of those arts associated with "magic realism," a name applied by Franz Roh in 1925 to

German painting that took up the mode of "metaphysical art" associated especially with De Chirico.

> De Chirico's metaphysical painting was the first to question this "apparent reality." Nietzsche's call for a poetic reappraisal of everyday objects and scenes to reveal the more profound meaning beneath superficial appearances is clearly reflected in the series of Italian squares painted in Paris, and is particularly crucial for an understanding of the Ferrarese metaphysical still lifes. Nietzsche described Turin, surely the most metaphysical of Italian towns, with its vast open squares and nineteenth-century frockcoated monuments, as the only suitable place for him, and his evocation of its atmosphere impressed De Chirico: "The innovation is a strange and profound poetry, infinitely mysterious and solitary, based on the mood of an autumn afternoon when the weather is clear and the shadows are longer than in summer, for the sun is beginning to be lower. Turin is the city where this extraordinary phenomenon is most apparent." (Tisdall, 9)

De Chirico valued Nietzsche for his ability "to gather those 'happy moments of the metaphysical'" (Carra, 20). The declared purpose of his paintings was "to find the daemon in everything.... The mysterious appearance concealed behind every object is revealed to the artist in certain magical or 'abnormal' moments of his creative contemplation" (20).

Although the term "Magic Realism" spread around the world and found an especially receptive cultural climate in Latin America, another term for the new mode – *Neue Sachlichkeit* (New Objectivity) – introduced by Gustav Hartlaub for an exhibition mounted also in 1925 – became the preferred phrase in Germany (Guenther, 34). As with Magic Realism, New Objectivity or New Sobriety involved an approach to "things" – not so much raw material "objects," the translators point out, but *Sache, thing* as fact or as fallen under a category or classification. Here is the chief relevance to choragraphy's focus on the categorial (metaphysical) image. Things in their categories have become "uncanny" in the epochal transition from literacy to electracy. "The goal of this post-World War I art was a new definition of the object, clinically dissected, coldly accentuated, microscopically delineated. Over-exposed, isolated, rendered from an uncustomary angle, the familiar became unusual, endowed with an *Unheimlichkeit* (uncanniness) which elicited fear and wonder. The juxtaposition of 'magic' and 'realism' reflected far more the monstrous and marvelous *Unheimlichkeit* within human beings and inherent in their modern technological surroundings of which both Freud and de Chirico wrote" (36). This "wrestling with the uncanny in the real" expresses a new sense of space and time that Anthony Vidler has called "warped space" (space measured to include subjectivity). In the context of the electrate apparatus, the uncanniness of objects is a symptom of a shift not only in technology and institutional practices but also of identity experience: "spirit" (the ghost-in-me of psyche) is moving again, from the confines of the individual body into the produced environment of object and image (what Scott Bukatman calls "terminal identity").

The choral map brings the objectal details of the scene into a relationship. The things accumulate until the pile becomes a heap, until the flash of the vortex(t) transforms the remainders into a category whose coherence is a mood, The artists teach us about what Heidegger might call the thinging of things. An intuition of the uncanny is the feeling of a new identity formation that is becoming for electracy what selfhood is for literacy.

That Franz Kafka is associated with this German context makes a direct link from the problematic of the object to learning how to map a chora. The link is Deleuze and Guattari's study of Kafka and their reading of his "new sobriety" as a "cartography."

> *The Castle* brings about another sort of progress by breaking away from that which was too spatial in *The Trial* in order to bring out what was already there but still too covered up by spatial figures: the series becomes intensive, the journey reveals itself as an intensity, the map is a map of intensities, and the moving frontiers are themselves thresholds (blocks of intensity). The whole first chapter of *The Castle* works in this mode, from threshold to threshold, from low intensities to high ones and vice versa, part of a cartography that is certainly not interior or subjective but that has definitely ceased to be spatial. The low intensity of the bent head, the high intensity of the head that straightens and the sound that takes flight, a passage from one scene to another by thresholds: a language that has become intense makes its contents take flight in place on this new map. (Deleuze and Guattari, 1986: 78)

The artist works neither subjectively nor objectively in the established sense of these terms, but "machinically": As a "bachelor machine," gathering an assemblage, forming connections, not as "impressions" or "expressions" of a "self" but as a vehicle or channel for a group, a collective, a community. The egents of the EmerAgency learn to map intensities in this way, in an attunement that is categorial rather than personal (or, the personal as category). The methods include the juxtaposition or superimposition of "blocks" (rather than "memories"), especially of a childhood block operating simultaneously in adult conduct (an updating of "transference"). "Memory yells 'Father! Mother!' but the childhood block is elsewhere, in the highest intensities that the child constructs with his sisters, his pal, his projects and his toys, and all the nonparental figures through which he deterritorializes his parents every chance he gets" (79).

PHOTOGRAPH: BARBARA JO REVELLE

Image metaphysics is aesthetic, and ontologizes affective capacities of the body in the way that literacy ontologized analytical capacities of the mind. Such blocks open new passages through information, help overcome impasses paralyzing or fixating adult conduct and set desire moving again along a line of flight, a pathic dimension of intensities. The key lesson of this theory is that an egent is a "minor consultant," channeling the very dimension of discourse in which values are circulated among the parties to a group subject. Information is stored and retrieved in a choral zone by means of affective intensities (manipulated with the devices of art as rhetoric).

> What Kafka suggests for literature is even more valid for cinema, in as much as it brings collective conditions together through itself. And this is in fact the last characteristic of a modern political cinema. The point of

> view of culture, is doubly colonized: colonized by stories that have come form elsewhere, but also by their own myths become impersonal entities at the service of the colonizer. The author must not, then, make himself into the ethnologist of his people, nor himself invent a fiction which would be one more private story: for every personal fiction, like every impersonal myth, is on the side of the "masters." ... There remains the possibility of the author providing himself with "intercessors," that is, of taking real and not fictional characters, but putting these very characters in the condition of "making up fiction," of "making legends," of "story-telling." The author takes a step towards his characters, but the characters take a step towards the author: double becoming. Story-telling is not an impersonal myth, but neither is it a personal fiction: it is a word in act, a speech-act through which the character continually crosses the boundary which would separate his private business from politics, and which *itself produces collective utterances.* (Deleuze, 1989: 222)

The egent (EmerAgency consultant) is at once the "author" and "characters" of this telling, and the "myths" in question are those that guide the ethics and politics of a coming global community. The desiring machine of writing operates directly with the real (in the manner of cinema – writing with objects and activity rather than with abstractions and symbols), with passage from one

PHOTOGRAPH: BARBARA JO REVELLE

contiguous thing to the next. Deleuze and Guattari replace "aesthetic" with "machinic" to name this *transversal* connective movement, which they call the logic of sense, and which in electracy is the inference procedure called "conduction," at work in flash reason.

> The points of connection between series or segments, the remarkable points and the singular points, seem in several respects to be aesthetic impressions. They are often sensible qualities, odors, lights, sounds, contacts, or free figures of the imagination, elements from a dream or a nightmare. They are connected to chance. For example, in the fragment "The Substitute," three points of connection intervene: the portrait of the king, the little bit of the phrase that the anarchist is supposed to have pronounced ("Hey, you up there, bandit!"), the popular song ("While the little lamp is burning"). They intervene to determine new couplings, and cause the proliferation of series; and the substitute notes that they can enter into innumerable polyvocal combinations, forming segments that are more or less near each other, more or less distant. But it would be a great error to refer the points of connection to the aesthetic impressions that subsist in them. *Everything Kafka does works to an exactly opposite end,* and this is the principle behind his antilyricism, his anti-aestheticism: "Grasp the world," instead of extracting impressions from it; work with objects, characters, events, in reality, and not in impressions. Kill metaphor. (70)

Guattari clarifies that the method continues a "certain" aesthetic state of mind: "The aesthetic power of feeling, although equal in principle with the other powers of thinking philosophically, knowing scientifically, acting politically, seems on the verge of occupying a privileged position within the collective Assemblages of enunciation of our era" (Guattari, 1995: 101). They are worried about the baggage associated with "aesthetic" in the post-Holocaust condition in which Hitler usurped the associative network of common sense (Deleuze, 1989: 264). The image crisis in its larger context may be understood in these terms: Hitler remains within the interior in the common sense of each modern person.

"Schizoanalytic cartography" shows how to intervene in any impasse, how to map the fixed borders and lines moving through them, in the relation of individuals to communities and institutions. "The most autistic psychical world is not in itself lacking in alterity. It is simply engaged in a constellation of Universes disconnected from the dominant assemblages of sociality. Lines can be thrown to the psychotic by mediations which will give consistency to certain of these components of Universes, or by the aggregation of other components which did not previously exist. (Through the introduction of materials of expression unknown to the subject, for example, relating to the plastic arts, video, music, theater or quite simply… cooking!)" (71). The mapping has a purpose, which is to find that autopoeitic umbilical point of connectivity that transforms chaos into assemblage (a good description of chora). "Worlding a complexion of sense always involves taking hold of a massive and immediate ensemble of contextual diversity, a fusion in an undifferentiated or rather de-differentiated, whole. A world is only constituted on the condition of being inhabited by an umbilical point – deconstructive, detotalizing and deterritorialising – from which a subjective positionality embodies itself" (80). This new cartography maps not (only) geography but the habitus, the socius, the category formation of the scientific worldview, by means of an alternative (electrate) "metaphysics": "On the one hand, formalist reasoning seeks clues from which to gain access to its interpretations, to a hermeneutic, on the other, an apparently mad reasoning starts from the notion that universality is to be found in the direction of singularity, and that singularity can become the authentic basis for a political and micro-political organization that is far more rational than what we have at the moment" (Guattari, 1984: 164).

The egent's testimonial operates with this categorical singularity. Guattari confirms the value of syncretizing poststructural epistemology with the oral practices of Vodou with respect to the heterological conditions of modernity.

> Archaic societies are better equipped than White, male, capitalistic subjectivities to produce a cartography of this multivalence of alterity. With regard to this, we could refer to Marc Auge's account of the heterogeneous registers relating to the fetish object Legba in African societies of the Fon.… The Legba is a handful of sand, a receptacle, but it's also the expression of a relation to others. One finds it at the door, at the market, in the village square, at crossroads. It can transmit messages, questions, answers. It is also a way of relating to the dead and to ancestors. It is both an individual and a class of individuals: a name and a noun.… Contemporary machinic assemblages have even less standard univocal referent than the subjectivity of archaic societies. But we are far less accustomed to the irreducible heterogeneity, or even the heterogenetic character, of their referential components. Capital, Energy, Information, the Signifier are so many categories which would have us believe in the ontological homogeneity of referents. In the context of a reductionist modernity, it is up to us to rediscover that for every promotion of a machinic intersection there corresponds a specific constellation of Universes of value from the moment a partial non-human enunciation has been instituted. (Guattari, 1995: 45–47)

The point of our trajectory, from "situation" to "machinic assemblages," is just to say that choragraphy derives from a certain modernist arts tradition of poetic encounters in the city a practice of information design. The artists have done for electracy what the philosophers did for literacy: invented the logic that is the basis of a skill-set for netizens in a digital civilization. Our task in this study is to show how this logic may be mounted and applied in a civic practice (the testimonial).

Those Haitian traders barely visible in "Crossroads"... An encounter.

CHAPTER 4

Encounter

Third Meanings

The egent is in Miami, staying at the Miami Inn, going out every morning with her cameras, practicing the drift.

The procedure for making images in the zone is based on Roland Barthes' theory of how to read the "third meaning" of a photograph. Barthes's way of reading (hermeneutics) is also a way of making (heuretics). The egent does not and could not plan or intend what Barthes called an obtuse meaning, neither to include nor exclude this dimension. "The detail which interests me is not, or not strictly, intentional, and probably must not be so; it occurs in the field of the photographed thing like a supplement that is at once inevitable and delightful; it does not necessarily attest to the photographer's art; it says only that the photographer was there, or else, still more simply, that he could not not photograph the partial object at the same time as the total object (how could Kertesz have 'separated' the dirt road from the violinist walking on it?). The Photographer's 'second sight' does not consist in 'seeing' but in being there" (Barthes, 1981: 47).

PHOTOGRAPH: BARBARA JO REVELLE

 | DOI:10.1163/9789004534643_005

Electracy is not only a camera, but also a thought practice (flash reason). To make a time-image adequate to the dromosphere the egent becomes a voyant. The consultation as a practice depends upon the egent's experience of recognition. Making is undertaken from a position of reception (receptacle). The material cause of the punctum that stings the viewer is some detail in the photograph. In a picture of a family of American blacks, the spectacle of their household interests Barthes, but he is stung by "the belt worn low by the sister," and above all "by her strapped pumps" (being out of fashion, they refer him to a specific date in time) (43). The key to the method, and to the obtuse as an experience, is in the relationship between the detail in the image and the memories and their associated emotions that the detail awakens in the viewer.

> However lightning-like it may be, the punctum has, more or less potentially, a power of expansion. This power is often metonymic. There is a photograph by Kertesz (1921) which shows a blind gypsy violinist being led by a boy; now what I see, by means of this 'thinking eye' which makes me add something to the photograph, is the dirt road; its texture gives me the certainty of being in Central Europe; I perceive the referent (here, the photography really transcends itself: is this not the sole proof of its art? To annihilate itself as medium, to be no longer a sign but the thing itself?), I recognize, with my whole body, the straggling villages I passed through on my long-ago travels in Hungary and Rumania. (Barthes, 1981: 45)

This recognition recalls the workings of a mode of experience called variously epiphany, attunement, extimacy (Lacan), whose purpose in the testimonial is to locate and map for the egent the categorial border between inside outside (of everything: logic, personal identity, nation – the ideological boundaries). The detail attunes Barthes to the setting, objects, characters of the scene in a way mapped by the inside-outside figure of a moebius strip – the extimate topology of the psyche. "The punctum, whether or not it is triggered, it is an addition: it is what I add to the photograph and what is nonetheless already there. To Lewis Hine's retarded children I add nothing with regard to the degenerescence of the profile: the code expresses this before I do, takes my place, does not allow me to speak; what I add – and what of course is already in the image – is the collar, the bandage.... Once there is a punctum, a blind field is created (is divined)" (57). The colloquial use of "divined" is reconnected with its ritual meaning in testimonial. The punctum, then, occurs when there is a match between a signifier in the scene (in the photograph, or in the zone), and a scene in the memory.

PHOTOGRAPH: BARBARA JO REVELLE

Haiku

Barthes' use of Haiku as an analogy for how a photograph produces a kind of satori brings photography into relation with theoria and illuminates how state of mind informs a practice. Our sense of how a theoros operated was influenced, that is, not only by the Greek and Hebrew versions of the institution, but by the "bhikku" or spiritual wanderers of Japan, exemplified in the life and work of Basho. Basho's masterpiece, *Back Roads to Far Towns*, is a travel journal recording in a prose poem style ("haibun") – punctuated by haiku – his nine-month journey (in 1689) with his companion, Sora, through the mountains and along the sea coast of the region between old Tokyo and Kyoto. "It was to be more a pilgrimage than a case of wandering scholarship: a sight not uncommon even in modern Japan, visiting from temple to temple, seeing old acquaintances, places famed in history or poetry or legend, touchstones for the life lived, the dying to come and what life continues" (Corman, 7). This relay clarifies the process of constructing a chora, as we are adding a stratum of art discourse over the initial documentation of site as history, news, tourism.

Haiku poetics is a relay for the poetic method of testimonial. Especially instructive is the talent of the Japanese for articulating their aesthetic categories, assigning names and concepts to the effects and devices that produce them. Basho helps bring into focus "state of mind" as the product of testimonial. We may learn from him the possibility of bringing into an image (a text of word and picture) the mood of a place. Basho called "slenderness" the ability "to immerse himself into the heart of an object or an incident and to catch the impersonal mood it shares with the universe" (Ueda, 1995: 160). The experience was that of attunement (but which Basho called "inspiration"): "If you get a flash of insight into an object, record it before it fades away in your mind." The method was to "let there not be a hair's breadth separating your mind from what you write. Quickly say what is in your mind; never hesitate at that moment" (162). The state of mind that Basho sought, as might be expected in the context of his Buddhism, had all the attributes of "ataraxy" – serenity, quiet, ascetic distancing from desires. His existential feeling about the life-world resonates with that of the shipwreck metaphor. "For Basho, sorrow was the word to describe life and the world at large; in his view this life was a 'demoniac world of the lusts' and mankind was 'drowning in a filthy ditch'" (155). The poet did not withdraw from ordinary life, but lived within it unperturbed, to the extent at least that he could sustain the state of mind he called "loneliness" – distinguished from "sorrow" by being an "impersonal emotion." This desired state of mind is "sabi," derived from "sabishi."

> Loneliness –
> Standing amid the blossoms,
> A cypress tree.

The effect of sabi was achieved by means of "shiori," referring to language that is flexible, supporting productive ambiguity. The verb meaning "to bend" (flexibility) is homophonous with the verb "to wither," and in Basho's poetics the two terms merged, to give the instruction that the sabi mood could be produced by the image of a "withering flower" (158).

God-Is-Able impounded, mattresses, buckets, bicycles. Haiti waiting.

John Frow clarified the links among Basho's poetics, tourism, photography, and the experience of recognition, declaring that "Basho's voyage is precisely a model of contemporary tourism":

> The writings of the ancient poets establish the formal essence of the tree, and all later seeing is governed by the possibility of conformity to this pattern. Just as the tourist guidebook stipulates an ideal core of interest in the sight, so the authority of a poetic tradition that constantly refashions

> the essence of the tree, its normative beauty constrains the visitor to a recognition of a conformity between the particular, more or less contingent shape of the tree and its ideal form. The poem by Kyohaku provides a second modeling of the form for Basho, and his own poem confirms (like the tourist's photograph) not an empirical act of seeing but the congruence of the sight with the idea of the sight. The poetic record thus promulgates a form of knowledge that can be recognized in and has a greater force than the appearances of the world. What the traveler sees is what is already given by the pattern. (Frow, 1991: 124–125)

For EmerAgency consulting, "empirical seeing" is the name of a blindness, an inadequacy of sight. In the context of cognitive psychology, or prototype theory, Frow's account shows how a poem and a photograph participate in the dynamic of schema formation (the mnemonic connection between concepts and images, abstract categories and particular embodiments). It was never a question of separating Miami from Myami (the architectural and the imagined city) but of materializing their shared border, in the manner of Robert Smithson documenting the "monuments" of his hometown in New Jersey. "Noon-day sunshine cinema-ized the site, turning the bridge and the river into an over-exposed picture. Photographing it with my instamatic 400 was like photographing a photograph. The sun became a monstrous light-bulb that projected a detached series of 'stills' through my instamatic into my eye. When I walked on the bridge, it was as though I was walking on an enormous photograph that was made of wood and steel, and underneath the river existed as an enormous movie film that showed nothing but a continuous blank" (Smithson, 1979: 52–3). Smithson's tour of Passaic, New Jersey, exemplifies the composite nature of theoria, making a third position from the paths or ways of theory and tourism. His example is all the more important as a relay because of the fact that Passaic was his hometown. The homology between technology and culture manifests itself in the use of the camera, the snapshot, which is precisely the equipment necessary for the traumatized witness.

As every tourist knows who prefers the photo to the site, the camera is a Chorus that has our epiphanies for us, saving them to allow for the time of understanding. The tourist trail, with its predetermined vantage points marked as "photo opportunities," institutionalizes what was once the privilege of mystics and then poets (and this vernacular mass habitus offers an opportunity for appropriation by theoria). Walter Benjamin commented on the way that technologies of mechanization such as the camera were related to new behaviors. "Of the countless movements of switching, inserting, pressing, and the like, the 'snapping' of the photographer has had the greatest consequences. A touch of the finger now sufficed to fix an event for an unlimited period of time. The camera gave the moment a posthumous shock.... Thus technology has subjected the human sensorium to a complex kind of training. There came a day when a new and urgent need for stimuli was met by the film. In a film, perception in the form of shocks was established as a formal principle" (Benjamin, 1969: 174–175). The camera is the equipment best coordinated with the extimacy of traumatized, alienated memory. Moreover, words are not excluded from this "cinema-ized" mode of production (the poststructural theory of "text" refers to just such a modality).

Revelle looked at the Miami River the way Barthes looked at the photograph by Kertesz. She did not try to photograph the third meaning (that is impossible by definition). Rather, she imaged the place, and the obtuse dimension emerged afterwards, when the group (the FRE*) read the record (the obtuse is an event of gaze).*

Dialectical Image

Electracy is the apparatus of imaging, just as literacy is the apparatus of writing. What is the genealogy of the third meaning (the experience of image reason)? Walter Benjamin anticipated Barthes's punctum in his "Small History of Photography." "No matter how artful the photographer, no matter how carefully posed his subject, the beholder feels an irresistible urge to search such a picture for the tiny spark of contingency, of the here and now with which reality has so to speak seared the subject, to find the inconspicuous spot where in the immediacy of that long-forgotten moment the future subsists so eloquently that we, looking back, may rediscover it" (Benjamin, in Gilloch, 113). Benjamin established the precedent for adapting photography – the lightning flash of the moment frozen in time in the still – and cinematography (montage), to the work of philosophical discourse. The method began with the *Denkbild* (thought-image) used to capture the feel and mood of cities such as Naples and Moscow. This flash must be put into a general practice, for a public sphere adequate to the dromosphere.

The Situationist drift through a neighborhood of a city, mapping its mood in pictures and text, is already there in Benjamin's application of the *flaneur's* wandering and the journalist's reportage to produce the *Denkbild*. The strategy was to attend to the ephemeral, marginal details of daily life in some part of a city, and to attempt to capture the peculiar specificity of the place (*genius loci,* spirit of place) in an anecdotal scene, such as this incident in Naples: "Some years ago a priest was drawn through the streets of Naples for indecent offenses. He was followed by a crowd hurling maledictions. At a corner a wedding procession appeared. The priest stands up and makes the sign of the cross and the cart's pursuers fall on their knees" (29). The *Denkbild* method perhaps took the photographic analogy too literally, in that it was limited to concrete descriptions without reflective commentary. Nonetheless, in the manner of the ideogram, the collection of details were gathered into coherence around a particular mood or feeling that Benjamin named. Naples, for example, gave him a feeling of "porosity," because of the way the private domestic and public street worlds flowed into each other, held apart only partially by an architectural divide (34).

As his work matured Benjamin developed the thought-image into the dialectical image, based on Eisenstein's cinematic montage, itself an adaptation of cinematic editing (from D. W. Griffith) and the photography of events and action, to intellectual discourse. Eisenstein was learning how to perform Marx's philosophical method of dialectical materialism in the medium of (silent) film. The other primary influence on Benjamin's image text was Proust. Gilloch declares Proust's "involuntary memory" to be the key to the dialectical image, and we would clarify, to the time-image. Deleuze considers the cinematic time-image as the means by which photography gains access to Proust's involuntary memory. "The direct time-image always gives us access to that Proustian dimension where people and things occupy a place in time which is incommensurable with the one they have in space. Proust indeed speaks in terms of cinema, time mounting its magic lantern on bodies and making the shots coexist in depth. It is this build-up, this emancipation of time, which ensures the rule of impossible continuity and aberrant movement. The postulate of 'the image in the present' is one of the most destructive for any understanding of cinema" (Deleuze, 1989: 39). This juxtaposition of past and present in a flash is necessary for decisions in Now-time. Eisenstein based his filmic discourse not only on the example of other filmmakers such as Griffith, but also on the Japanese writing system, the kanji, similar to Pound's use of the Chinese ideogram as a relay for the invention of an image category. "By what route is it possible to attain a heightened graphicness combined with a realization of the Marxist method? The first stop along the path will be to carry the montage principle over into his story. That is, to build up the large constructions out of the smallest, precisely fashioned structural elements. Indeed to detect the crystal of the total event in the analysis of the small, individual moment" (Benjamin, in Gilloch, 112).

In his memoir of Berlin Benjamin practiced the method of involuntary memory in a directly Proustian way, using the objects and places of his childhood to evoke personal memories. In the Arcades Project the method is applied to Paris as an image not of a biography but of a social history. In either case, the dialectical image produces insight through a juxtaposition of past and present circumstances. "'It is not that the past casts its light on the present or the present casts its light on the past: rather, an image is that in which the Then and the Now come into a constellation like a flash of lightning. In other words: image is dialectics at a standstill' (Benjamin). The dialectical image is a pause, a moment of interruption and illumination, in which past and present recognize each other across the void which separates them" (113). Gilloch cites Ralf Konnersmann on Benjamin's metaphor of historiography as (photographic) image. "The metaphor of the photographic snapshot encapsulates and illustrates several of those attributes which characterize the conditions and modes of this historiography: the transience of the chance which presents itself; the suddenness with which the motif appears; the momentariness of the truth which is to be established; the fleetingness of the spatio-temporal constellation in which one must act; the visualization of the past as an image which receives its illumination from references to the present" (Konersmann, in Gilloch, 113). The suddenness of epiphany, being an event outside of time, gains through aspect some perspective on the instant Now.

"Crossroads" is a dialectical image.

Back of Beyond

The point for understanding how images form electrate categories is that the personal private and collective public histories are brought into explicit relationship (correspondence). It is not that the egent suspends her own formation to record in some empirical or neutral manner the feeling of the place, but that there is an "attunement" between two situations (Then and Now, There and Here). The dialectical image activates (necessarily "involuntarily") the past of both the egent and the zone in a way that reveals the propensity, the direction, of the forces at work in the zone. This is the counterintuitive quality of warped space: the singular instance generates a holistic coherence through shared feeling of mood (mood as the equivalent of Kant's "common sense"). How this generalizing of the concrete occurs may be seen by considering further the nature of involuntary memory. The effect of this mnemonics is to help subjects recover the conditions of the situation in which their identity was constructed, prior to "the age of reason."

In Proust the mood thus recovered when he tasted the madeleine dipped in tea was one of happiness. "No sooner had the warm liquid, and the crumbs with it, touched my palate than a shudder ran through my whole body and I stopped, intent upon the extraordinary changes that were taking place. An exquisite pleasure had invaded my senses" (Proust, 34). He begins to reflect. "Undoubtedly what is thus palpitating in the depths of my being must be the image, the visual memory which, being linked to that taste has tried to follow it into my conscious mind" (35). Finally the past scene comes to mind.

> And once I had recognized the taste of the crumb of madeleine soaked in her decoction of lime-flowers which my aunt used to give me (although I did not yet know and must long postpone the discovery of why this memory made me so happy) immediately the old gray house upon the street, where her room was, rose up like the scenery of a theater to attach itself to the little pavilion, opening on to the garden, which had been built out behind it for my parents (the isolated panel which until that moment had been all that I could see); and with the house the town, from morning to night and in all weathers, the Square where I was sent before luncheon, the streets along which I used to run errands, the country roads we took when it was fine. And just as the Japanese amuse themselves by filling a porcelain bowl with water and steeping in it little crumbs of paper which until then are without character or form, but, the moment they become wet, stretch themselves and bend, take on color and distinctive shape, become flowers or houses or people, permanent and recognizable, so in that moment all the flowers in our garden and in M. Swann's park, and the water-lilies on the Vivonne and the good folk of the village and their little dwellings and the parish church and the whole of Combray and of its surroundings, taking their proper shapes and growing solid, sprang into being, town and gardens alike, from my cup of tea. (36)

The times of Then and Now appearing together in the dialectical image are both equally saturated with preexisting, default moods. Here is the purpose of chorography as map: to record and track and render accessible the moods already circulating in a zone. These moods, anchored in objects, places, events, people, carry values that guide action and thought. Such a map is of at least two regions at once, and through this alignment a third time and place (*chronotope*) *become legible* (*not yet, elsewhere*).

To fully unpack Proust's memorial scene required considerable thunder of text – the whole of *The Remembrance of Things Past*, in which Proust analyzed his own sensibility in relationship to what amounts to a dramatized sociology of class relations in the France of his era. Benjamin's feeling about being raised in an upper middle class home was less happy. What bothered him in retrospect was the bourgeois "illusion of security," of the permanence of their world (Gilloch,

76). The mood for Benjamin of his grandmother's house was not that of a happy garden but a murder scene in a detective novel. "The furniture style of the second half of the nineteenth century has received its only adequate description, and analysis, in a certain type of detective novel at the dynamic center of which stands the horror of apartments. The arrangement of the furniture is at the same time the site plan of deadly traps, and the suite of rooms prescribes the fleeing victim's path" (Benjamin, in Gilloch, 81). "On this sofa the aunt cannot but be murdered," Benjamin observed, associating the middle-class furnishings with the stagecraft of the baroque *Trauerspiel* and its primary emblem, the corpse and its related idea of *vanitas*, death and ruin. Alienated from this repressive atmosphere, Benjamin as adult was attracted aesthetically, ethically, and politically to the "unrespectable" regions of the city – proletarian neighborhoods and the bohemian hangouts of the red-light districts. As a child these districts were forbidden, off-limits, designated as "back of beyond" (85).

Here is a source of that expectation of safety noted in the Florida image crisis.

In his writings about his childhood in Berlin Benjamin recorded what the Italian poet and filmmaker Pier Paolo Pasolini called "the pedagogy of things." The egent's sensibility (identity, worldview, habitus) is formed first of all by an experience of things bearing values and behaviors, which are interiorized, memorized in the body and spirit.

> Our first memories are visual ones. In memory life becomes a silent film. We all have in our minds an image which is the first, or one of the first, in our lives. That image is a sign, or to be exact, a linguistic sign.... The first image of my life is a white, transparent blind, which hangs – without moving, I believe – from a window which looks out on to a somewhat sad and dark lane. That blind terrifies me and fills me with anguish: not as something threatening and unpleasant but as something cosmic. In that blind the spirit of the middle-class house in Bologna where I was born is summed up and takes bodily form. Indeed the images which compete with the blind for chronological primacy are a room with an alcove (where my grandmother slept), heavy "proper" furniture, a carriage in the street which I wanted to climb into. These images are less painful than that of the blind, yet in them too there is concentrated that element of the cosmic which constitutes the petty bourgeois spirit of the world into which I was born. But if in the objects and things the images of which have remained firmly in my memory (like those of an indelible dream) there is precipitated and concentrated the whole world of "memories,"

> which is recalled by those images in a single instant – if, that is to say, those objects and those things are containers in which is stored a universe which I can extract and look at, then, at the same time, these objects and things are also something other than a container.... Their communication was essentially instructional. They taught me where I had been born, in what world I lived, and above all how to think about my birth and my life. (Pasolini, 1983: 28–29)

Interpellation (*Hailing*)

The punctum is an uncanny experience of retrospection: we recognize the forgotten yet familiar. This pedagogy of things is one channel of the process of identity construction known as interpellation. One hypothesis of the consultation is that the feeling of recognition associated with interpellation (the event of hailing that is the hypericon of Althusser's theory of ideology: when the cop yells at the departing crowd, "Hey you!," who turns around?) could, if supported, augmented, enhanced by the prosthesis of internet technology and the testimonial practice, be transformed into a capacity of a different sort. The psycho-marxist theory of identification – "interpellation" – provides a reference point for how discourse formations operate. The exchange is made through the multiple meanings of "recognition," related to the ideological fable of intersubjectivity as a Master-Slave relationship. The two meanings (my ability to identify something previously seen; my demand for acknowledgment) are based on my "position" in society.

The relevance of "discourse" to the new consultancy is that "individuals and interest groups will simply generate a social problem out of their own interests, with or without the data from objective reality.... It is the claims-making activity of various actors – journalists, politicians, social workers, union organizers – that defines a social problem" (Croucher, 1997: 14). Testimonial is not so much a language of its own, but a way to bring into relationship all the discourses (the institutions) in which our lives are conducted. Althusser borrowed his notion of a recognition scene, in which a person experiences the real as a surprising encounter (*tuche*) from Lacan's psychoanalytic theory (*tuche* as the point of collapse between the psychic and material orders) which in turn drew upon the tradition of epiphany – a tradition connecting the surrealists to the Magi (Cohen, 1993: 153). The Florida image crisis was a matter of how discourses interact with states of mind. This interaction is mediated in terms of the atmospheres, the moods, evoked by images. The "feelings" one has of being American or Chinese, Protestant or Catholic, Female or Male, or the sense of belonging to any given identity category are carried by atmosphere. There is more or less one predominant mood that is the default for each identity category, determined in each case by the norms of the hegemonic order in a given society (Caucasian Protestant Bourgeois, or Arab Muslim Proletarian, for example). Advertising displays the fundamental trait of electracy in its nearly pure capacity to manipulate atmospheres. Ralph Lauren, for example, designed clothing that evokes "the whole atmosphere of the good life," as established by classic cinema: "One imagines that his Paris is the Paris of *Gigi*, his Wild West that of *High Noon*, his Main Line Philadelphia the house where Katherine Hepburn lived in *The Philadelphia Story*" (Holly Brubach, in Santner, 184).

The experience of interpellation as a direct revelation of belief offered a potential point of departure for the testimonial. The generalization of the epiphany to politics revises the experience of the punctum to be a description of what happens when interpellated subjects take notice of the upholstery tack (quilting point) in their psyche. The image (quilting point) alludes to Jacques Lacan's metaphor for the way the figure of the other (scapegoat) sutures or achieves (illusory) closure in a given ideological system. Or, we might say that when individuals recognize themselves (identify with) the subject positions prepared for them in advance by the dominant ideology within the discourses of the Symbolic order, they are accepting default or readymade (dead) epiphanies. The "point" is to discover the place from which one is seen in the field of identity. The dominant discourse operates not only by power (externally) but also as knowledge, internally. Individuals (minorities and majorities alike) are

PHOTOGRAPH: BARBARA JO REVELLE

subjected "by the power of inner compulsion and subjective con-formation to the norm," which was the lesson that Stuart Hall drew from Frantz Fanon's *Black Skin, White Masks*. "I was responsible at the same time for my body, for my race, for my ancestors," Fanon wrote. "I subjected myself to an objective examination, I discovered my blackness, my ethnic characteristics; and I was battered down by tom-toms, cannibalism, intellectual deficiency, fetishism, racial defects, slave-ships, and above all else, above all: 'Sho' good eatin'" (Fanon, in Jordan and Weedon, 303).

Chora maps the default moods at work in the configuration of a region. What are the moods organizing the Miami River? Search the zone, or the images of the zone, for a triggering detail.

Epiphany

The punctum is a relay that takes us to the core (heart) of the testimonial. The method of the dialectical image, that is, depends upon a specific kind of experience, physically felt during the consultation. This feeling expresses a certain kind of intelligence whose operations it is the purpose of the EmerAgency to

PHOTOGRAPH: BARBARA JO REVELLE

enhance and augment. The goal is to combine a practice with a technology (testimonial on the internet) that supports the feeling that Barthes associates with satori. The hypothesis is that there is a continuum, a family resemblance, such that interpellation or "hailing" (the identifications with "parent" or authority figures formed as part of the construction of identity within the various institutions of the popcycle) may transform into satori. The "jubilation" manifested by infants recognizing their own image in a mirror is the basis for every experience of "Aha!" (eureka). To say that testimonial is to electracy what tragedy was to literacy is to suggest that anagnorisis and peripeteia, performed by the actors in theater, are performed in testimonial by netizens online. The family of terms, arranged more or less in an order of ascending "insight," include: interpellation (misrecognition), recognition, the uncanny, epiphany, anagnorisis, eureka, satori. We will perhaps have to add a new term to this family, a term from the creole vocabulary (possession?).

What is the mood (attunement) of the modern city? To what extent is this attunement writable (scriptible) as well as readable? The poetic recording of a city environment using epiphany is a commonplace of modernist poetry. Indeed, one way to consider our experiment is as a "popularization" of modernist poetics, or postmodernizing (in that testimonial is an appropriation, decontextualizing and remotivating, of an extant practice), since our approach as consultants is to take something invented within the specialized disciplines of Arts and Letters and move it into the public realm, to make a practice for the general public. Our premise is that the modernist epiphany lends itself to an electrate method, to open a dimension of experience that digital technology supports (some parts of epiphany are relevant to electracy, and some are not). This dissemination of a practice from its origin into other institutions and discourses is a given of the relationship between the sciences and the community (our name for this circulation of ideas and practices is "popcycle"). The calculus that Newton invented is now taught in high school, and is still relevant despite the continuing evolution of Physics and Mathematics. "Poetry" has this same potential functionality for general education electracy.

Gesture

In a testimonial the "flash of spirit" arises in the juxtaposition of a poetic image between two situations (individual and collective – the constructed subject and the discourse formation) in the consultation. To undertake her drift, her psychogeographic mapping, of the Miami River, Barbara Revelle settled in a bed and breakfast place and became acquainted with the denizens of the zone – a heterogeneous group representing a cross-section of urban types, from Coast Guard officers to the homeless. An example of using story to bring together private and social worlds in the mode of attunement is Rilke's *The Notebooks of Malte Laurids Brigge*. Brigge, the poet's alter ego, undergoes a version of Baudelaire's mood of spleen while wandering the streets of Paris. The story or anecdote describes Brigge's encounter on the Boulevard Saint-Michel with a poor fellow who is an object of bemused derision for the crowd. The fellow, one of the "castaways of life" whose "shipwreck" Brigge witnessed, suffers from St. Vitus's Dance. The effect on Brigge (who has been reading Baudelaire's poems) of witnessing the scene in which the sufferer struggles to suppress his symptoms only finally to give in to the spasms of the illness, is a kind of epiphany. The scene returns his gaze, meaning that a correspondence forms between the "dancer" in the street and Brigge's inner world. Brigge identifies with the humiliated dancer, and the outer scene becomes a map to the poet's inner reality. The experience involves the same immersion in the object spoken of in Haiku poetics.

James Joyce expanded upon the use of the poetic epiphany as a device for rendering the mood of a city (Dublin). His fiction shares the quality with much other modernist fiction of being an allegory for which no interpretive key exists. "When the material of experience assails us without our possessing its interpretive framework – when we notice that the codes of interpretation can be different, more open, flexible, and full of possibilities, and yet we still don't have the key for using them – then experience must show itself directly in the word" (Eco, 1982: 37). For Joyce this meant that "the form of the story becomes a revelatory image of an entire situation and is, using the categories of Stephen Dedalus, an epiphany. But it is not an epiphany-vision; it is an epiphany-structure" (37). Each story in *Dubliners* is described as one vast epiphany: "key moments that arise in a realistic context. They consist of common facts or phrases but acquire the value of a moral symbol, a denunciation of a certain emptiness of existence. The vision of the old dead priest in the first story, the squalid inanity of Corley with his smile of triumph while showing the little coin in "Two Gallants," the final crying of Chandler in "A Little Cloud," and the solitude of Duffy in "A Painful Case" are all brief moments that turn moral situations into metaphors as the result of an accent placed imperceptibly upon them by the narrator" (25). These moments resonate with Benjamin's *Denkbild* anecdotes.

A similar effect is achieved in the photograph or film "still," as Barthes argued, referring to the tradition in the arts of the "tableau vivant," of representing "the pregnant moment" – the gesture that sums up all (Barthes, 1977: 73). The punctum triggered by recognizing a detail in the image (the emergence of a match between a signifier and a memory), may not touch immediately, but, like Wordsworth's "spots of time" (one of the key moments in the secularizing of the epiphany), only as an after-effect, recollected in tranquility.

> Odd contradiction: a floating flash. Nothing surprising, then, if sometimes, despite its clarity, the punctum should be revealed only after the fact, when the photograph is no longer in front of me and I think back on it. I may know better a photograph I remember than a photograph I am looking at, as if direct vision oriented its language wrongly, engaging it in an effort of description which will always miss its point of effect, the punctum. Reading Van der Zee's photograph, I thought I had discerned what moved me: the strapped pumps of the black woman in her Sunday best; but this photograph has worked within me, and later on I realized that the real punctum was the necklace she was wearing; for it was this same necklace (a slender ribbon of braided gold) which I had seen worn by someone in my own family, and which, once she died, remained shut up in a family box of old jewelry (this sister of my father never married, lived with her mother as an old maid, and I had always been saddened whenever I thought of her dreary life). I had just realized that however immediate and incisive it was, the punctum could accommodate a certain latency (but never any scrutiny). (Barthes, 1981: 53)

Brecht's concept of the "social gest" is a culmination of this tradition, and is, according to Barthes, "the most intelligent concept ever produced in dramatic theory" (73). "[The social gest] is a gesture or set of gestures (but never a gesticulation) in which a whole social situation can be read. Not every gest is social: there is nothing social in the movements a man makes in order to brush off a fly; but if this same man, poorly dressed, is struggling against guard-dogs, the gest becomes social" (74). Collectively the moral scene of *Dubliners* is a vision of the city in a state of "paralysis" (Eco, 1982: 33). The "poses" that Barthes inventoried in the scenario of "the lover's discourse" become in cinema the seeds around which crystallize image categories.

> "Give me a body then": this is the formula of philosophical reversal. The body is no longer the obstacle that separates thought from itself, that which it has to overcome to reach thinking. It is on the contrary that which it plunges into or must plunge into, in order to reach the unthought, that is life. Not that the body thinks, but, obstinate and stubborn, it forces us to think, and forces us to think what is concealed from thought, life. Life will no longer be made to appear before the categories of thought; thought will be thrown into the categories of life. The categories of life are precisely the attitudes of the body, its postures. (Deleuze, 1989: 189)

Attunement

A further contribution to understanding the categorial capacities of image is Heidegger 's notion of attunement, which goes back to *Timaeus* (which in turn was informed by the Pythagorean notion of the "music of the spheres"). In his history of the term "Stimmung," Leo Spitzer noted the importance of *Timaeus* for figuring the world soul as musical. The importance of music in classical education through the middle ages was based on the idea of morality as a tuning of the individual soul to this world harmony (Spitzer, 34). Heidegger revisited this tradition as part of his return to the Greeks, designating "Stimmung" as one of the existentials grounding one's being in the world. It named the ground of feeling that lets us know where we are "at," how we are doing, how things are with us or stand with us (in us), how we find ourselves to be (*Befindlichkeit* – "situatedness" or "state of mind") (Ballard, 1991: 33).

Translated as "atmosphere," "ambiance," "mood," "Stimmung" (whose root is "Stimme" meaning "voice," in the sense intended in studies of style – "tone") calls attention to the attunement of modern people as "dread," according to Heidegger, replacing the classical Greek attunement of "wonder" (the feeling that gave rise to philosophy). Most importantly, a mood is not private, but is fundamentally a revelation of being-together, *Mit-Sein*, of a belonging, an intersubjectivity (hence a politics and an ethics). "Attunements are not side-effects,

PHOTOGRAPH: BARBARA JO REVELLE

but are something which in advance determines our being with one another. It seems as though an attunement is in each case already there, so to speak, like an atmosphere in which we first immerse ourselves in each case and which then attunes us through and through. It does not merely seem so, it is so.... An attunement is a way, not merely a form or a mode, but a way – in the sense of a melody that does not merely hover over the so-called proper being at hand of man, but that sets the tone for such being, i.e., attunes and determines the manner and way of his being" (Heidegger, 1995: 67). The egent asks of a zone (the Miami River) the question of "Befindlichkeit" on behalf of a collectivity: "which attunement are we to awaken for ourselves today? We can answer this question only if we know our situation well enough to gather from it which fundamental attunement pervades us" (69).

Heidegger showed that mood is the effect of a merger of the outer and inner worlds – understood in our context as including both time and space. He spoke of "attunement" rather than correspondence to name the way in which things may be known through moods, as in this example of a book that is "boring." "The book must make itself felt, not as an inducing cause, but rather as that which attunes us. This is where the question lies. If the book is boring, then this thing outside the soul has in itself something of the possible, perhaps even suppressed attunement that is in us. Thus, although it is inside, the attunement plays around the thing outside at the same time.... It does not cause the

PHOTOGRAPH: BARBARA JO REVELLE

boredom, yet nor does it receive it merely as something attributed by the subject. In short: boredom – and thus ultimately every attunement – is a hybrid, partly objective, partly subjective" (Heidegger, 1995: 88).

Rilke's "Weltinnenraum" was one source for Heidegger's understanding of how Being manifested itself to humans (his notion of the "opening" or clearing was borrowed from Rilke). In a scene of central impotence to Heidegger's attunement, Malte Laurids Brigge in his walks around Paris came upon the wreckage or ruin of a house that had been torn down, and the object returning his gaze inspired fear (was a shock).

> They were houses that no longer existed. Houses that were torn down from top to bottom. What was there was the other houses, the ones that had stood along side them, tall neighboring houses. They were obviously in danger of collapsing after everything next to them had been removed, for a whole framework of long tarred poles was rammed aslant between the ground of the rubble-strewn lot and the exposed wall [...] But most unforgettable were the walls themselves. The tenacious life of these rooms refused to let itself be trampled down. It was still there; it clung to the nails that had remained; it stood on the handsbreadth remnant of the floor; it had crept together there among the onsets of the corners where there was still a tiny bit of interior space. You could see that it was in the paint, which it had changed slowly year by year; from blue to an unpleasant green, from green to gray, and from yellow to an old decayed white that was now rotting away. (Rilke, 1958: 46–47)

Later when Brigge reflects upon the experience, he used the term "recognition." "I've been talking all along about this wall. You'll say that I stood in front of it for a long time; but I'll take an oath that I began to run as soon as I recognized the wall. For that's what 's terrible – that I recognized it. I recognize all of it here, and that's why it goes right into me: it's at home in me (48).

The time-image, manifested in the ruin of a tenement, or in any entropic scene of wear and decline, any manifestation of destruction or dissolution, makes time perceptible and hence through it death, the measure and aporia of human life. The feeling aroused by this glimpse into time the Japanese call wabi-sabi.

Images gather and hold heterogeneous attributes not by means of essence but holistic atmosphere. The term "recognition" connects epiphany with an experience fundamental to tragedy, as in this comment on Joyce's epiphanies. "Joyce's internal drama of everyday life is driven by moments of 'gnosis,' revelatory instants in which characters acquire knowledge they did not have before and whose subsequent action is directed by that knowledge. Thus, the dramatic process involving epiphany is at least analogous to the Aristotelian idea that peripeteia and anagnorisis should occur simultaneously" (Weir, 1996: 30). The value of the modernist poem as a relay for testimonial is due, in other words, to this principle of correspondence, aura, recognition in which the scene and the spectator come into relationship. In modernist poetics usually dated from Baudelaire (but anticipated in the Romantics) "metaphor and image are the primary carriers of information: the perceived world is a metaphor for the self, that what you tell about what you see reveals more about you than about the object you are ostensibly describing" (Dobyns, 1997: 72). Rilke worked more explicitly perhaps than anyone else with this exchange between the outer and inner worlds, in which his attending to the details of his environment functioned as thinking. This correspondence is known as "Weltinnenraum" "One space spreads through all creatures equally – / inner-world-space. Birds quietly flying go/ flying through us. Oh, I that want to grow,/ the tree I look outside at grows in me! / It stands in me, that house I look for still, / in me that shelter I have not possessed. / I, the now well-beloved: on my breast/ this fair world's image clings and weeps her fill" (Jayne, in Baron, 199). This experience of extimacy introduces measure (border, limit) into a zone of disorder, and is the basis for chora as digital ontology.

The God-Is-Able in me. These are not just theories, but instructions.

"Lichtung"

Citing the passage from *Malte Laurids Brigge* telling about Brigge's recognition during his wanderings in the city of the place of the torn-down house, Heidegger stated: "Poetry, creative literature, is nothing but the elementary emergence into words, the becoming-uncovered, of existence as being-in-the-world. For the others who before it were blind, the world first becomes visible by what is thus spoken. We may listen to a quotation from Rainer Maria Rilke as testimony on this point" (Heidegger, 1988: 12). In Heidegger's reading testimony is a reception not an expression. "What Rilke reads here in his sentences from the exposed wall is not imagined into the wall, but, quite to the contrary, the description is possible only as an interpretation and elucidation of what is 'actually' in this wall, which leaps forth from it in our natural comportmental relationship to it" (13).

The description of the destroyed house gives an image of the ontological clearing within which Dasein (human Being) manifests itself. The effect is that of significance. "Clearing away is release of the places at which a god appears, the places from which the gods have disappeared… Secular spaces are always the privation of often very remote sacred spaces. Clearing away is release of places" (10). Like chora (sacred space), Heidegger's "place" sorts and orders what is. "Place always opens a region in which it gathers the thing in their belonging together" (10). This poetic way of gathering is what is at issue in our experiment: the Miami River as a chora (a category), a new kind of (w)hole. What makes this experiment electrate is that the image supporting the revelation or epiphany is a photograph (among other things). The modernist poetic image, in short, is the logic of flash reason, and the operating rhetoric for deliberative politics in the dromosphere.

This clearing (image category) is Heidegger's alternative to Western metaphysics, to the category formation of ontological concepts, according to which "things" and other "entities" are defined in terms of essences and accidents. Heidegger "steps back" from this order to notice the very emergence of distinction as such, the interval of differentiation of presence and absence, and the coordination across different dimensions made possible by analogy. He returns to the origins of Greek philosophy to find another mode of gathering (logos) that he refers to as the four-fold, (*Geviert*). The four-fold names the epistemology of correspondences (nearly universal in pre-modern civilizations), the correlational thinking of macrocosm, microcosm (as above, so below). How sky and earth, divine ones and mortals, participate in the site (Ort) of world is given in the figure or scene of a crossroads illuminated at night by a lightning flash. "In this turning, the clearing belonging to the essence of Being suddenly clears itself and lights up. This sudden self-lighting is the lightning-flash. It brings itself into its own brightness, which it itself both brings along and brings in' (Heidegger). With the analogy of the 'clearing' – yet is it still an analogy in the sense of a relationship obtaining between a true meaning and an image evoking it in concrete fashion? – we reach within the *Ur-Sache*, the 'original tenor' of Heideggerian thought. It has abandoned the claim to elucidate the world in terms of causal connections or transcendental relationships. Simply, it allows the thing to appear in its frailty, without keeping any reserve in back" (Janicaud and Mattei, 139).

As Jean-Francois Mattei indicates, Heidegger draws upon the same poetic (mythic) discourse Plato used in *Timaeus* to evoke how chora (space, receptacle) allowed being and becoming to interact. To explain Heidegger's thinking of the four-fold gathering, that is, Mattei turns to Plato and his commentators, such as Proclus on the *Republic*. "Three terms, 'demonic location,' 'crossroads,' and 'meadow,' designate the same location in the cosmos. The demonic location is in the middle of the sky and earth, such that the crossroads and meadow are there too. Now, in the middle of these two, there is nothing but ether.... Consequently, demonic location, meadow, and crossroads are nothing but the middle of the entire sky according to Plato.' The chiasmus is indeed the center of the world: its four branches bind together the microcosm (the soul) and the macrocosm (the universe) in order to ensure the correspondence of the totality. The primordial symbol of the X – in the shape of the Greek cross, the reverted cross, or the cross of Saint Andrew – immediately expresses the perennity of a cosmic structure or archetype to which Platonic thought is probably not the only one in attesting" (101–102).

The scene of a country path in a forest, opening into a clearing from where may be glimpsed at night a particular star, is a site of memory of Heidegger's childhood around his hometown of Messkirch, and an emblem of his cosmology (148). "Crossroads" evokes this same tradition of thought revived by Heidegger, and proposes to make it the basis of a practice of electrate thought and public policy consultation, on behalf of collective well-being. The challenge for contemporary correspondences is that the music of the spheres now is dissonance. Time is out of joint, none of the dimensions of the "great chain of being" align to produce similitude. Correspondences between microcosm and macrocosm no longer are there to be discovered, but must be composed, designed, as matters of collective choice. Why an ontology of aesthetic experience? Because in our transcendent epoch (literacy) we created the possibility of annihilating this very ground, the everyday material world that sustains life.

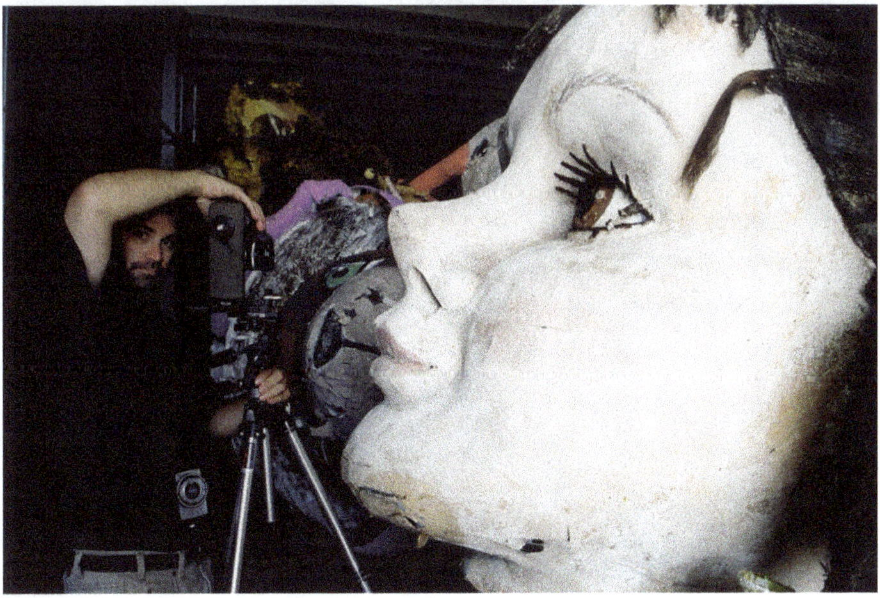

PHOTOGRAPH: BARBARA JO REVELLE

A lightning storm over Biscayne Bay exposes in a flash the vessels of the Haitian traders moored along the banks of the Miami River. The database (lightning become electricity) makes it possible to unfold the whole of the four-fold working in this moment.

PHOTOGRAPH: BARBARA JO REVELLE

CHAPTER 5

Memory

Our goal is to design and test an image logic (assuming that the technological prosthesis of cyberspace within which this "flash reason" becomes functional is taking care of itself). So far we have the equipment (photography), an image and its associated context (the "Crossroads" picture), an occasion for making the image (the EmerAgency consultation on public policy), the place in which the image was made (the Miami River), a method for mapping the area (drift), an aesthetics for selecting the primary image (punctum or third meaning), a formal structure (poetic epiphany), a philosophical rationale (Stimmung). Choragraphy gathers up every thread of the tradition of topics, the places of memory, from Aristotle's dialectical topics through memory palaces to the modern Situationist drift. There is one more dimension of this process, to indicate the challenge of designing an Internet civic sphere, to transduce a region of "Miami" into a chora named "Miautre."

Miautre is in the pedagogical tradition of the memory palace. The ground or "site" – the receptacle (chora) – holding together the disparate levels or registers of the mnemonic system – is a geography familiar to the maker (the medieval student who had to memorize large quantities of information for the purposes of public oratory). The first level of the scene was a familiar street of a hometown, or even one's own residence (in classical practice). Distributed through a representation of this location were sets of strong or memorable images paired with the information being memorized from a discourse such as that of law, religion, government, or any matter to be stored for later quick retrieval, in a debate for example. The image was generated associatively, based on some "aesthetic" link with the information (a punning or iconic connection). The modern epiphany continues this mnemonics, adjusted yet again to a different episteme. "Wordsworth's spots of time have obvious associations with mnemonic schemes, for they evoke the past through a juxtaposition of images and places. The imagery of the spots are emotionally attached to his deepest memories of past states of mind, and these images in turn are situated within landscapes that are unforgettable in their awe or enchantment. All his life he had been a traveler, and his excursions into his own past were shaped by his memories of these travels. As in the classical art of memory, the association of concrete images with particular places was essential to his method" (Hutton, 1993: 56).

The "sacralization" process by which a given place becomes a "tourist attraction" follows rules of formation that are the public, collective equivalent of

 | DOI:10.1163/9789004534643_006

the mnemonic techniques that dominated literate pedagogy up to the age of print. Indeed, the tourist attraction is a secular survival of the pilgrimage, with the tour of the Holy Land being a relay for how to bring Miami and Myami together in a chora on the information highway. Again, the medieval example, as researched by Halbwachs, provides the relay. "Halbwach's thesis is that the biblical Holy Land was an imaginary landscape conjured up during the Middle Ages in Europe and superimposed on the actual terrain of Palestine. Beginning in the fourth century, European pilgrims constructed a Holy Land of religious shrines that faithfully reflected their imagined conceptions of Jesus' sojourn on earth. The Holy Land, therefore, was not a discovery but a localization of a mind-set they carried with them on their journeys" (Hutton, 1993: 80). As indicated by the contemporary Israeli/Palestinian conflict, the same landscape may support incompatible mind-sets (active images and associated information). This observation applies to any region, including the Miami River.

Reflecting the fact that a pilgrimage is simultaneously an outer and an inner journey, the architecture marking the sacred site, the goal of the pilgrimage, "often created a microcosm of the pilgrimage. An internalized world was created in which the pilgrim's journey is reenacted in the architectural setting – the spiritual path and place" (Barrie, 1996: 28). This microcosm often was included in sacred buildings far removed from the destination, for believers unable to make the journey. Thus the 660-foot circumambulation found in Chartres Cathedral, said to recreate the pilgrimage to Jerusalem (Barrie, 30), is a relay for the relation of the Internet chora to the Miami River zone. The "paths and places" of architecture translate readily into the "links and nodes" of the web. While sacred architecture shares (through pilgrimage) the myth of the hero that structures most Hollywood features, we must forego its "clear path to a point of decision" (55), just as we suspend the closures of melodramatic narrative.

Choral design learns from any practice concerned with the exploration of place or space. Andre Breton's *Nadja*, a semi-autobiographical account of wandering the streets of Paris late at night with a mysterious beautiful woman (who turned out to be mad), may be read as an attempt to reconstruct the once familiar pedagogy of using the cityscape as a memory system. At certain places in the city the couple experience strong sensations of the uncanny. Margaret Cohen uses tourist guides of the period to reconstruct the historical events and social conditions that determined the mood of these places.

> Strolling through *Nadja's* prologue with the guides reveals Breton to experience his uncanny sensations at sites where ghosts of past insurrectional

> activity cluster, sites frequented by bohemians both in the present and in the past. In addition, Breton's narrator associates his uncanny sensations with contemporary aspects of the site where the obscured past persists in disfigured form. At the Place Maubert Breton fixates on the statue of Etienne Dolet, the only visible reminder of "certain riots"; at the Pantheon he turns to Jean-Jacques Rousseau. What particularly frightens Breton in the latter case is the superimposition of pieces of chopped wood on his skull. Once we remember that the Pantheon is a Parisian monument associated with revolutionary violence, the association of hacked wood with heads conjures up the guillotine. When Breton discusses the attractions of the Boulevards and the Porte Saint-Denis, he can find no visible detail on which to fixate. It is important to note, however, that in his very failure to locate the source of his uncanny sensation, Breton confirms, albeit negatively, that the haunting power of these sites may come from their relation to temporal passage and spatial location. (Cohen, 1993: 98)

Breton's wanderings anticipate (and influeced) the Situationist drift. We may imagine his electrate counterpart using mobile computing with augmented reality capabilities to investigate the historical turbulence intuited at each location.

Diagramming Enjoyment

To do a choral map of a zone one must locate the points and lines of intensities (to use terminology from Deleuze and Guattari). The egent tuning a zone brings a childhood block into line with local circumstances in the real to situate the desiring machine at work in the community (correspondence between macro- and micro cosm). In relation to the materiality of a zone she is in the position that Lacan described in the anecdote of his encounter with an empty tin of sardines. Lacan tells the story to illuminate his theory of the gaze. The incident occurred on a day when Lacan, a young intellectual in his twenties with romantic ideas about adventure and the nobility of the physical life, had joined the crew of a fishing boat. "One day, then, as we were waiting for the moment to pull in the nets, an individual known as Petit-Jean, that's what we called him – like all his family, he died very young from tuberculosis, which at that time was a constant threat to the whole of that social class – this Petit-Jean, pointed out to me something floating on the surface of the waves. It was a small can, a sardine can. It floated there in the sun, a witness to the canning industry, which we, in fact, were supposed to supply. It glittered in the sun and Petit-Jean said to me – You see that can? do you see it? Well, it doesn't see you!" (Lacan, 1978: 95).

The joke was at Lacan's expense and in his lecture he reflects upon why he did not find it amusing. The incident served as a kind of map of his situation, his positioning that was "out of place" in relation to the institution formation implicit in the sardine tin. "If I am told a story like that one, it is because I, at that moment – as I appeared to those fellows who were earning their livings with great difficulty, in the struggle with what for them was a pitiless nature – looked like nothing on earth. In short, I was rather out of place in the picture" (96). Specifically, the theoretical point included the idea of the subject being located in a field within which he himself is a stain of opacity (what blocks his view is precisely himself): he has no access to the position from which he is observed. "This is something that introduces what was elided in the geometrical relation – the depth of field, with all it ambiguity and variability, which is in no way mastered by me. It is rather it that grasps me, solicits me at every moment, and makes of the landscape something other than a landscape, something other than what I have called the picture" (96).

The stain does not block singular vision in flash reason, because the image is constituted not through perspective but event (it maps an ecology). The "gleam of light" that, as Lacan said, lay at the heart of his story, places the theory in relation to the tradition of epiphany. The sardine tin is a gest, producing a punctum. "For Bohme, the mystic way started with a contemplation of simple, material objects which, at a certain point, might, as it were, burst into

PHOTOGRAPH: BARBARA JO REVELLE

flames in an epiphany which revealed the true nature of the invisible. His own vocation had been decided when, being still a young man, gazing at a tin pot struck by the rays of the sun, he was suddenly vouchsafed a vision that became, like Borges's Alef, a privileged moment in which the light of God present in all things suddenly disclosed itself" (Eco, 1995: 183). The sobriety of choragraphy, of course, dictates that the common objects constituting the connecting lines of the map will reveal their intensities as metaphorical flame. Lacan demystified the mystical flame with a diagram of the subject that accounts for the magical power of ordinary objects constellated into desiring machines. "The three angles of the triangle stand for the three fundamental dimensions which, according to Lacan, structure the human universe: the Real (the 'hard,' traumatic reality which resists symbolization), the Symbolic (the field of language, of symbolic structure and communication), and the Imaginary (the domain of images with which we identify, and which capture our attention). 'J' in the middle of the triangle designates *Jouissance*, the abyss of traumatic/excessive enjoyment" (Zizek, 1997: 175).

Movement within a choral zone is generated by the differential forces represented in this diagram. Slavoj Zizek explains the functions of the different objects in the diagram by applying them to several stories by Patricia Highsmith. In anticipation of the narrative to come in the next part of this guide, the following description maps Zizek's reading of Lacan's diagram of the subject onto Revelle's Miami "Testimonial" (with the Miami application in italics). It is important to note that according to the diagram the subject is a vortex (whirlpool) formed by the operations of interior and exterior forces (drive and desire). Lacan's diagram, that is, is an analytical representation of extimacy, of correspondence between the conditions of the zone and the habitus (deep memory) of the egent. What will the egent recognize?

> The three vectors in this schema do not indicate a relation of causality: I – S does not mean "the imaginary determines the symbolic," it marks rather the process of the symbolization of the imaginary. The object small *a* is thus the "hole in the real" that sets symbolization in motion (*the "mattress"* for example: the screen for the projection of our fantasy narrations); capital phi, the "imaginarization of the real," is a certain image that materializes the nauseous enjoyment (*the destroyed, scattered, vandalized goods in the burglarized warehouse,* for example); and finally S(A), the signifier of the lack in the big Other (the symbolic order), of its inconsistency, the mark of the fact that "the Other (as a closed consistent totality) doesn't exist," is the little bit of the real functioning as

> the signifier of the ultimate senselessness of the (symbolic) universe (*the magnetic "sculpture" built by Frank, the former taxi driver*). The abyss in the middle (the balloon encircling the letter J – *jouissance*) is of course the whirlpool of enjoyment threatening to swallow us all; the pothole exerting its fatal attraction (*like the impounded boat of the Haitian trader*). The three objects on the sides of the triangle are perhaps nothing but the three ways to maintain a kind of distance toward this traumatic central abyss. (Zizek, 1991: 135)

The drift allows the subject to respond freely to the intersubjective forces of the psyche. Choragraphy maps this vortex (plaque tournante), this circulation or passage from object to object around the whirlpool of enjoyment. That it is precisely "enjoyment," in the psychoanalytic sense of jouissance (bliss), is important, since the project of electracy is to ontologize this dimension of embodied satisfaction as the basis for a metaphysics of well-being. Psychoanalysis is the proper relay for this new subject formation, in that it is a theory of the human being focusing on desire and drive (embodiment). Electracy requires a visceral metaphysics. For Revelle the mattresses distributed through the zone are the object small *a* – "a quite ordinary, everyday object that, as soon as it is 'elevated to the status of the Thing,' starts to function as a kind of screen, an empty space on which the subject projects the fantasies that support [her] desire, a surplus of the real that propels us to narrate again and again our first traumatic encounters with *jouissance*" (133). Here is a theoretical explanation for the mystery or magic associated with objects giving rise to the uncanny feeling of metaphysical art. Deleuze and Guattari credit Lacan with initiating the theory of desiring machines (the subject is not confined to the body but is spread across a field of alterity) but they want to go beyond the limitations of his linguistic frame, bodily organs and functions, and familial institution. They extend the object small *a* by means of chaos theory. "This object-subject of desire, like strange attractors in chaos theory, serves as an anchorage point within a phase space (here, a Universe of reference) without ever being identical to itself, in permanent flight on a fractal line. In this respect it is not only fractal geometry that must be evoked, but fractal ontology" (Guattari, 1995: 94–95).

The egent's task in mapping the zone is to notice and register the pull and push of attraction/repulsion. Zizek's explanation of the turn in late Lacan suggests that the *sinthome* (symptom, the fourth register of the subject diagram), functions in this fractal way. "Lacan undertook to delineate the outlines of a certain 'free-floating' space in which signifiers find themselves prior to their discursive binding, to their articulation. This is the space of a certain 'prehistory' preceding the 'story' of the social bond, i.e., of a certain psychotic kernel evading the discursive network.... This One is of course precisely the One of *jouis-sens*, of the signifier insofar as it is not yet enchained but rather freely floating, permeated with enjoyment; it is this enjoyment that prevents it from being articulated into a chain. To indicate the dimension of this One, Lacan coined the neologism *le sinthome*. This point functions as the ultimate support of the subject's consistency, the point of 'thou art that,' the point marking the dimension of 'what is in the subject more than himself' and what he therefore 'loves more than himself" (Zizek, 1991: 132). Choragraphy learns from these theories how to include "jouis-sens" in the mix of other senses and meanings–good sense, common sense, logic. The sinthome, then, this sensational uniqueness of the person, is what makes it possible for an egent to function as category.

From Koenigsberg to Miami

The larger importance of Lacan for our project, and for choragraphy and electracy in general, concerns his place in the history of topology. To understand the ambition of the virtual city (the online tour of which constitutes a practice of reasoning) requires a review of its place in a certain history. The history begins with Euler's invention of topology, and passes through Lacan's appropriation of this mathematics for his *topologerie* (this neologism intended to indicate that Lacan's usage is not literal and is specific to the discourse of psychoanalysis) – his exploration of topological figures such as the moebius strip, the cross-cap, and knots to research the nature of the psyche. The next step (Miami image as collective memory) is justified by the origins of topology in a city tour. This trajectory is fundamental to electrate metaphysics, showing the "etymology" so to speak of how a city zone becomes an information space.

The philosophical implications of the drift as method is suggested by an association with Immanuel Kant, who made his hometown into the "Capital of Philosophy."

> Immanuel Kant is so identified with Koenigsberg that a sketch of this city is essential to a correct knowledge of the life of her most famous son. Here he was born and educated, here he taught and died; and this city, with its immediate vicinity, was the scene of all his labors, hardships, and triumphs. Its social, religious, and intellectual condition exerted a potent influence on his character and views; but he gave the city a fame such as it had never before enjoyed, and has for ever associated its name with one

> of the most important epochs in philosophy, so that for his sake it was called "The Capital of Philosophy" and also "The City of Pure Reason." Koenigsberg is a frontier city of Germany [now Kalinigrad, in Russia], being situated in the northeastern corner of Prussia, near the Russian border. The city is built on undulating ground, in an attractive region, and its position is favorable for commerce. It is situated at the mouth of the Pregel, a river which forms an important means of communication with the interior of the province and also with Poland, though for its mercantile importance it is mainly indebted to its location on a bay of the Baltic. Its extensive commerce brought the city into communication with numerous seaports of Europe, Asia, Africa, and America, as well as with the whole province of Prussia and the adjoining countries. (Stuckenberg, 1986: 1)

One of the anecdotal icons of Kant's life is the walk he took every afternoon.

> "It is difficult to write the history of the life of Immanuel Kant," noted the poet Heine, "for he had neither life nor history." He lived the mechanically ordered and almost abstract life of a bachelor, in a quiet, retired little street of Koenigsberg. I do not believe that the large clock of the cathedral did its daily work with less passion and with greater regularity than its countryman, Immanuel Kant. To rise, drink coffee, write, deliver lectures, eat, take walks, everything had its appointed time; and the neighbors knew that it was exactly half-past three when Kant, in his grey coat and with the Spanish reed in his hand, stepped out of doors and walked towards the mall Linden Avenue, which is still called after him, "The Philosopher's Walk." Eight times he walked up and down there, at all seasons of the year; and when the weather was unfavorable or the gray clouds portended rain, his old servant, Lampe, might be seen wandering anxiously behind him, with a long umbrella under his arm, like a picture of Providence. (162)

Kant's walk was as legendary as his ideas were profound. "In summer he walked very slowly, so as not to perspire; if he noticed that he was about to do so, he at once stopped, because he thought that his constitution required that he should by all means avoid perspiration. And it is stated as a remarkable fact that even in the hottest weather he never perspired" (161). "In order to walk more firmly, he adopted a peculiar method of stepping: he carried his foot to the ground, not forward, and obliquely, but perpendicularly, and with a kind of stamp, so as to secure a larger basis, by setting down the entire sole at once" (De Quincy, in Cutrofello, 1994: 53). Although the anecdote does not say so, we may suppose that Kant indulged himself in trying to solve the local puzzle of the Koenigsberg bridges. In any case, he emblematizes a certain relationship between the pace of walking and a style of thought.

We are tracking, in other words, a certain figure of a walk through the places of memory as a guide to the design of an online civic sphere. We want to keep in mind the allegorical link between the individual body and the geography of the city, as in William Carlos Williams' *Paterson*, an important example of the application of Pound's ideogram (vortex poetics) to a specific location. He says "a man like a city and a woman like a flower." Today we might say *a man or woman like a city in flower.*

> *Walking –*
> *cramped arena has been left clear at the base*
> *of the observation tower near the urinals. This*
> *is the Lord's line: Several broken benches*
> *drawn up in a curving row against the shrubbery*
> *face the flat ground, benches on which*
> *a few children have been propped by the others*
> *against their running off*
>
> *Three middle aged men with iron smiles*
> *stand behind the benches – backing (watching)*
> *the kids, the kids and several women – and*
> *holding,*
> *a cornet, clarinet and trombone,*
> *severally, in their hands, at rest*
> (Williams, 1995: 63)

Williams used the walk as an organizing device, a structure, to control the release of information: the mind's eye, it is said, moves through thought at the pace of a walk. Of course at times "Thought clambers up, / snail like, upon the wet rocks," especially when contemplating "Things, things unmentionable, / the sink with the waste farina in it and / lumps of rancid meat, milk-bottle-tops: have / here a tranquillity and loveliness / Have here (in his thoughts)/ a complement tranquil and chaste" (38).

Say it, no idea but in things.

Deleuze and Guattari's *Anti-Oedipus* provides another suggestive reference continuing the sequence joining topology to psychology.

> A schizophrenic out for a walk is a better model than a neurotic lying on the analyst's couch. A breath of fresh air, a relationship with the outside world. Lenz's stroll, for example, as reconstructed by Buchner. This walk outdoors is different from the moments when Lenz finds himself closeted with his pastor, who forces him to situate himself socially, in relationship to the god of established religion, in relationship to his father, to his mother. While taking a stroll outdoors, on the other hand, he is in the mountains, amid falling snowflakes, with other gods or without any gods at all, without a family, without a father or a mother, with nature. Everything is a machine. Celestial machines, the stars or rainbows in the sky, alpine machines – all of them connected to those of his body. The continual whir of machines. "He thought that it must be a feeling of endless bliss to be in contact with the profound life of every form, to have a soul for rocks, metals, water, and plants, to take into himself, as in a dream, every element of nature, like flowers that breathe with the waxing and waning of the moon." To be a chlorophyll- or a photosynthesis-machine, or at least slip his body into such machines as one part among the others. (Deleuze and Guattari, 1977: 2)

How is reasoning conducted with choras rather than topics? Choral thinking is like taking a walk in a scene of correspondences. All these philosophical and literary sources provide "instructions" for the assignment: how to map a chora? Listen for the whirring of machines! But we understood that "machine" describes any flow and its segmentation, replacing "structure" and even "sign" in this poststructural theory. Bliss? Jouissance perhaps. It is not so pleasant to open oneself to the line of attraction-repulsion tracked by desiring machines.

Games of Invention

> We picture a whole branch of mathematics being born out of a game. This game was a puzzle in connection with the bridges at Koenigsberg. The River Pregel, which runs through Koenigsberg, has two islands. Those two islands are joined to each other and to the shores by means of seven bridges. The puzzle is whether it is possible, starting at any point, to work out a walk so that we get back to our starting point having walked across all the bridges, but only once over each bridge. The problem would not be altered at all if the bridges leading to the same island or shore converged at

the same points (this would merely cut out the walks on the banks of the river). The problem can now be reformulated by asking whether this figure can be drawn with one stroke of a pencil without lifting the pencil off the paper in such a way that no part of the figure should b e drawn twice and that at the end we should get back to the starting point. (Peter, 1976: 35–6)

Euler was able to solve the puzzle when he realized that it had nothing to do with geometry.

He realized that the exact layout of the islands and bridges was irrelevant. What was important was the way the bridges connect, that is to say, the network formed by the bridges. The actual layout of the river, islands, and bridges – the geometry of the problem – is irrelevant. In Euler's network, the bridges are represented by edges, the two banks and the two islands are represented by vertices. In terms of the network, the problem asks if there is a path that follows each edge exactly once. Euler argued as follows. Consider the vertices of the network. Any vertex that is not a starting or a finishing point of such a path must have an even number of edges meeting there, since those edges can be paired off into "path-in-path-out" pairs. But, in the bridges network, all four vertices have an odd number of vertices that meet there. Hence there can be no such path. In consequence, there can be no tour of the bridges of Koenigsberg that crosses each bridge exactly once. (Devlin, 1997: 175)

Euler's work has served as a point of departure for discussions of network theory in general, with application to the nodes and links of the Internet in particular, and this fact is the bridge connecting our register of the apparatus (concerned with rhetoric) with the technological register: Networks inside and out. "In retrospect, Euler's unintended message is very simple: Graphs or networks have properties, hidden in their construction, that limit or enhance our ability to do things with them. For more than two centuries the layout of Konigsberg's graph limited its citizens' ability to solve their coffeehouse problem. But a change in the layout, the addition of only one extra link, suddenly removed their constraint. In many ways Euler's result symbolizes an important message of this book: The construction and structure of graphs or networks is the key to understanding the complex world around us. Small changes in the topology, affecting only a few of the nodes or links, can open up hidden doors, allowing new possibilities to emerge" (Barabasi, 12). Choragraphy works in this spirit, as a mapping that constructs as much as it records the routes of information.

Cosmological Cross-Cap

Lacan argued that topology is not simply a metaphorical way of expressing the concept of structure; it is structure itself. The importance as a precedent for choragraphy is to note how Lacan appropriated topology to figure the operations of the psyche.

> He emphasized that topology privileges the function of the cut, since the cut is what distinguishes a discontinuous transformation from a continuous one. Both kinds of transformation play a role in psychoanalytic treatment. As an example of a continuous transformation, Lacan referred to the moebius strip; just as one passes from one side to the other by following the strip around continuously, so the subject can traverse the fantasy without making a mythical leap from inside to outside. As an example of a discontinuous transformation, Lacan also refers to the moebius strip, which when cut down the middle is transformed into a single loop with very different topological properties; it now has two sides instead of one. Just as the cut operates a discontinuous transformation in the moebius strip, so an effective interpretation preferred by the analyst modifies the structure of the analysand's discourse in a radical way. (Evans, 1996: 208)

The torus is another one of the figures that Lacan used. "The torus in its simplest form is a ring, a three-dimensional object formed by taking a cylinder and joining the two ends together. The topology of the torus illustrates certain features of the structure of the subject: one important feature of the torus is that its center of gravity falls outside its volume, just as the center of the subject is outside himself; he is decentered, eccentric. Another property of the torus is that its peripheral exteriority and its central exteriority constitute only one single region. This illustrates the way that psychoanalysis problematises the distinction between "inside" and "outside" (extimacy) (Evans, 208–209). This property of *extimacy* is a major feature of object small *a* and hence is central to the passage into Miautre, since what distinguish the inventional approach to consulting is the assumption that the "things" of a policy problem are not conceptual things, but display the extimacy of warped space, at once outside and inside the subjects of knowledge (*Problems B Us*).

The cross-cap is another topological type that Lacan used to show extimacy. "To add a crosscap to a sphere," in topological theory, "you cut a hole and sew on a moebius band. The entire edge of the moebius band must be sewn to the circular hole; in three-dimensional space this can only be done if you allow the surface to intersect itself. To sew on the entire edge without such self-intersection you need to work in four-dimensional space" (Devlin, 186). Here is what Lacan does with this figure.

> I tried to draw on the blackboard the topology of the subject according to a sign that I once called the interior 8. This is certainly reminiscent of Euler's famous circles, except, as you will see, that Euler was concerned with a surface that could actually be made. The edge is continuous, except that at one point it does not proceed without being concealed by the surface that has previously unfolded itself. This drawing, seen from a certain perspective, may seem to represent two intersecting fields. I have placed the libido at the point at which the lobe defined as field of the development of the unconscious covers and conceals the other lobe, that of sexual reality. The libido, then, would be that which belongs to both – the point of intersection, as one says in logic. But this is precisely what it does not mean. For this sector at which the fields appear to overlap is, if you see the true profile of the surface, a void. This surface belongs to another whose topology I have described to my pupils at various times, and which is called the cross-cap, in other words, the mitre. (Lacan, 1978: 155–156)

That Lacan includes in the significance of the cross-cap its embodiment of its shape in the "mitre" or "miter" (the official headdress of a bishop in the Western church, as well as a certain kind of joint) indicates the difference between topology and *topologerie*. The point of the demonstration in part is to deconstruct the categorial border inside-outside with respect to the human subject. The cross-cap figure lends a theoretical dimension to the effect explored by poets at least since the beginnings of modernism in Baudelaire's "correspondences," Rilke's *Weltinnenraum*, or T. S. Eliot's objective correlative – the experience of recognizing some "thing" found in the world as an equivalent of one's inner life. That objects produce this uncanny effect is a symptom of modern alienation, the conditions created in the industrial city of commodity capitalism. One no longer recognizes one's own agency, Marx explained, one's responsibility for how the world is, except through these intuitions of exitimacy. In fact, the interior 8 of the cross-cap alludes to the same chiasmatic figure of Greek cosmology that interested Heidegger. Plato's discussion of macrocosm microcosm correspondences in the *Timaeus* resonates with Lacan's extimate topologerie.

> It is well-known that the demiurge of the *Timaeus* used harmonics to shape the soul of the world: first he took "the indivisible substance that

> always consists in the invariable," the eidos of the Same, and the "divisible substance that is in the bodies," the eidos of the Other. Then he mixed them together with an intermediary substance obtained from the previous two according to a given proportion so as to obtain a series of harmonic numbers comparable to a sort of gauge or scale that he sliced in two, lengthwise. "This entire compound he divided lengthwise into two parts which he joined together at the center like the letter X, and bent them into a circular form, connection them with themselves and each other at the point opposite to their original meeting point, and comprehending them then in a uniform revolution upon the same axis, he made the one the outer and the other the inner circle. [36b-c] (Janicaud and Mattei, 99)

Lacan also used that part of topology dealing with knots to show the operations of the subject. The commentators remind us that what counted for Lacan in his manipulation of these *ficelles* or loops of string in his seminars was the holes thus formed, the holes figuring the orifices of the body. And if Euler showed a certain impossibility in the path across the bridges of Koenigsberg, Lacan followed suit in his account of the path of the subject in its aim to achieve satisfaction.

> When we look at it more closely, we see that something new comes into play – the category of the impossible. In the foundations of the Freudian conception, this category is an absolutely radical one. The path of the subject – to use the term in relation to which, alone, satisfaction may be situated – the path of the subject passes between the two walls of the impossible. This function of the impossible is not to be approached without prudence, like any function that is presented in a negative form. I would simply like to suggest to you that the best way of approaching these notions is not to take them by negation. This method would bring us here to the question of the possible, and the impossible is not necessarily the contrary of the possible, or, since the opposite of the possible is certainly the real, we would be led to define the real as the impossible. (Lacan, 1978: 167)

Rhythms of Place

Movement through time and space has been an analogy for thought from the beginnings of philosophy. The "impossibilities" figure "aporias" of reason, and resonate with the "impossibility" of taking a walk along the banks of the Miami

River itself. This latter impossibility has like the others certain theoretical and epistemological implications. The larger ambition of Miautre is to initiate a mode of knowledge that leads one day to Miami as "the city of choragraphy," commemorating the contribution its river made to the logic of electracy. The Koenigsberg game was relevant not only to the geography of that city but through the invention of a new mathematics to any network whatsoever, including the connectivity of the Internet. The implications of choragraphy similarly are not confined to Miami. Like Kant's walks around his city, generalized we might say by Euler into topology, the "impossible walk" (the drift) along the Miami River may be generalized into a practice introducing into the nodes and links of the World Wide Web the capacity to think with a group subject (the Internet as a prosthesis of the unconscious mind). The difference between literacy and electracy is a difference in (machinic) rhythms, a different groove. An electrate Kant learns salsa.

> The regularities that the culture of the Caribbean shows begin from its intention to reread (rewrite) the march of Nature in terms of rhythms "of a certain kind." I'll give an example: Let's suppose that we beat upon a drum with a single blow and set its skin to vibrating. Let's suppose that this sound stretches until it forms something like a salami. Well, here comes the interruptive action of the Caribbean machine; it starts slicing pieces of sound in an unforeseen, improbable, and finally impossible way.... The Caribbean rhythm is in fact a metarhythm which can be arrived at through any system of signs, whether it be dance, music, language, text or body language. Let's say that one begins to walk and all of a sudden he realizes that he is walking "well," that is, not just with his feet, but also with other parts of his body, which nonetheless adjusts admirably to the rhythm of his steps. It is quite possible that the walker in this circumstance might experience a mild and happy sensation of well-being... What has happened is that the center of the rhythmic ensemble formed by the footsteps has been displaced, and now it runs from muscle to muscle, stopping here and there and illuminating in intermittent succession, like a firefly, each rhythmic focus of the body. (Benitez-Rojo, 18–19)

Antonio Benitez-Rojo sums up this "certain kind of way" in a word – "improvisation." He anticipates and confirms the syncretic possibility of the consultancy in his study of how "the certain kind of way" of the Caribbean rhythm is manifested in literature. The point recalls us to the specific dimension we need to ontologize (render categorial), which is that of the bodies capacity for sensation, its experience of itself, and its particular manner of performance, its style.

> If this double seduction reaches the intensity of "a certain kind of way," both the text and the reader will transcend their statistical limits and will drift toward the decentered center of the paradoxical.... I mean by this that the space of "a certain kind of way" is explained by poststructuralist thought as episteme – for example, Derrida's notion of differance – while Caribbean discourse, as well as being capable of occupying it in theoretical terms, floods it with a poetic and vital stream navigated by Eros and Dionysus, by Oshun and Elegua, by the Great Mother of the Arawaks and the Virgen de la Caridad del Cobre, all of them defusing violence, the blind violence with which the Caribbean social dynamics collide, the violence organized by slavery, despotic colonialism, and the Plantation. (23)

A metaphorical mise-en-scene of the dance-step way is useful for working at the edge of problems, on the line marking the inside-outside of anything whatsoever. This certain-kind-of-way is the "way" of method, the "steps" of reasoning, that are becoming less sober, more corporeal. To add dance movement to the abstract "movement along a path," it is to include well-being: it does

PHOTOGRAPH: BARBARA JO REVELLE

not suffice to move only, but to move well (figuratively speaking, regarding the inference path of conduction).

> The crossing of borders always announces itself according to the movement of a certain step [pas] – and of the step that crosses a line. An indivisible line. And one always assumes the institution of such an indivisibility. Customs, police, visa or passport, passenger identification – all of that is established upon the institution of the indivisible, the institution therefore of the step that is related to it, whether the step crosses it or not. Consequently, where the figure of the step is refused to intuition, where the identity or indivisibility of a line is compromised, the identity to oneself and therefore the possible identification of an intangible edge – the crossing of the line – becomes a problem. There is a problem as soon as the edge-line is threatened. And it is threatened from its first tracing. This tracing can only institute the line by dividing it intrinsically into two sides. There is a problem as soon as this intrinsic division divides the relation to itself of the border and therefore divides the being-one-self of anything. (Derrida, 1993: 11)

The consultation unfolds this problematic of borders and thresholds. What is at stake is identity at every level and scale: the interior border of an individual

PHOTOGRAPH: BARBARA JO REVELLE

identity (the feeling of being a differentiated self; the split subject – conscious and unconscious); the border between two political states or nations; the border distinguishing between two concepts (39–40). Personal, political, and logical identity are braided into one aporia whose problematic is the future of democracy in an era of teletechnologies. Digital technologies have opened new spaces of relationship on a global scale and put in question the established modes and experiences of "belonging." The location of the river as a site of permeability (porosity) across the registers of all classifications available within the apparatus of literacy contributes to its status as an electrate category. The objective correlative mapped in the testimonial marks the singular location of this border.

Such is the movement of invention, as Guattari pointed out with respect to the origins of the steam engine in an Ancient toy. Euler transposed a tour of Koenigsberg into topology; Lacan transposed topology into the path of the subject (dynamics of drive and desire); the FRE proposes to extend this topo-psyche into a logic for an Internet database and interface such that a tour of Miautre is a mode of reasoning useful to decision making in a virtual civic sphere. The geographical material crossroads of the river port in Miautre become a memory palace of a global wisdom.

PART 2

Myami – Narrative

It is just possible that photography is the prophecy of a human memory yet to be socially and politically achieved.

JOHN BERGER (*About Looking*)

CHAPTER 6

Allegory

"Crossroads" is a "decisive moment" in a situation. It marks the hub or turntable, a vortex through which at least four stories pass, or asymptotically converge. "To judge by Cartier-Bresson's own photographs, the decisive moment was not when meanings suddenly and unequivocally came to the surface of reality; it was rather a moment when meaning was made available, when the image became unreliable, when it moved out of its own field of reference.... The 'decisive moment' is the interface between reference (the indexical/iconic) and significance (the iconic/symbolic) – an intimation, or augury, or emerging design in appearances – and this interface is opened up by the least prejudicial of titles, the indexical (time and place of taking)" (Scott, 63). The camera (photography in general) makes the electrate apparatus technically adequate to the instantaneity of Now. Cartier-Bresson's most famous book of photographs was entitled "L'image à la Sauvette." The French phrase means to "hawk something on the street with one eye watching out for police," alluding to the technique of the candid shot "stolen" from the situation. Cartier-Bresson referred to a seventeenth-century Cardinal who said "there is nothing in the world that does not have a decisive moment" (Cheroux, 96).

The principle, related to a long tradition in the visual arts of concern with capturing the essence of a situation in one scene, is that the world is dynamic, in flux, but that at certain moment everything aligns in a way that reveals the underlying tendency of things, the entelechy or causal directedness of reality. These patterns are not apparent in themselves, but are captured in the optical unconscious of the camera, managed by the artist attuned to the temporal moment of kairos, the instant of opportunity. For Cartier-Bresson, the clues to this moment conformed with Euclidean geometry and the Classical designs of the Golden Section. This aesthetics is updated in the context of electracy to include the topological forms of complexity theory. The FRE consultation on the river selected this scene to represent the "decisive moment" of the group mystory. The mystory genre was used to correlate the four-fold registers of the river condition. Mystory writes an epiphany.

The ambition of Miautre as online civic sphere is to create the conditions favoring a collective or community epiphany.

 | DOI:10.1163/9789004534643_007

Epiphany – the poetic means for expressing an uncanny encounter with some "thing" – is impossible in principle in our time, epoch, episteme (overseeing the death of the author) or only accessible negatively, according to Fredric Jameson.

> Stimmung – much stronger than the English "mood" in its designation of moments when a landscape seems charged with alien meaning (Jullien Gracq), when a glimpse of sordid wallpaper chokes us with anxiety or a framed and distant vista fills us with an equally unaccountable elation – is the very element of what Frye, following Joyce, terms the romance "epiphany." The latter term, however, is misleading, precisely to the degree to which it suggests that in the secularized and reified world of modern capitalism, epiphany is possible as a positive event, as the revelation of presence. But if epiphany itself is a mirage, then the most authentic vocation of romance in our time would not be that reinvention of the providential vision invoked and foretold by Frye, but rather its capacity, by absence and by the silence of the form itself, to express that ideology of desacralization by which modern thinkers from Weber to the Frankfurt School have sought to convey their sense of the radical impoverishment and constriction of modern life. So the great expressions of the modern fantastic, the last unrecognizable avatars of romance as a mode, draw their magical power from an unsentimental loyalty to those henceforth abandoned clearings across which higher and lower worlds once passed. (Jameson, 1981: 135)

The experience of revelation is replaced in a secularized civilization by an intuition of the uncanny – a symptom of encounter with the event of our own alienated agency. Walter Benjamin believed that children naturally used a kind of mimetic (correlational) cosmology to understand the world, but lost this possibility as adults. The abandoned clearings of bygone correspondences (the impossibility of epiphany in a disenchanted world) symptomize for Jameson the fundamental obstacle to problem formulation and solution in the reified conditions of modernity: "the growing contradiction between lived experience and structure, or between a phenomenological description of the life of an individual and a more properly structural model of the conditions of existence of that experience" (Jameson, 1988: 349). To extrapolate from Jameson's examples of the disconnect between global realities and local experience, we would say that the events recounted in the news datelined "Miami" are not caused by anything local, but these local breakdowns have their origins in a global structure of multinational corporations and the entire history of colonialism, (are part of an unintelligible, unattainable "sprawl" that disperses any ordered whole or totality). Even if these "global realities are inaccessible to any individual subject or consciousness," Jameson adds, meaning that they are "unrepresentable" and may "never emerge into the presence of perception," they may nonetheless "find figures" that make the "absent cause" known "in distorted and symbolic ways" (350). The challenge, or rather the opportunity, for testimonial, working à la Sauvette, is that, within these conditions, according to Jameson, "we can say that if individual experience is authentic, then it cannot be true; and that if a scientific or cognitive model of the same content is true, then it escapes individual experience."

Jameson calls for a "cognitive mapping" – a new collective logic, poetics, aesthetics that would reconnect the dimensions of essence and appearance, truth and experience (but certainly not in the sense of an unmediated, direct, transparent self-presence). When he mentioned "allegory" and the play of figuration as means of bringing into contact the different levels and scales of reality he established a context for mystory. He explicitly invoked medieval allegory as a relay for the discursive practice missing from modern society. The epiphany, whatever it might have been as an experience, depended formally on allegory. His project, modified to emphasize the affective dimension of the mapping, sets an agenda for the EmerAgency.

Mystory

The updating of epiphany and allegory as the basis for testimonial consulting is a development and extension of Gregory Ulmer's work with "mystory" as an electrate genre (Ulmer, 1989). Mystory as a genre helps map what Guattari calls a "transversalist subjectivity, one which permits us to understand both its idiosyncratic territorialised couplings (Existential Territories) and its opening onto value systems (Incorporeal Universes) with their social and cultural implications."

> Should we keep the semiotic productions of the mass media informatics, telematics and robotics separate from psychological subjectivity? I don't think so. Just as social machines can be grouped under the general title of Collective Equipment, technological machines of information and communication operate at the heart of human subjectivity, not only within its memory and intelligence, but within its sensibility, affects and unconscious fantasms. Recognition of these machinic dimensions of subjectivation leads us to insist, in our attempt at redefinition,

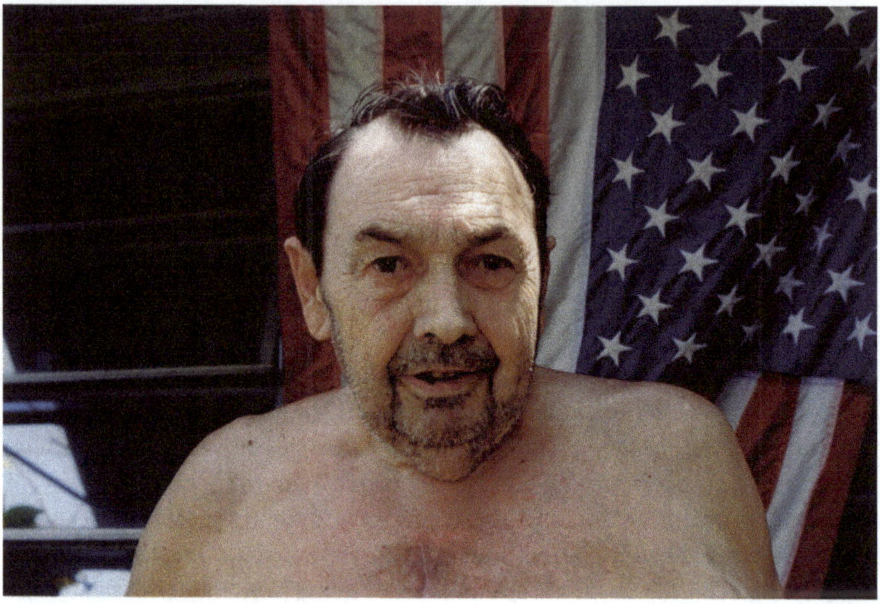

PHOTOGRAPH: BARBARA JO REVELLE

> on the heterogeneity of the components leading to the production of subjectivity. Thus one finds in it: 1. Signifying semiological components which appear in the family, education, the environment, religion, art, sport… 2. Elements constructed by the media industry, the cinema, etc., 3. A-signifying semiological dimensions that trigger informational sign machines, and that function in parallel or independently of the fact that they produce and convey significations and denotations, and thus escape from strictly linguistic axiomatics. (Guattari, 1995: 4)

To make a mystory is to map the path of connections that traverse these heterogeneous discourses, to locate the "existential refrain" that brings into alignment or attunement the institutions of Family, Entertainment, School, and Discipline (Career Specialization). Entertainment as an institution is understood in the broadest terms, as one of the core discourses of the "popcycle" – the matrix of languages and administrative orders that every citizen learns, and that serve to "interpellate" (hail) or socialize people into the cultural order of their community (the nation-state within literacy). The popcycle is so named to reflect the circulation of ideas or "memes" through all the institutions, with any of the discourses being a potential source for materials used in any of the other discourses.

PHOTOGRAPH: BARBARA JO REVELLE

An egent documents the consultation in a mystory.

Family
The discourse of Family operates by means of:
natural language
oral forms (anecdote, proverb, joke)
commonsense reasoning
self-evident proof
Through Family experience a person comes to embody a specific version of the ideological categories of identity formation (race, ethnicity, religion, gender, sexuality, class, nationality).
Entertainment
The discourse of Entertainment operates by means of:
pop culture
media forms (ad, sitcom, news, documentary)
mytho-logic
fashion as proof
Through experience of Entertainment a person internalizes a set of norms (VALS – values and lifestyles) reflecting the preferred choices among all the possibilities available within the ideological identity categories (the

hegemony in the United States of white, Anglo-Saxon, Protestant, masculine, heterosexual, bourgeois, American...). These stories bridge the gap between the official ideals and values authorized by School and the actual experience of life in Family.

School

The discourse of School operates by means of:

official history

textbooks

cultural literacy

proof by authority

The institution of School instills the practices of literacy through training in analytical reason in general, and indoctrination in the version of history approved by the local and national community.

Career (Discipline)

The discourse of Career operates by means of:

disciplinary knowledge

treatise

(formal) logic

proof by argumentation

Through the institution of Career one becomes an expert in a specialized field of knowledge. The method of the field embraces (or simulates) the scientific state of mind – objectivity and systematic doubt – without acknowledging that the dynamics of the "entry into language" that operate in the other discourses by means of identification and introjection (internalization of tutor figures as imagoes) also occurs in this discourse.

Every first-world educated citizen is native to these institutions, and the circulation of information, ideas, and values through these different logics constitutes an economy of creativity in the modern world. The history of invention shows that the greatest problem solvers are those who intuitively learned how to think with the whole popcycle. Or, to use Guattari's terms, creativity involves thinking with that same group subjectivity that also threatens a fall into fascism (the dangerous zone articulated formerly in tragedy). This multi-tasking thought may be described as bringing to bear on discipline puzzles or enigmas the resources of problem solving figured imaginatively in the researchers' experience of the discourses supposedly irrelevant to the disciplinary paradigm. Inventors have had to be their own Champollions, discovering how in each case to decipher the Rosetta Stone of their popcycle – how to translate across the different languages and logics internalized through interpellation to form their cognitive maps, their psychogeography. The mystory is a cognitive map locating the maker's position in each of the popcycle institutions. A goal of the mystory is, in Nietzsche's phrase, *to find that secret point at which the aphorism of thought intersects with the anecdote of life*. Such a *crossroads* exists virtually in every person's biography, marking the border or destiny of identity. Mapping this intersection is one way of providing testimony as a witness in the creation of public policy. Through mystory egents experience the EmerAgency slogan: *Problems B Us*.

An apparatus must be able to coordinate in one discourse all the registers of human experience. In the medieval world this correspondence was represented in allegory, with anagogy in the form of Christian cosmology providing the coherence. In postmodernity we returned to allegory, but the coherence principle organizing the disparate orders of experience comes now from the singular trait marking an individual's style of thought, expression, being – the *sinthome* (symptom). Mystory is a version of this modern allegory, exchanging Lacan's four exigencies for the discourses of the popcycle. It is important to note in our appropriation of allegory for testimonial that the renewal of this mode within postmodernism reflects a dissatisfaction with the literal account of the world established within the literate apparatus. The impulse to allegorize the reports and recommendations of empirical consulting is a symptom of a paradigm shift just as it was at the beginnings of Philosophy, when the Greeks allegorized their myths (or when the Christians allegorized the Old Testament).

The fullest theoretical elaboration in our time of an allegorical quartenary was made by Jacques Lacan (hence, no doubt, his interest in Joyce). Human experience in Lacanian psychoanalysis is situated at the intersection of three orders – the Real, the Symbolic, the Imaginary – supplemented by a forth order, the Symptom. The object *a* (or "@," "a" inside "O") is said to exist precisely at the point of intersection or overlap (hence its power to produce effects of uncanny "illumination"). We put our own spin on these "exigencies" (Leupin, 1991), to bring them into our adaptation of allegory as the method for generating testimonial as a consulting practice:

- The Real, in this adaptation, corresponds to the level of Anagogy (but in Lacan it is impossible to define, lies outside formalization, and is known only through inference, by its effects, which often produce shock, trauma, as in the sinking of the Titanic);
- The Symbolic is the Literal (in Lacan, the historical collective dimension of social institutions, law, language, and all other systems, equivalent to the Superego in Freud);
- The Imaginary is the level of "Allegory" in this narrower hermeneutic sense (in Lacan the production of projections and identifications created through intersubjective specularity – the Freudian Ego);

- The Symptom is the Moral level, based in what is specific, even unique, to the existence of an individual.

Testimonial

Mystory performs a *testimony,* the mode that Elie Wiesel said defines our era. The registers or levels of our episteme are different from those organizing medieval experience, obviously, but their nature and function may be understood by analogy with the medieval example. In mystory the allegorical scheme is appropriated not as a means of interpretation but of invention, production. The expectation is not to live an epiphany, but to write one, by juxtaposing a modern equivalent of the allegorical levels. This use of allegory identifies mystory as part of the postmodern condition. The contemporary quartenary or square consists of four discourses that together make up a "popcycle" (referring to the circulation or cycling of units of signification through the different discourses, from popular to specialized and back) (Ulmer, 1994).

The relevance of mystory as a basis for testimonial is that it is a holistic practice, designed originally as a way to compose simultaneously in four (more or less) discourses. What one testifies to in the first place is the inventory of identifications or quilting points constituting one's feeling of being a unified "self." Mystory is a map of its maker's psychogeography, locating those discourse positions already occupied by the subject. The effect of "encounter" or "recognition" is triggered when the egent feels the draw or pull of one of these positions. The popcycle that Ulmer worked out in *Heuretics* remotivates an allegorical quartenary. The four levels of allegory serve as the four elements (Earth, Air, Fire, and Water) into which Plato's chora sorted chaos.

Mystory as Allegory

Anagogy = Science (Career, Discipline).

The collective meaning of history is determined now not by Christian Theology but by whatever world view is embodied within the specialized knowledge that one acquires as an expert in some given career field. This expertise may be disciplinary in the academic or professional sense – physics, law – or it may be a craft or trade (plumbing, football). This knowledge is the means by which one earns one's livelihood (work), but the knowledge of an avocation may be used instead.

Moral = Family (biography of the Individual).

PHOTOGRAPH: BARBARA JO REVELLE

The individual is considered in terms of his/her family upbringing, with the language being the one learned in the home (English, French, Creole) and the discursive regime being the habits and customs specific to that family, as governed by such things as ethnicity, race, gender and the like.

Allegory = Entertainment (commercialized information and story). The interpretive "key" comes not from the "imitation of Christ" but from the identification with certain favorite celebrity stars. The discourse learned is that of cultural mythology encountered in popular genres (Westerns, film noir, romances) carried through the media – television, cinema, newspapers, magazines, advertising. The entry into this discourse starts in early childhood due to the nearly ubiquitous presence of television in the home.

Literal = School (K-12).

The historical fact that grounds the figuration is not the story of Israel as codified in the text of the Old Testament, but the history of one's nation, state, or community – the official version – as codified in textbooks taught in schools K-12. This history represents the memory of the collectivity.

"Crossroads": The Allegory

Egents do not file a report, but a testimonial, using the mystory procedure. The image generated mystorically is categorial – it extracts and holds the mood of

PHOTOGRAPH: BARBARA JO REVELLE

the zone, and this mood is what makes the image capacious, capable of gathering into a set the whole (hole) of the zone (to chart its determining propensity). The established means by which a picture holds a narrative are either that it represents the "pregnant" or decisive moment – the crisis, peripateia or turning point of a narrative; – or that it offers what is typical in terms of the everyday reality of a situation. While both of these possibilities are observable in "Crossroads," neither is necessary since the method does not depend upon the form of what is visible in the image. The image functions rather as a "souvenir" of the triggering detail (punctum). It is to the testimonial what a film still is to a movie. The network of associations snapped in this shot is enhanced by the database technologies of electracy.

To put into a database the flash of insight captured in "Crossroads" involves unspooling the four threads ("ficelles") of allegory. What is the effect of juxtaposing these four scenes? The *sinthome* that links them together – the event of attunement – is "singular," a specific "situation," and yet, as *Whatever* (Agamben), has categorial power. If our purpose were to write a screenplay, to make a movie, then these tracks could be integrated in one narrative. "Crossroads" is a virtual film still evoking the diegesis (imaginary space and time) of Myami (a narrative diegesis is what makes a particular story recognizable regardless of the medium in which it is represented). "Myami" is "Miami" in the media spectacle. The dynamic movement of a narrative from beginning to end is managed by the contracts governing the three modes of exchange that constitute society: economics, kinship, and communication. Events become potential stories (become worth recounting) when the contracts that regulate the exchanges are violated (the enactment of such vices as stealing, cheating, lying).

These exchanges and their violations are at work in the scene and moment indexed by the photograph. The mystory registers each exchange, juxtaposing or blocking them into an assemblage, in order to foreground a different dynamic that is also at work in the scene: contingency, luck, (it goes by many names, each with its own nuance), chance, coincidence, synchronicity. Beyond the ordering powers of argument, narrative, and figure, there is a disruptive, chaotic operator. An "omen," for example, is defined as the conjunction of a personal problem, an event, and chance. In electrate metaphysics there are no essences, but only "accidents." "This happens on the road, out at the crossroads, where novel experience arises contingently and so novel speech must also arise. Plato thought not only that Hermes invented language but that he did so in relation to 'bargaining,' which implies that a prime site of linguistic invention is the marketplace, another place where we are likely to meet strangers with strange goods, and, crossroads-wise, find ourselves forced to articulate newly. Moreover, the market at the crossroads may be a metaphor for metaphor itself, or for any original speech, the linguistic flowers that sprout at the crossroads of the mind. The mind articulates newly where there is true coincidence, where roads parallel and roads contrary suddenly converge" (Hyde, 1998: 299).

The Miami River is a port, a marketplace, literally a crossroads. The lines of our potential narrative are left untied in order to allow a figure – a metaphor or metonym – to emerge, an image, whose "event" or happening blocks the discourses into a pattern, a constellation, a vortex. The point in any case is not to tell one story, but to show the ground and limits of story and concept. We assume the cultural schema, the familiarity with narrative form, so that it matters little whether we imagine "Crossroads" as a still from a movie that has been disassembled, or one that cannot be made. What are the stories?

Literal Level: History (the story of the Miami River).

Literally "Crossroads" shows a ship docked next to a warehouse on the Miami River, City of Miami, Florida. It is not the ship we anticipated in our table work (it is not the *Rose-Marie Express* but the *M/V Deliverance Express*). The cargo is not immigrants but used, second-hand items: bicycles, plastic buckets, clothing, mattresses. It is piled around the warehouse, and some has been loaded onto the vessel. Atop a pile of mattresses two crewmen are seated.

What cannot be seen but must be inferred is the Problem at work in each track or register. One of the selection principles determining what to include in each thread is this schema of Problem-Solution that frames the modern understanding of "Até" (the blindness that produces calamity in tragedy). In this case the ship is impounded, caught in Operation Safety Net – the "Caribbean Code" – conducted by the United States Coast Guard, directed at "substandard shipping," vessels under 500 gross tons. The nature of the problem depends upon one's point of view or position in the situation. For the Coast Guard it is "irresponsible owners" trying to extend the economic life of old ships. The established policy solution is to eliminate such shipping through inspection and impounding. For the Haitians the problem is to make enough money to feed dependents back home, a goal thwarted by the opponent – the Coast Guard. Here is a situation of the kind explored in tragedy: a conflict of two goods.

Allegorical Level: Entertainment

The place where the photograph was taken (the Miami River zone) is animated by the exchange of communication, media, language, in the sense of the circulation of stories and information that motivate or discourage tourist travel and immigration. "Problems" in the Media quarter are constructed as spectacles to be consumed as entertainment. Although we designated this track as the Media exchange, an immediate Problem structuring it concerns tourism, the visitors who come to the location of a "spectacle." Sites of problems of all kinds have long been destinations favored by tourists. One of the most popular stops on the "Murder, Mystery and Mayhem Tour" of the crime scenes in Miami, for example, is the gothic Mediterranean mansion on Ocean Drive known as the Amsterdam Palace, former home of Gianni Versace. The steps and sidewalk in front of the mansion, the place where Versace was murdered by Andrew Cunanan, is a shrine for mourners and a "photo opportunity" for the many visitors who pose for snapshots. The best representative of this register is the television series *Miami Vice*, with its pastel updating of the noir cop drama. *Miami Vice* is a kind of postmodern Eiffel Tower, holding a place for the city's identity in an international cognitive map.

The policy question in this quarter has to with "murder" to the extent that crime statistics influence plans for the gentrification of downtown Miami. The Myami of neonoir narratives, like much American entertainment, foregrounds murder as the essemce of "problem." A practical concern for mediated Miami is the physical impossibility of moving visitors continuously through the river district. "Few things can dampen an urban experience more than the realization that each stop on an itinerary requires an automobile and a parking space. Until downtown can offer an assortment of attractions within walking distance of each other or in the reach of a short ride on a convenient public shuttle, business persons and visitors alike will have no patience with the idea of spending time downtown." So states the Master Plan of the Miami River Coordinating Committee. Solution? A River Walk, including urban renewal projects of various kinds in addition to the board walk: condominiums, restaurants, pollution controls, and the like. Thus this thread embodies the river as an object of study for the expertise of urban planners and architects.

Moral Level: Family **(*Personal Memory*)**

The photograph was taken sometime in June, 1998, by Barbara Jo Revelle, an artist and member of the FRE, during her five-week stay in a tourist Inn next to the river. Her personal trajectory is inscribed within the exchange of kinship, exogamy, tracing a circulation away from her father through a marriage and divorce to her present relationship with Ron.

PHOTOGRAPH: BARBARA JO REVELLE

Revelle's relationship with Ron is the motivation for the burning question of her consultation, which is framed syncretically as empirical and oracular. The Problem orienting the exchange is love – manifested in this case as jealousy, which motivates the intense and repeated fights between Barbara and Ron. He accused her of wanting to start an affair with some of the men she met during her stay. His justification was a genuine (and justified) fear for her safety. The default Solution is counseling, couples therapy. Her story is simultaneously that of a querent and theoros, working in the "middle voice" which is the "mood" of theoria (the tour of the problem out there is a reading of the egent's Family situation).

Anagogic Level: Career (Disciplinary Knowledge)

The quarter of Expertise is absent from most narratives, since the discourse of disciplinary knowledge avoids story and prefers science, argument, expository and argumentative writing: the treatise format. The structural account of circulating objects, of exchanges governed by contracts, nonetheless applies equally well to knowledge, especially to the institution of consulting that is the target of our deconstruction. Knowledge circulates between the academy and the community – the agencies, commissions, and councils overseeing policy formation in the City of Miami, for example. The Problem considered within this institution is every problem and any problem that comes to the attention of the public. The default Solution is science (Natural and Social Sciences) in the form of instrumental reason, empirical applied knowledge.

PHOTOGRAPH: BARBARA JO REVELLE

The difficulty for the FRE *is that our expertise (Arts and Letters) is excluded from this exchange. The narrative obstacle for the* FRE *is that its expertise is refused, remains unsolicited, rejected as irrelevant to the policy formation addressed to the zone. It is important to note, at the same time, that the kind of knowledge we have to offer does not conflict with the Problem-Solution terms operating in each quarter of the allegory. Rather, what we have to offer is a different category, a different framing of events altogether, from the problem-solution narrative of conventional consulting. Nor does the EmerAgency replace conventional consultants, but supplements them with egents, citizen witnesses, working through a new medium – the Internet.*

Impasse

Four threads, four circulations, four "props" run through the mystory: used domestic goods; tourists; Barbara Revelle herself; knowledge. It is not that all this information is "in" the photograph, but that it all may be thought together through this image, designed as interface for a Myami "mirror world." Why four? The *Geviert*. The number is symbolic, not instrumental. Each circuit in its own way is "blocked" (frozen, immobilized, paralyzed): the narrative exchange is interrupted, at an impasse. The block in each case involves a safety crisis, a calculation of risk. The snapshot records this situation at a specific moment in time: it is one "throw of the dice," just as in divination a consultation samples the querent's situation at a given moment, with one shuffle of the deck, one casting of shells, stones, coins. The reading reveals the play of forces at work in the situation at that moment, their "propensity" or tendency, the possibilities or potentials for change, the timing of that change, its possible direction. At the same time the reading confirms the general condition of the modern city, THE city (image-city), already established for Baudelaire's Paris, Joyce's Dublin, Eliot's London.

A situation is not (only) problem, or project. Mystory could be described as an intersection of problem and project in a zone of disorder. A deconstructive approach to consulting puts in question the state of mind that conceives of the world always in terms of problem/solution and individual intention. "Event" is the convergence of these possible framings. "Problema can signify projection or protection, that which one poses or throws in front of oneself, either as the projection of a project, of a task to accomplish, or as the protection created by a substitute, a prosthesis that we put forth in order to represent, replace, shelter, or dissimulate ourselves, or so as to hide something unavowable – like a shield (problema also means shield, clothing as barrier or guard-barrier) behind which one guards oneself in secret or in shelter in case of danger. Every border is problematic in these two senses" (Derrida, 1993: 11–12).

Derrida extends his etymological analysis into an allegorical exposition of the effect of problem as shield by means of the scene in which the ghost of Hamlet's father, a "revenant," appears in full armor on the ramparts of Elsinor.

The definition involves the fusion of a series of terms: advice, advise, adviser, advisory, visor, alluding to the experience of cultural artifacts or phenomena that "return our gaze." "To feel ourselves seen by a look which it will always be impossible to cross, that is the visor effect on the basis of which we inherit from the law. Since we do not see the one who sees us, and who makes the law, who delivers the injunction; since we do not see the one who orders 'swear,' we cannot identify it in all certainty, we must fall back on its voice" (Derrida, 1994: 7). Such is the experience of stain in a field of the gaze. Myami spectacle configured through mystory into chora (Miautre) enables citizens to experience their effect as stain (the unintended calamity).

"Problems" in this account look at the consultants. The catastrophe outside somehow corresponds to an inside, in a feeling of obligation – the demand of the ghost (revenge), objectifying a force or orientation in experience. It is the question of ethics as a specific kind of experience (feeling, emotion). To produce a testimonial with respect to a problem requires that the consultant be included within the matrix of the problematic (rather than as an external objective source of solutions), in order to undergo (and not just understand) one's agency. Advice is specific to each ad-visor, depending upon who or what is recognized behind the visor. Problem as revenant covers multiple, heterogeneous heritages, and approaches to solutions are based on the ad-visors' experiences outside of disciplinary expertise, in the habitus of life. The consultation moves from Miami-vice, through ad-vice, to virtue.

The difference between problem and aporia is that problem concerns that for which there is a determinable, calculable outcome that may be produced according to a program (possibility), as opposed to the aporia whose outcome remains outside the reach of prediction whether calculated or divined (impossibility). Problems work with concepts, while aporias work with events. Event is by definition that which may neither be predicted nor divined, but which is absolutely unforeseeable, completely surprising. While concepts permit planned programs, events require testimony, oaths, faith, confidence, conjuring. Problems engage knowledge (epistemology), while aporias engage responsibility (ethics). Problems concern law, while aporias concern justice. The two dimensions are not separable, but are in tension, wrapping the theoria in a double bind (Derrida, 1997: 67–69). To tune a zone is to look for the impasses to be found there. The impasses, knots, or blockages filling or suturing a hole are not to be resolved or untied but experienced, rendered, made tangible, even enhanced, heightened, brought into thought and perception, "toured," the assumption being that these blockages are otherwise inaccessible to a traumatized public sphere. There is no concinnitas (no harmony) in a zone, but dissonance, derangement.

Each story of the Miautre mystorical testimonial is outlined separately in the following levels. Perhaps one might see them integrated some evening, watching TV *with remote in hand: an artist's biography, an episode of* Miami Vice, *a documentary on Caribbean trade, all threads probed forensically as crime scenes (in order to remain comprehensible to the distracted viewer). Zapping is an inference path internalized through a viewing style. You are distracted, worried, anxious, (modern), about problems in your world, so you don't notice how the* TV *schedule coheres in a constellation. That is how it often is. You could pose your worry as a burning question to your city as oracle. That is what Barbara Jo Revelle did, in the summer of 1998, and then she drifted through the designated zone to receive an indication. Miami, like Delphi, neither conceals nor reveals, but intimates. Each egent will have asked: what is that for me, in my own place and time?*

CHAPTER 7

Mystory 1: Family

Divination

With the "table work" completed – a review of the media image of Miami (Myami), the next step was to tour the designated zone, the Miami River. The FRE delegated the official tour to Barbara Jo Revelle, acting as theoros (theory tourist). As a creative photographer, Revelle's task was to measure the mood or atmosphere of the zone. Her procedure was documentary, including interviews with anyone she met and a journal recording her experiences, memories, and reflections related to the tour. These documents (word and picture images and recordings) provided the link between Miami and Myami. In this way we underwent Miami "for ourselves" (theoria).

Given the decision, based on the table work, to focus on the "haitian" as the other of the tourist, representing the range of attraction-repulsion, it was important to take into account post-Enlightenment warnings about the limitations of science – its colonizing ideological construction of the other –, and to make explicit this potential abuse in our own disciplinary practices. A primary function of a choral map is to register the default moods already circulating in a zone, organizing events and judgments, predetermining in principle "what there is" – what features are selected as "problem" and what counts as a plausible "solution." These defaults include the traditions within which the FRE itself works. The historical research showed that our civilization is continuing the process of syncretism, adding an Afro-Caribbean worldview to the post-Augustinian jew-greek synthesis. Our intention was to experiment with a hybrid of poststructural theory and the divinatory practices of the Afro-Caribbean diaspora (specifically Vodou and Santeria). The larger goal is to design Miautre as an internet oracle, to undertake a pratice of global wisdom. This hybrid is undertaken having in mind Bruno Latour's reminder that "we have never been modern." His concern overlaps with our experiment in transforming literate category formation.

> But how can the principle for classifying entities be changed? How can the illegitimate multitudes be given a representation, a lineage, a civil status? How can this *terra incognita* that is nevertheless so familiar to us be explored? How can we go from the world of objects or that of subjects to what I have called quasi-objects or quasi-subjects? How can we move from transcendent/immanent Nature to a nature that is still just as real, but

 | DOI:10.1163/9789004534643_008

PHOTOGRAPH: BARBARA JO REVELLE

> extracted from the scientific laboratory and then transformed into external reality? How can we shift from immanent/transcendent Society toward collectives of humans and nonhumans? How can we go from the transcendent/immanent crossed-out God to the God of origins who should perhaps be called the God below? How are we to gain access to networks, those beings whose topology is so odd and whose ontology is even more unusual, beings that possess both the capacity to connect and the capacity to divide that is, the capacity to produce both time and space? How are we to conceptualize the Middle Kingdom? As I have said, we have to trace both the modern dimension and the nonmodern dimension, we have to deploy the latitude and longitude that will allow us to draw maps adapted both to the work of mediation and to the work of purification. (Latour, 77)

A quasi-object resembles conceptually a transitional object for adults, reflecting a "warped" condition that includes passion within materiality. The "Great Divide" separating West from East is a construction, Latour argues, a projection of an inner split by means of which the West invented the "exotic other." He calls for a categorization whose topology is more adequate to the continuous networked relationship between the local and global dimensions. His "we" is an invitation, and not just to "Americans." Roland Barthes, who anticipated our crossing with divination in his theory of text, provided a point of departure for our hybrid "consultation."

The Starred Text

> We shall therefore star the text, separating, in the manner of a minor earthquake, the blocks of signification of which reading grasps only the smooth surface. The tutor signifier will be cut up into a series of brief, contiguous fragments, which we shall call lexias. It will suffice that the lexia be the best possible space in which we can observe meanings; its dimension, empirically determined, estimated, will depend on the density of connotations… The text, in its mass, is comparable to a sky, at once flat and smooth, deep, without edges and without landmarks; like the soothsayer drawing on it with the tip of his staff an imaginary rectangle wherein to consult, according to certain principles, the flight of birds, the commentator traces through the text certain zones of reading, in order to observe therein the migration of meanings, the outcropping of codes, the passage of citations. (Barthes, 1974: 13–14)

PHOTOGRAPH: BARBARA JO REVELLE

We designated the zone (run-down, mixed-use industrial area) of the Miami River as a text-scene of divination, which we would "consult" by surveying the movements, positions, identity, and "cries" of the denizens – Haitians, Coast Guard, Italian taxi driver, the homeless, the developers, and the rest – in relation to the personal situation of an egent. Divination provided an interface metaphor for the tour of the river, communicating instantly the strategy for transforming indifferent circumstances into a *situation*. Divination is an analogy for our Copernican revolution in consulting, in which we did not try to explain the problem; it explained us. *Problems B Us*. The hybridity of the method was contained within the pun on "consultation" that linked the practices of science and religion. In our case, Barbara Revelle, acting as querent, posed a burning question (a question at once personal and collective, the one that everyone asks, hence general, and yet the answer is unique in each instance) and then her tour of the zone, when distributed through the four-fold matrix of the mystory, produced a number of signifiers that constituted in its pattern an answer to her question, showing the *topologerie* organizing the zone.

The querent poses a Burning Question prior to the tour of the zone, thus configuring this unknown place as "forest of signs," an allegoresis of her project.

Barbara Jo Revelle walked out into a humid sunny summer morning on the Miami River, day after day, for five weeks during May and June of 1998, performing as querent (inquirer) for the FRE theoria. Her burning question was formulated as a hybrid between the general one borrowed from Heidegger's attunement ("how do things stand with me?" – *Befindlichkeit*), and the specific one based on divination. What was her question? "Should I stay in my love relationship?" or more basically: "What is my situation with Ron?" One does not ask such a question casually, nor will the answer come to one not in crisis.

Revelle's consultation sampled the situation of the zone at one historical moment. Her use of a personal question was posed within the frame of her status as theoros, one member of a theoria. The FRE acted as diviner to her querent, providing a cultural form within which her question and the image record could come together to provide an answer. Whatever the value of this answer to Revelle personally (and only she may decide that), *the image-answer serves also to reveal the attunement* (*the mood*) *of the zone*. The image produced in the process of the consultation is an electrate category, gathering the heterogeneous entities found in the zone into a scene (a mise-en-scene which is also a mise-en-abyme) that is to electracy what the concept is to literacy. The claim (which is falsifiable, open to proof) is that it is possible to use this image not only to think and reason and feel with respect to the Miami River, but with respect to other matters as well. The poetic categorial image has abstractive, logical, classifying power, in the context of a digital prosthesis.

The iconic or emblematic image was selected not by Revelle alone, but by the FRE using the mystory form to sort the data (in this case, nineteen hours of videotape and many rolls of film as well as a journal) into an affective map that produced a pattern. "Cognitive map" is the term Fredric Jameson used (extending the meaning from its source in architecture to name a citizen's mental image of her environment) to name a new aesthetic capable of connecting personal experience with expert knowledge about global and institutional realities. The choral affective map takes this development another step. The triggering detail (the punctum) appeared as a signifier, a certain quasi-object, that repeated in the discourse of each register of the popcycle. The title of the photo-emblem is "Crossroads."

Dialectical Image

The egent as querent shifts the state of mind of the consultation from instrumental sufficient reason to emotional burning question. This syncretic strategy solved not only the problem of "compassion fatigue" that Miautre must overcome as an internet civic sphere, but also motivated the formal necessity dictated by our method to compose a dialectical image (a time-image). To configure a chora, to prepare it as the interface for a mirror-world database managing deliberative reason in the dromosphere, the egent needs to objectify and render accessible to design the pre-judgments determining her decisions. In short, chora is an augmented prudence (virtue of good judgment). Mystory inherently produces dialectical images, in that it juxtaposes the egents childhood memories of Family experience with the other quarters of the popcycle. The FRE group mystory was experimental in that Revelle was contributing her Family childhood scene not to a private mystory but on behalf of a group. Our confidence in this expansion of the practice was based on the precedent of the Wooster Group (actors ensemble improvisation), who generated theatrical works by having the group improvise from scenes of personal experience provided by one member of the group – Spalding Gray (in the Rhode Island trilogy). A further adjustment in the original individual mystory was to shift the School quarter from the history of the egent's community to the history of the site addressed in the EmerAgency consultation. The other quarters of Entertainment and Career were similarly adjusted to reflect the group and the target of the consultation.

This first Level documents the childhood scene of memory that provided the source of the "Then" to juxtapose with the "Now" of the Miami River. Prudence requires instant assessment of past-future Now, a requirement objectified in the dromosphere. Prior to her tour of the zone Revelle produced a series of short sketches recalling her childhood living near Pleasant Lake in the State of New York. The purpose of the sketches is to accumulate sufficient details to establish the tone, the atmosphere, of the memorial scene. Each egent will have asked "what is that for me, in my own case?"

Back of Beyond

From the notebooks of Barbara Jo Revelle

Skeeversville 1: The Upholsterer's Basement

I am looking at this map I drew hurriedly of Pleasant Lake, my side of the lake with my parent's big safe home they called "Lost Lodge" and the other side of the lake called "Skeeversville" by my Dad. And the Swamp behind our house into which my Dad dumped kerosene – to kill the mosquitoes, but instead he killed all the little spring peeper frogs and created our own silent spring – and nothing to eat the mosquitoes, which survived in alarming numbers, huge clouds of mosquitoes. I'll try to divide the main memories about the place where I grew up and my attendant memories into parts:

Linda McDonald's house – with the basement where her Daddy did upholstering when he wasn't drunk, and where the kids from Skeeversville played sexual and scary games in the dark amidst the mattresses, old chairs, and moldering upholstering material. Two brothers from Skeeversville, Ricky and Bobby Eason – were often in on those games. There was one we called *Perfume in the Black Hole*. I can't quite remember the perfume part, but I think we divided into two groups, boys and girls, and went to parts of the basement where we couldn't see the other group. The girls passed around a perfume bottle, and put some on in turn, and the idea was that where you put the perfume was related to the "secret you were required to remember and tell to the rest of the group." We went around the circle, and when a player couldn't think of any more secrets to tell, she "lost." The punishment was that she had to crawl into a mildewed and spider-infested space created by a folded-over, broken-open mattress. This was the "black hole." Then someone from the boys side of the basement, chosen through some more manly ritual we never got to witness but were told had to do with their "privates," would crawl inside and try to guess the losing girl's identity by feeling her body.

Nancy Moyer and I talking earnestly and often about whether we could be pregnant because of these game or because of our practice of getting into sleeping bags with boys inside their "forts" built in the woods which surrounded Skeeversville. (We all had forts. Mine I lined with milkweed silk, a summer-long project.) Nancy and I were eleven and both had started menstruating the summer we thought ourselves pregnant. We didn't know what intercourse was, except one day we had seen two of the Irish setters my parents raised for extra cash, stuck together. One was crying and so we tried to pull them apart.

Skeeversville 2: The Stupp's

The Stupp's is the house right next door to Linda McDonald's house. Stupp's is the most low-down shack over there–and there were lots of Stupp kids, but the one who I remember best is George Stupp who was older and sexy, and Mrs. Stupp who, legend had it, was so "stupid" she cooked a frozen turkey with all the giblets still frozen in it inside the paper bag, and tried to serve it to her family. And there was Odd-Eye Aldridge who hung out with the Stupp boys, a dangerous and edgy kid with a wall-eye who allegedly raped Barbara Eason, my friend Bobby Eason's older sister. This is the same house that I photographed when I went back with Ron last year – with the graffiti on it that says "house of pain." My Daddy said if anyone raped his wife or daughter he would go kill the guy himself.

Skeeversville 3: Bobby Eason

In the slightly better section of Skeeversville, (but where I was still forbidden to go) was my best friend Bobby Eason's house – still covered today with the same brown shingles. Bobby was five years older than me, so fifteen when we began our friendship. Of course my parents frowned on this friendship, but in spite of that we swam to the Island, Snake Island, half a mile up Pleasant Lake, every day one summer. We imagined ourselves in training to be long distance swimmers. Bobby had a dog he found and kept in a pen in the woods well away from his house and we both used to sneak it food – a black dog he claimed was a rare black German shepherd and called Saber Jet. One summer I used to meet Bobby every evening by Saber's pen and bring scraps from dinner. One day it just disappeared.

At our meetings we talked about our shared passions – Bridy Murphy and reincarnations, flying saucers, and death and whether God could bring things back from the dead and whether or not prayer worked. We put live bugs in jars and buried them and did other religious experiments. Later when Bobby went AWOL from the paratroopers during the Vietnam War he sent wildly decorated packages from Pakistan or Shri Lanka or wherever he was hiding out from his AWOL punishment – lumpy, fat offerings covered with quotations from Ferlingetti or Ginsberg or with "get out of Vietnam" slogans, packages and letters filled with drawings, decals, soil samples, animal bones, and a bit of broken china, a doll's hand. One came wrapped in fur. He was a "mail art correspondent" though he never knew it and neither of us had heard of Allan Kaprow or Ray Johnson. Later Bobby drowned off the coast of Santa Barbara, SCUBA diving for sea urchins. There was a deaf mute on the boat and Bobby's body was cremated before any of his family or the woman he lived with saw it. There was supposed to be an inquest to see if there was foul play, but the parents were too old and ill to follow through and the woman, not being his legal wife, apparently had no rights.

Skeeversville 4: Lost Lodge

Lost Lodge is the remodeled summer cabin on Pleasant Lake where I grew up and George is the stuffed tiger's head over the stone fireplace. Teddy Roosevelt is supposed to have shot it. Family legend has it that this information came to us from a piece of old newspaper stuffed inside George's head. He was handed down in the family on my father's side. In many dreams I am trapped in the house–Lost Lodge–and there are tigers roaming around outside. Sometimes I am holding a tube, like a toilet paper tube, with my hands over the ends, and inside is an embryo and I am trying to protect it from the tigers. They are quiet and they skulk around. Sometimes I forget about them being outside and lean out the window and fall out and am eaten. They say if you die in a dream you really die, but it must not be true because I die in some version of this dream all the time. So now Daddy has an African game room and goes to Africa on safari several times a year trying to fill in his collection with this or that rare animal, but back when I was a child there was just George and a few deer heads, some stuffed ducks, and there was the stuffed parrot, Polly, on the drift wood, who nobody but my brother could bear to look at.

Skeeversville 5: Stuffed Parrot

Mom and Daddy are going to ride their motorcycle to Laconia, New Hampshire for some biker's week motor cycle races and Jackie my younger brother and I are staying with the McDonald's: Linda McDonald with the drunken upholsterer father and her brother Alan, her mother Mary with a huge mole on her forehead and her little sister Bonnie. I am eight. So all five kids have to sleep in the same bed – Linda, her younger brother Alan, her sister Bonnie who still wets the bed, Jackie and me, and it is hot and smells of urine and I can't sleep and I am longing for my mother and feel abandoned. I have a little black kitty named Snowball at home, but McDonald's won't let me bring it to their house. My parents have told me if we leave it outside it will stay home and I can go down with Mary each day and feed it. They say it will be OK. Mary has to go around the lake to our house and feed all the Irish setters anyway. The Irish setters are in kennels, and we raise them for money. So the first thing that happens is that we go there, Mary with the mole, and me, and it is missing–my baby Snowball. I look all over the woods and the swamp all day, calling for it, but it has vanished.

Mary also goes to feed the parrot, Polly, who is Mama's beloved. This parrot calls to be covered at night by saying it is a "Sleepy Sleepy Poll," and it does a routine like an auctioneer, at the end of which it cries out triumphantly "Sold, American!" I think Mary forgets to give it water or it gets too hot because one day it is sick and laying on the bottom of its cage in all its green and white marbled shit. Mary brings it up the hill to McDonald's house and we try to feed it water with an eye dropper. It doesn't work. Polly breaks the dropper with her beak. She dies. I start crying and tell Mary to call my Mama and tell her to come back, but Mary won't. Instead she gives us money for ice-cream and we walk barefoot on the dirt road further around the lake than I have ever been, to the little store where they sell beer and milk and newspapers. We buy red popsicles which melt all over our hands. As we walk back Carrie Nepage, an older girl who has big breasts, invites us into her house to see what her family has just acquired. On the wall is a carnival plaster replica of a macaw. It has green, red and yellow in just the same places as Polly – first experience of synchronicity, or a malevolent god, so I decide to run away. I cannot sleep in that piss-smelling mattress one more night with everyone's clammy flesh and no way not to touch them, and Bonnie kicking me and all the pain in my chest from Polly dying and losing Snowball. When it gets dusk I slip away unobserved and take the path around the lake to my parents' empty house. It is locked so I go into the kennel and curl up in one of the dog houses with Duchess and her pups. I hear Mary and, later, her drunken husband Mac calling and calling for me for hours, even coming near the kennel, but I stay still and they don't find me till morning when they come to feed the dogs.

When finally Mary tells us that Mama and Daddy have come home I rush down the hill and there is Mama sitting on the steps in front of the green painted house with the Lost Lodge sign. Her beautiful curly-haired head is down and she is crying about Polly and not even happy to see me. About a month later Daddy tells my brother and me that he has a surprise and we can't go in the dining room until supper. When we do go in we must keep our eyes closed, and open them only at his command. There, on a free standing piece of drift-wood, is Polly, stuffed, with glass eyes peering out of her yellow head.

End of Revelle's notebook.

The Burning Question: Ron?

The querent's burning question is posed within the Family quarter of the mystory, in the present. These samplings from Revelle's journal composed during the consultation document the context of this dilemma.

From the journals of Barbara Jo Revelle, written during the consultation, May–June, 1998.

Address Unknown (*Unsent Letter*)

Dear Dana,

How does it go with you my dear old friend, my first true love? I am living on the Miami River for five weeks, doing a project in which I pay a lot of attention to things like chance encounters, odd notes I find on the ground, my dreams – making these things into directives for where to go on the river each day, how to spend my time. You were in a dream last night so I thought I should write to you and see if you are still on the planet – we cannot safely make assumptions like that about each other anymore. Anyway, I think we should keep up our habit of checking in with each other every decade or so, don't you?

You appearing in a dream always stands for the search for – or trying to get back to – "the loved one." I am sure I must have told you this before since it has been part of my unconscious symbol system for probably thirty years. In this dream I am lying on an old mattress by the mouth of the Miami River, actually I think I saw it yesterday filming at the site of some apartment buildings which are being torn down. This mattress has an intricate flower pattern and is all pink and yellow in the late afternoon light. The whole scene is suffused with saturated reds and yellows and oranges like in an early Kodachrome just after color snapshots were invented – no blues or greens. There has been a picnic and I have a taste in my mouth like bland Spanish rice or maybe nuts and there is someone with me–a serious looking man with a beautiful back, crouching down intently looking at some evidence of evicted apartment dweller's lives – some child's drawing he has found in the rubble. I don't recognize him but I know he is the one I love and I ask him where he is from. He says: Provincetown, so I know he is you – Dana – (the original loved one) but he doesn't look right – doesn't look like you or anyone I know. We get in an old wood canoe with peeling white paint – (the canoe trip on the Mississippi River we planned in 1968 but never made because you fell in love with someone else just before our departure?).

We go a long way up the river to a town that I almost recognize. It is like Chichicastenango, a village in the mountains of Guatemala where Alex (my ex) and I went the year Katie was born–with a smoky market near a church with a medicine man doing some kind of ceremony with purple flower petals and Mayans wearing exquisitely complex embroidered huipilles with parrots, deer, fish – like some peyote hallucinations. No tourists. No one speaking English or even Spanish. I like this strange dream man I am with – he seems to love me fiercely – but not being able to recognize him when I look at his face is unsettling, so there is both pleasure and suffering, and this feeling is connected to a quality of the light which is yellow and pulsing, flickering, shimmering, beating, I can't quite think of a word for the way it was.

I wake up struggling to stay in the dream. I want to be in the exotic village with the man, don't want to be alone in my little room in the Miami River Inn where it is too hot, too blue, too green, too humid, where the reality of my love relationship with Ron, who has left in a rage in the middle of the night to drive back to Gainesville, is too much work. Since you and I last corresponded I got a divorce and now I am with this jealous, moody, journal-keeping Irishman who has, as he puts it, many dark nights of the soul. He was a helicopter rescue pilot in Alaska for many years. He lost some lives he thinks he should have saved, had to bring back bodies in all states of decomposition, and has an overdeveloped sense of mortality (what a shameful thing death is). I'd say he has experienced more than the usual share of losing, dying, trying to forget.

Ron and I knew each other as children. He loved me when I was a girl of twelve. We went to a Halloween party together. I didn't remember this, but he did in uncanny detail, and there it was when I looked it up in my first red plastic "ponytail diary" c. 1959. I ditched him for another boy on the Halloween hay ride, and thus became his first unrequited love. I didn't see him for thirty some years, then he found out I was divorced in a chance encounter with my brother three years ago, called me, came to see me, we fell in love, and now here I am dreaming about you, which always signifies the re-emergence of the question: How should I live my life? – with whom? – coming up again and again even though I should have settled it years ago. I keep wondering if I'm supposed to be with this man or, perhaps, someone else; somewhere else. This restlessness has a level of absurdity, of course, in a 51 year old woman, but here it is, and it feeds right into his pain and jealousy, so we fight for our lives when we fight, and we fight all the time.

Of the thousands of things I could tell you, what should I actually write about? I've been here about two weeks working on this project about a little five-mile river with the nasty reputation – *Miami Vice*, drug dealing, toxic water, and much more. I drifted into it not knowing where it was going. But now it has a very particular shape. There are just two rules: (1) Each day spend an hour writing about my waking life, dreams, what I read, what is said to me, whatever happens; and (2) be responsive – go out and photograph and video-tape whoever or whatever comes up in the writing. If I read about a baby girl's body rolling out of a sewer pipe into the river and it strikes me, then I go to that site, even if it happened in 1908. I just go and see who's out there – I go to places that attract or scare me and wander around like I'm looking for clues to something. All of this work is condensed into a question – the "burning question" that one might ask of a diviner. And the river is supposed to give me an answer.

The river has this kind of composition for me. It connects all that is unconnectable. Like Borges's Chinese Encyclopedia, the river constitutes an unthinkable space. I guess this letter has gone on long enough. I wonder how you are, sometimes. How are your children? How is your wife? How do you spend your days? What are you passionate about? What do you think about being this old? My 52nd birthday is in 9 days. Did we know when we were suntanned lifeguards on Race Point Beach we'd ever be 52? Unimaginable. "I hate old poet-men…" wrote some young poet but the years came on him anyway.

Mapping

I've figured out about mapping my own childhood – growing up on Pleasant Lake – onto the Miami River – or why I am, of all the things one could choose to think about on the river, most attracted to the Haitian boats, docks and warehouses and the sanctions against the people. When I grew up on my little Upstate New York lake I was forbidden by my father to cross (transgress) to the other side–which he called "Skeeversville," meaning shanty town, as it was inhabited by poor people who lived all year in un-insulated little cottages intended by their builders only as summer camps. There were drownings over there, dog packs, rapes…. I was told the people on the other side of the lake were dangerous, stupid, and immoral. Linda McDonald had a basement crammed with broken furniture and bedding, as her father made a living doing upholstering when he wasn't drunk. This basement is where the kids from "across the lake" hung out in the winter, killed rats, and did sexual things hidden in the stuffing and the debris. My side of the lake, where all the kids on the honor roll lived, was order, sanity, safety, reason and was absolutely boring. I was wildly

PHOTOGRAPH: BARBARA JO REVELLE

attracted to everything across the lake, had secret friends and even a dog there, to which I sneaked food daily for over a year.

Ron comes down to stay with me in Miami sometimes. Sex gets into the consultation. Ron and I go out and videotape a group of Haitians loading a boat and afterwards he comments that the man he was talking to was wearing his pants so low you could see his pubic hair. Later Ron writes me a fantasy in which I am attracted to this black guy and invite him over to my little apartment to be interviewed, and in which the Haitian man gets into my bed while I am in the bathroom. When I come out he is naked under the sheets – his clothes in a pile on the floor. In Ron's text the myth of black men having huge cocks does not prove entirely accurate, and his fantasy is written using words like "his erect manhood" – an expression I had associated with Victorian romantic novels, until we go into a porno store in Miami Beach and I notice that they make vibrators molded from some porno star's "actual erect manhood." You can purchase these in "natural" (pink), black, or electric green with sparkles.

Ron

Forgetting to Suffer

BARBARA: In my dream you were writing words all over my body – trying to take off my clothes so you could finish your sentences. We were in a restaurant, something like that outdoor restaurant with the shoe.

PHOTOGRAPH: BARBARA JO REVELLE

RON: The shoe? Oh you mean that gigantic shoe that was once a boat or something – used for a wedding in Venice.

B: Yeah, and fish scales on the restaurant, silver, remember? That yuppie restaurant by the river with the plastic humanoid statues – you know with the big eyes.

R: I don't remember it that way. I just remember suffering?

B: Suffering?

R: From the heat

B: I guess I was just so taken by the sights I forgot to suffer. Remember we walked from the restaurant to where they were bulldozing those apartments in front of the Sheridan – wandered around in the rubble, ruins, finding that painting of the family?

R: Yeah, lots of bits of those peoples lives. Whatever happened to that stuff we collected?

B: You had it – in a yellow folder.

R: No you took the best stuff to photograph for your slide show.

B: No, I only took that notice saying how the Nicaraguan woman should come in and get her bronzed baby shoes and those photos that looked like illegal aliens hiding in drainage tunnels.

R: You took the eviction notices too, the school kid's notebooks about Nada or something.

B: We got in that fight because you thought I threw the pink baby chair in the river to photograph.

R: You DID throw it in, I saw it in the debris earlier. Then it was mysteriously in the river. You'd do anything for a picture.

B: I would not.

R: Well then how come you kissed Frank?

B: I just kissed him on the cheek for Christ sake, after hours of hearing his life story and taping – don't start that stupid jealousy shit. God you are the worst. You knew I was going to do this thing down here – going out every day and meeting whoever and videoing them.

R: Yeah, well, how come there are no women in this art piece? Aren't there any woman around the Miami River? How come everyone you just happen to meet is an eligible, black man?

B: An eligible black man? Eligible for what Ron? Eligible to fuck? Is that what you mean? Eligible to fuck a fifty-two year old woman? I'm an old woman, Ron. I just talk to these people–its part of my project. You make me crazy.

R: Oh yeah, you just talk to them. Then you follow them to their pathetic little rooms and get on their beds and photograph them. I saw the pictures, remember?

B: Frank's room is so small I had to stand on the bed to take pictures of his magnet sculptures–on his dresser. You have a sick fucking detective mind. I don't believe you are bringing this up now.

R: You had to stand on his bed. Not the floor; his bed!

B: Yeah his little dog – it was on the floor – sick on the floor. He said it had parvo.

R: Oh of course. It all makes perfect sense. And I suppose when Winston invited to you his party – totally innocently right? He never imagined something might get going between you two?

B: Winston? Oh now its Winston too? Ron, he's in this thirties. He just wants to tell me how the Coast Guard is screwing the Haitians over. That's all.

R: Yeah well I know you weren't home till late that night of the party because I called and called. Were you with Winston?

B: I told you, it didn't start till way late. Winston didn't even show up until after eleven.

R: Yeah Right, Barbara, for his own party?

B: Look Ron, get out of my life at this level. I don't want this fucking surveillance. I won't account to you for all my waking hours. Winston Murdock is not interested in me as a woman and I will not do this idiotic, defensive conversation over and over.

R: God you are so naive. Look: you're white, he's black, and your exotic with your movie cameras paying all kinds of attention to them and everything. Don't you see that makes you desirable? Like you think white women come up and ask Winston who burned up his boat and can she interview him about the injustice of it all every day?

B: So what?

R: So you're playing the race card.

B: What the fuck does playing the race card mean Ron? Do you think in twenty-seven years of doing this–like teaching photo issues and cultural theory and all that–I haven't thought about exploitation, about the complexities of taking photographs across class and race barriers, haven't a clue what I'm doing?

R: Well O.K. What ARE you doing then? Wandering up and down the river all day and night, hanging out with the homeless, speaking your stupid version of Spanish 'hola, que passo?' – your eight words of creole? Waiting till some guy hits on you in some seedy restaurant or bar, writing letters for people…

B: I'm just trying to open up to the unknown–to let it come in and stir things up…

R: Yeah like drug dealers with guns–like going into Haitians boats and wandering around in filthy warehouses where you don't even have permission. It's dangerous – psychotic.

B: People need to open up to what's out there – even if it's impious – forbidden. I don't get much feeling of danger anyhow.... When I was a twenty-three year old girl I was in danger, maybe. Now I'm invisible. You just see me as a girl still, cause you knew me when I was in the sixth grade.

R: Yeah, middle-aged women don't ever get raped, right?

B: Ron we're not talking rape here. Rape is not what you're worried about.

R: Yeah well I don't believe you really went back to write that letter for Simon Libyan or whatever that guy's name was, to send it to the Coast Guard. You went back because you were attracted to him–I saw the video. That's the kind of guy you fantasize about.

B: That's the kind of guy you fantasize about, Ron.

R: Well I don't want you out there kissing men

B: Any other rules today Ron? You need to give out some more rules and I better write them down.

R: I'm not making rules. But if I can't call you by the time it gets dark – and I don't even know...

B: Where'd they teach you to put rules on everyone, Commander Hoover? You learn that stuff in the Coast Guard? Nineteen years and that rule shit just gets in to your bones.

R: The Coast Guard is not the bad guy here Barbara,

B: Yeah the Coast Guard is the bad guy, because the Coast Guard is the stupid fucker who says shit like, the boat is un-seaworthy, so I'm doing the Haitians a favor not to let them back out on the river, cause they might drown out there trying to take food back to Haiti. I can't help it if the people in Haiti are starving. I can't help it if the boat owners are stuck in the river with no way to get the thousands of dollars I just told them to put in their boats or they cant leave. I don't care if their cargo is rotting in the warehouses. It's not my problem. The boats aren't safe–the boats get seized. That's the rule. I'm just doing my job.

R: Well if a cop gives you a ticket for bald tires you can't say, gee, officer, I haven't got the money – I'm too poor. It's not logical.

Bug Dream

In the middle of the night I wake and realize that my eyes are already open. I am lying on my belly with my arms pinned under my thighs and my head turned so my eyes are fixed on R's sleeping form. I will myself to move but my body does not respond. I feel something odd about my head, like there's an insect in my ear, but at first I am not positive and I lie there a long time trying to think what's wrong. Then it starts to get active and its wings are beating inside my head and it is tearing itself apart. Ron gets up – I still can't move – and without either of us saying anything he calmly pours hot water – boiling hot

PHOTOGRAPH: BARBARA JO REVELLE

water – in my ear. He tells me it won't hurt and it doesn't, and the thing dies and comes out in these pieces which look like parts of some tiny parrot. And I am thinking, oh that's great but can I hear any more?

Then there is a skip in what I remember and he is writing on my body – strange words that all start with "A" like, "anathema," "apotheosis" – words Ron never uses. And words in Spanish or maybe French some other foreign language – but Ron only speaks English. Then they change into pictures with colors, like intricate huichol paintings, then back into words in different typefaces – one that looks like skeletons. And I could make out that there were pairs – opposites like dark and light, love and hate, and anger. What was that word with anger?

What do I know about Ron, really? Not a whole lot for two years. Fifty-one. Once Coast Guard (how strange is that?). A pilot. A bush pilot. Helicopter pilot in Alaska, lots of tight situations: saves, near saves, and failures – some frozen or decomposing bodies. I know that he's married, that he writes in a journal, that we fight on a daily basis, that he never quite gets around to getting divorced from Tara, that he looked me up after mine, thirty-seven years after I dumped him for the taller Robbie Robinson – on that hay ride in 1959. We were twelve. He says that he loves me – that he has always loved me; that I am the love of his life. We drive around the river, looking hard at the map, losing the river again and again, trying to understand the relationship of the map to the roads, bridges, fences, warehouses, trucks, boatyards, cranes, a black man with shopping carts full of old clothes, a white woman with a green ski cap shouting at a Cuban man, piles of wire and mattresses and lumber and tires and metal scraps – undifferentiated, colorful, useless looking stuff which is being loaded on boats.

What Is Desire?

I ache. I am homesick for Katie and for my attempt at a home with Zelda and Violet and her. What is this place Miami and how can I turn it into something I can live in? Is that the question? Did I just endure Jamaica. I should have brought my Jamaica journals. I both want and dread not having Ron here. I remember this feeling of being in a foreign country fundamentally alone and feeling unloved and not looking for love, just human companionship. Only sex is offered and it's scary and hard to be expansive because it is a dangerous environment with no real friends, no community. "But friends are everywhere to him who behaves himself well... and a prophet is not esteemed in his own country" (Burton?). The future-past connection. The longing for something missing from the past? A deferment of experience in the direction of origin. Hyperbole–what does exaggeration, as a mode of signification, exaggerate? Gap between the signifier and the signified. Narrative as a means of inscribing this gap again and again.

PHOTOGRAPH: BARBARA JO REVELLE

Attraction, craving, longing, desire. They don't all mean quite the same thing. They have something to do with my feelings for Katie – my feeling of guilt about her childhood, my desire to give her a magic childhood. Some joy. I call Katie. Self conscious beyond anything I have seen so far. Tells me her heart beats and her face gets red if she is asked to talk in school. My mother, today on the phone, says the same thing – says that she was terrified like that in school; wet her pants on a stage, couldn't sing her small solo part of a song in church, that the organ played over and over the refrain, waiting for her to sing her one line, that finally she ran off the stage in tears. Could never talk in groups, and still can't. Has to read her presentation when she narrates her slides at her women pilot's 99 group – and I know this feeling too, although some say YOU? Why you talk all the time! Katie tells me, tonight as I talk to her on the phone, that she is not alone and that the boy is the same one she had a date with two days ago. We talk in code. I say, are you alone. She says no. I say are you with a boy or a girl, say one if a boy and two if a girl. She knows, maybe she knows, I share her joy if she can get out of herself, out of her pain and loneliness, in any way.

"Writing is the eruption of meaning in a mark that before that eruption had no name." The longing mark. The moment when a child sees that a mark can represent something. Helen Keller and the moment she understood the connection between the shape of the mark traced on her hand and the water she desired. Gap between signifier and signified being closed. Being understood. Being seen. Being heard. The relief of it. The longing for it. The pain when it doesn't happen – the particular pain so connected with trying to get what you desire. I tell Ron that I am not attracted to the Irish. This disclosure comes after seeing the movie *The Butcher Boy*, which traces the life of an Irish boy whose childhood is so grim, desperate and loveless that the viewer is made to empathize with a child who kills a woman, hacks up her body, writes the word PIG on her walls with her blood. I am thinking of the lugubrious, blunt drunkenness of the town's people and the futile, pathetic weakness of the parents – the father in particular – when I say that I'm not attracted to the Irish. I mean something so totally different by that utterance than Ron receives – he being so closely identified with the pain of the Irish oppression and, evidently, the pain of the child with an alcoholic father and a crazy mother; my being so totally identified with the anguish of living with Alex who drank to put out feeling, to distance himself from his wife and daughter. Pathetic, weak, self-indulgent drinking. Drinking vodka till he passes out in the sub-zero Chicago night and would have frozen if I hadn't found him and dragged him up the steps into the apartment.

Hunger: Saturday, 5 AM

During the course of several conversations yesterday, after he "stumbled on" my computer journal – and mostly on the phone from Ft. Lauderdale after he packed and left – Ron said "Barbara I love you so deeply, I want to come to a point where we share everything – our stream of consciousness, our whole interior life –total honesty." I say "it's impossible – no one wants to know every nuance of everything about their lover – its destructive." I say, "you can't handle it – I know you can't handle this radical truth thing you are asking for. You can't handle it if there is criticism of you in it or if it is about other men or our sex life…" "How do you know?" he said. "You never try me. I know I did some bad things before from jealousy, human frailty, insecurity, whatever you want to call it – I've had that issue in most of my life – but I've WORKED on myself. I know you are faithful – its the editing of your thoughts I don't want. That's what makes me crazy – not the CONTENT of your thoughts – just the idea that you have secrets in your thoughts you aren't telling me about. I want us to TAKE THE RISK OF TOTAL HONESTY. Maybe then I can tell you fully some of the stuff I have been afraid to tell you – that I have been holding back. Otherwise I don't believe we can ever have real intimacy."

"But it's not like that," I say. "I just want to write in my journal what ever comes into my head – stream of consciousness. I don't want to have to censor myself because you might read it – misunderstand it, take it literally. I would write differently, not explore my dreams or whatever, my attempts to access my unconscious. The stuff that comes out in free writing is not the full considered, final figuration of the ultimate truth anyway. You just try things out there. Run things up the flag pole. It would be like writing with the thought-police looking over your shoulder. It isn't necessary. We would just hurt each other unnecessarily and then have to spend hours and hours trying to sort it all out and undo the damage."

Finally, reluctantly, but with some hope in my heart or throat or wherever in the body hope resides – "OK, I'll try it. Here in Miami will be a radical experiment in total honesty." To myself I am thinking, well, why not? Maybe I haven't trusted him and that really IS why he is so insecure. Maybe it is not just the gap between the signifier and the signified – between what I mean by my words and what he receives, through his suspicious, fragile frame of reference, learned back in our childhood when he was a skinny, tender boy dumped by the twelve year old coquette I was then for the taller, more mysterious Robbie Robinson on the hay ride. Maybe I CAN trust a man deeply with whatever rushes through my mind. Why not give this radical truth thing a try if he believes he loves me that much? Maybe I have been wrong and it is, against all my instinct, the right step to make toward deeper intimacy.

I say OK. He drives back to Miami. We hug in the kitchen and it feels like home. When I try out a true utterance, it already seems he has forgotten this new deal and we fight and the pain closes me down again. I try to read and write. Then he says, that night, after I offer to read him what I have just written, "why don't you let me read it to you?" So, in the spirit of the new experiment in radical honesty, I let him. Then he says, can I read your other writing from down here in Miami – (the "stumbled on" stuff written after the Dream with Dana)?" Pause. I think: well, he's sure to misunderstand and be hurt, and he's exhausted and this probably isn't a good time – but OK, he says I need to trust him. This is what he wants.

And it all comes crashing down – all turns out to be just a dirty little trick to satisfy his curious, suspicious mind. There is nothing there in that writing that betrays him – nothing that is a show stopper, no deep dark secret, just general musings that include thoughts about other men. He doesn't do anything different with his hurt; just the same old tape loop: "I hope you find what you are looking for, Barbara," he says. "I feel like I have been lied to for two years. You don't know anything about love and commitment, you are sick, I see now that I have been wasting my life with you, that you can't love." And what happened to all the day's fine sounding protestations that I should trust him – that we can work through anything if we are just totally honest with each other? Vanished.

PHOTOGRAPH: BARBARA JO REVELLE

As meaningful as pounding shit down a rat hole–crushed on the side of the river. He just packs up and leaves again. The rage of this betrayal hits me like nothing I have felt before. The deepest betrayal of all – the biggest lie anyone has ever told me: "You can trust me with anything," he says to me, oh so earnestly – looking into my eyes.

The wooden Haitian boat – the owners with hope in their mouths, trusting against all logic that the wood vessel will be seaworthy enough to make it to the promised land and back–and my God–after terrifying days at sea, it arrives!. Hope seems to win out – the boat makes it to the Miami River against great odds! Only there, just when they thought they had reached a safe place finally – to be pronounced "derelict" – the boat dragged up on the banks of the river in the night and crushed by big machinery. And they are stranded here on the Miami River and can't even go home. So much for the promises of America: "give me your tired, your poor, your hungry masses yearning to be free. You have arrived. You can rest here." So much for the promises of Ron – "Give me your thoughts, trust me, I will love you and never leave you." Haul away the wreck. Leave no evidence. Nice clean river. I must remember this the rest of my life.

End of Revelle's journal. The mystory is accumulating, one layer after the other, forming a block activated at the moment of decision.

CHAPTER 8

Mystory 2: Entertainment

Continuing the unfolding of the allegorical narrative passing through "Crossroads," the next register of the mystory to consider is Entertainment. If the actual geography of the Miami River is the places in Miautre's memory theater, then the "active agents" associated with locations along the mnemonic route are supplied initially by Entertainment discourse. This register concerns how Miami is represented in the media, which is distinct enough from the actuality of the place to deserve its own name – "Myami." In order to discern the specific quality of the mystory – of Revelle's attunement of the zone –, it was necessary first to determine the "default" moods already circulating through Myami. Indeed, the encompassing relation of region (space) to place in the Classical vocabulary continues in figurative use of the spatial terms to name the operations of digital memory. Chora as category encompasses and brings into relationship the collection of topics (topoi) found in the four registers of the popcycle relevant to the scene of consultation. In short, when Revelle undertook her drift, the scene she entered was already constituted and underway. This is the condition that Heidegger refers to as "thrownness." There is a default wisdom running through the scene like the current of the river.

PHOTOGRAPH: BARBARA JO REVELLE

 | DOI:10.1163/9789004534643_009

A goal of the mystory attunement that produces a cognitive map of the city is to locate and correlate the external conditions of Miami/Myami within, in the interior (spirit) of the consultant (querent), in order to objectify the virtue of prudence. When citizens log on to a chora, they should be able to experience at once the default decision structuring the scene. Before borders and boundaries of places and categories may be negotiated or adjusted, they must be located and mapped. The method is that of the poetic encounter, which is a kind of opthamology for the mind's eye. Revelle poses her burning question: what is my situation with Ron? She then tours the zone to gather the materials with which to write an epiphany. The question will be answered by a pattern (constellation, ideogram) that flashes into appearance through the repetition of signifiers across the four registers of the popcycle. This flash moves through information at the speed of the dromosphere. Meanwhile, at this stage of invention (inventio), the task is to locate Revelle's place – the position prepared for her in advance (whether or not she recognizes it) – in Myami.

Myami

The spectacle of Miami is "dialogical," meaning that its nature and meaning are contested, occupied by multiple opposing discourses vying for control of its signification. To distinguish Miami-as-image, the discursive Miami, from the "thing itself," we used the name "Myami." Our file of clippings tracked over a period of several years the events considered newsworthy enough to put on the wire, datelined MIAMI (journalism as first responder):

- As a quadriplegic boy watched, three teens stole his VCR, video games and \$200 his mother had saved for his 12th birthday party.
- A suspicious tool salesman with a \$50,000 debt believed a hit man was going to kill him. So when Richard Deville, a United Airlines pilot and avid tool collector, showed up at Seminole Tools on August 6, Gerald Fishbone shot him dead.
- Three men were found shot to death at a cockfighting arena, but police don't know who killed them or why.
- A 23-year-old man pleaded guilty Wednesday to cutting off a dead man's hand as part of a voodoo-type ritual.
- The Port of Miami looked like a success, but appearances can be deceiving. Federal prosecutors contend three of the top people at the port were crooks, using port money like a personal piggy bank for everything from yacht maintenance to Disney World trips.

- One of Miami's most savage drug gangs, The Boobie Boys, blamed by police for 35 killings, including a 5-year-old boy, was dismantled Friday with a roundup targeting a suspended detective. The gang's favorite method of attack was drive-by shootings.
- The U.S. Army Corps of Engineers says it will help pay for most of the cleanup of the Miami River, a waterway so polluted it has been called a 'cesspool'.
- Unless architects redesign their sketches, a $100 million twin tower condominium soon will be built where archaeologists discovered what they believe are prehistoric ruins, a settlement of the Tequesta Indians. Archeologists speculate that the circle, made up of dozens of holes carved into stone, could have been used to measure time or keep track of the seasons.
- When business recruiter Frank Nero told a group of Japanese businessmen that he was from Miami, one formed his fingers to make an imaginary gun and said: 'Miami Vice, boom, boom, boom.'

How is the scene of interest represented in popular culture, or entertainment forms and media?

Noir

News is not the only genre forming the image (reputation) of Miami in the public sphere. Myami reflects the same paradoxical influence on the image of the city that Mike Davis found in the representations of Los Angeles in hard-boiled novels and film noir cinema.

> Together they radically reworked the metaphorical figure of the city, using the crisis of the middle class (rarely the workers or the poor) to expose how the dream had become nightmare.... It is hard to exaggerate the damage which noir's dystopianization of Los Angeles, together with the exiles' [European intellectuals living and working in L.A.] denunciation of its counterfeit urbanity, inflicted upon the accumulated ideological capital of the region's boosters. Noir, often in illicit alliance with San Francisco or New York elitism, made Los Angeles the city that American intellectuals love to hate (although, paradoxically, this seems only to increase its fascination for postwar European intellectuals). As Richard Lehan has emphasized, "probably no city in the Western world has a more negative image".... It is hard to avoid the conclusion that the paramount axis of cultural conflict in Los Angeles has always been about the construction/interpretation of the city myth, which enters the material landscape as a design for speculation and domination. (Davis, 20–21)

The same dynamic conflict between promotional public relations and noir narratives informs the city myth of Myami. A history of the promoters lists the cast of characters as follows: "Hamilton Disston, who brought Florida to life with a million dollars he didn't have; Henry Plant, who opened Florida's West Coast and helped win a war; Henry Flagler, who opened the East Coast and built the eighth wonder of the world; Bion Barnett, who proved bankers should take chances; Mrs. Potter Palmer, who used snob appeal to start a city; D. P. Davis, who sold island properties before he built the islands; Barron Collier, who made millions in advertising and poured them all into Florida; Carl Fisher, who created Miami Beach but who couldn't crack New York; George Merrick, who built Coral Gables but thought he failed. Their like will not be seen again; there are laws these days which would have handcuffed them – literally" (Harner, 1973: 11). Despite or because of the hucksters' propaganda, the promotional Myami oscillates between "Magic City" and "Paradise Lost" (Croucher, 1997: 25).

Miami, most notably in the works of Elmore Leonard and Charles Willeford, and in the Television series *Miami Vice*, and *CSI-Miami*, has received some of the same treatment as Los Angeles, albeit belatedly, or in a post- or neo-noir modality of the genre. A book about "the mythical America of crime writers" includes some discussion of the Miami River setting. The Interviewer, John Williams, spoke with James Hall, author of the hard-boiled *Squall Line*, as they rode in Hall's boat on the bay near the river's mouth.

> Hall points out the opening of the Miami River to me: "I took a trip down there the other day; it's where the ships come from the Caribbean. I was looking at these ships that are ready to go back and they're stacked with the strangest stuff... stolen bicycles, plastic buckets.... All these things you can't believe another culture places a value on, that they'll all be saying in Haiti, 'Oh, great, a ten-gallon bucket.' [...] Ships flying the flag of Honduras and Haiti are coming in to Miami, bringing in Colombian cocaine, and then returning to Haiti loaded up with items as diverse as school buses, rice, beans, stolen bicycles, all hidden under tarpaulins, stolen from pest-control firms. None of this strange cargo is listed on any manifest, and all disappears on arrival in Haiti. The sinister part is the identity of the people running this Miami River traffic: none other than Baby Doc's infamous Tonton Macoutes. Now minus their trademark sunglasses, they are currently upping the craziness stakes in Miami drug wars. A guy from the Sheriff's Office, quoted in the HERALD says, 'The Macoutes don't think they can die. They think they're immune to bullets. They do drug deals based on info from voodoo priests.'" Echoes here of *Squall Line*. (Williams, 1993: 33)

A State of Mind

Sheila Croucher, in *Imagining Miami*, summed up the issue: "Miami is a city without true substance, a state of mind instead of a state of being" (Croucher, 1997: 2). As James Donald observed, every city is a state of mind that owes as much to discourse as to people and place. "[The imagined city] has been learned as much from novels, pictures and half-remembered films as from diligent walks round the capital cities of Europe. It embodies perspectives, images, and narratives that migrate across popular fiction, modernist aesthetics, the sociology of urban culture, and techniques for acting on the city" (Donald, 1999: 7). We had identified three genres, all within the institution and discourse of entertainment, as contributing most to the complex image of Myami: news, advertising, noir narrative. The insight at this stage of research was the possibility that an understanding of Myami might be generalized into an understanding of the logic and rhetoric of electracy as such. Entertainment (understood as the design of information according to the demands of profitability) is the new institution of the emergent apparatus: it is to electracy what school was to literacy – the site of the invention of the practices capable of disseminating to the general population the practical benefits of a language technology.

PHOTOGRAPH: BARBARA JO REVELLE

The second insight concerned the way each mode constructed the image of the city to create an atmosphere, a feeling, an ambiance – a state of mind:

Miami-in-me. Every city is a state of mind, but not the same state of mind. Film noir has been defined precisely as being not a genre but a tone or mood. The lesson was that our zone came pre-coded, with a conflicting set of default moods already inscribed within it.

> *Miami Vice* has many antecedents, but most significant among them is the American cinematic genre known as film noir – the source of many of the program's thematic, narrative, and stylistic elements. [...] Paul Schrader has argued bluntly, "film noir is not a genre... It is not defined, as are the Western and gangster genres, by conventions of setting and conflict, but rather by the more subtle qualities of tone and mood." [...] Noir visual style is catalogued lucidly by J.A. Place and L. S. Peterson. They argue that film noir is fundamentally "anti-traditional" in its visual style – that it consistently violates the code of classical filmmaking that had evolved through the 1930s, listing noir violations as: low key (high contrast) lighting; imbalanced lighting; Night-for-night; deep focus; wide angle focal length; dissymmetrical mise-en-scene; extreme low and high angles; foreground obstructions. They argue that these stylistic elements are as responsible for the film's 'meaning' as are conventional components of plot and theme. "The characteristic noir moods of claustrophobia, paranoia, despair, and nihilism constitute a world view that is

PHOTOGRAPH: BARBARA JO REVELLE

> expressed not through the films' terse, elliptical dialogue, nor through their confusing, often insoluble plots, but through their remarkable style. In effect, unbalanced composition equals an unstable world view. [...] Noir thematics are assumed to be the dark side of the American dream, a negative image of the 1940s status quo. Men are the ostensible heroes of most films noir. Most commonly they are men with an indiscretion in their past and unpleasantness in their future toward which the present rapidly carries them. The noir protagonist is alienated from a combustible, hostile world, driven by obsessions transcending morality and causality. The obsessive noir protagonist is drawn into a destiny he cannot escape; he is impelled toward his fate by exterior forces beyond his power and interior forces beyond his control. (Butler, 1996: 290–291)

Noir (and neo-noir) is a report on a modern state of mind circulated globally through the spectacle. Like advertising, it is a theory of (un)happiness. The struggle between booster advertising and noir is between two states of mind: nostalgia versus nihilism. Our measurement or sampling of the mood of the zone would have to take these default moods into account. Our concern was not to try to suspend these schemas, but to include them within the design. We understood this inquiry to be related to consulting in the terms outlined by Croucher, having to do with the subjective nature of public problems. The question empirical consultants do not address is, "why and how certain images, issues, and identities capture the public imagination and others do not." We took up in turn her focus "on the processes that transform private perceptions into public images" (Croucher, 1997: 17, 18). The long-range goal was to learn how to revise the default public images and states of mind with alternative moods: to ask if it is possible to address policy formation at the level of state of mind (to create situations).

A distributed testimonial, interlinking a multitude of individual consultations of the zone, has as its goal the democratizing of this passage from private to public and back. The struggle for Myami (the image and its associated state of mind) exists at this point of convergence or overlap of discourses. "Miami is a composite of images none of which are devoid of material referents but all of which are constructed through processes distinct from the content or conditions they appear to reflect. These images can be traced to the tug and pull of vested interests and community politics" (23). We needed to correlate this point with a specific geographical setting in order to perform as theoria.

The advantage of this association of Miami with a narrative genre defined more by mood than plot is that "mood" is the quality of an image that in electracy has the capacity to become categorial. The egent knows to revise the title for her own testimonial: [City] Virtue.

Spleen

One motivation for the formation of the EmerAgency was Paul Virilio's warning about the spectacle and the Internet, associated with the disappearance (becoming invisible, the illegibility) of the city and the loss of the public sphere. Our method of the poetic encounter takes its point of departure from the beginnings of modernism and the crisis of the agrarian order brought on by the rise of the city associated with the industrial revolution. One way to consider a possible response to the present community crisis, the demise of the city form, is to review what the response was to the emergence of the city. The nightmare corruptions of the noir city, in fact, continue the ambivalent attraction-repulsion that informs Baudelaire's *Paris Spleen* that initiated the modernist image. Spleen persists within the default mood of the noir city. For the purposes of research on the default moods of Myami, the important feature of Baudelaire's founding attunement of the modern city is the feeling generated by the crowd, the mass of people that converged on the city, thrown together in a new experience of anonymity and deracination, leading ultimately to the definitive modern condition of alienation (objectification, reification). The initial crisis caused by fear of "the crowd" is still with us today. "As for the Third World masses, nothing will circumscribe them. How can we absorb them? The moderns raise the question in anguish. How can they all be modernized? We might have done it; we thought we could do it; we can no longer believe it possible" (Latour, 77).

The modern attunement, Heidegger said, is angst. That is the default mood of modern people. Walter Benjamin organized his study of Paris as "capital of the nineteenth century" around Baudelaire's use of poetry to absorb the city of trauma.

> "Paris as Heaven, Paris as Hell": it is in the motif of the crowd that this dual identity finds its quintessential expression. The image of the crowd is a monad in which Benjamin's contradictory responses to modernity are crystallized. On the one hand, destructive or genuine experience and diversity, the crowd rigorously regiments the individual and denies his or her autonomy. The crowd is the negation of individuality and the self. On the other hand, despite its brutalizing tendencies, the crowd retains a certain allure for the outcast who wishes to remain undetected. The anonymity afforded by the crowd denies yet surreptitiously facilitates,

> the possibility of the idiosyncratic.... Benjamin's ambivalent response to the urban crowd was superseded by a growing pessimism. In his later writings on Baudelaire and Paris, an increasing emphasis is given to the dehumanizing tendencies at work in the crowd: towards conformity, uniformity, anonymity and passivity. The metropolitan crowd emerges in a new light: namely, as a threatening, undifferentiated mass. The mass is the final extinction of the bourgeois individual. (Gilloch, 146)

The modern city created its own crisis of seeing and looking, of intelligibility and legibility. Benjamin found in Baudelaire the first response to this crisis of the emergent city.

> Thus in order to see properly – to rediscover a "stranger relationship" which can no longer be based on religion or tradition – it is necessary to interrupt through a shock the temporal alienation of seriality, of the ever-different and ever-the-same. Benjamin would find this shock again in surrealist or Brechtian montage, as the practice of any "modern" avant-garde. But it is already poetic principle in Baudelaire." ... This interruption that shock produces in the flow of time is nothing other than spleen. "Spleen is that feeling which corresponds to catastrophe in permanence." In this sombre tete-a-tete of the subject with himself, the time of seriality, "progress" and the early culture industry is interrupted or immobilized. It becomes memory, quintessence of history: spleen inaugurates a time that is "outside history, as is that of the involuntary memory." This is why, unlike in the baroque where death and the abyss were objectivized, death is internalized in Baudelairean spleen.... To have the corpse inside oneself – a radically disturbing novelty which demolishes the acquired certainties of the "subject" – is also to do violence to oneself, to become victim and executioner, seer and seen. (Buci-Glucksmann, 76–77)

The modern writer who sets himself against the forces of progress and industry that produced the city enters a duel, according to Baudelaire, "in which he screams with fright before he is vanquished" (75).

Screams in the night.

Inventing Myths

The crisis of the modern city created by industrialization and expressed in the crowd, the anxiety and shock of the masses, was one of the defining qualities not only of European experience but of American identity as well. Here is a point of coherence focusing the EmerAgency consultation. The crowd in this context was composed of the "dangerous classes" – the workers (proletariat), immigrants, and freed slaves – and viewed as a threat to the hegemony of the ruling WASP elite. To legitimate their privilege, to reconcile the contradictions introduced into the American belief system by the forces of capitalism and technology, WASP culture invented the myth of the Frontier. The history of this invention is traced by Richard Slotkin in three volumes, covering nearly the entire history of the United States from the colonial period through the Vietnam war (Slotkin, 1992).

Any hermeneutics may become a heuretics. Slotkin's history of the invention of one myth may become a poetics for the making of another myth (a mythography). Indeed, the point of the EmerAgency intervention in policy formation is at this level of mythography, keeping in mind the principle of the apparatus – that the apparatus creates the categories that determine what is real; the categories (metaphysics) organize the belief system; the values select what counts as problem and acceptable solution; policies implement this selection. To address the policies operating in Miami requires nothing less than the invention of a new category (image metaphysics) and a new myth that the category makes functional. Part of the point of Slotkin's study is that the Frontier myth is obsolete, anachronistic, and counterproductive as a guide for action, whether political or ethical, a view that is shared by most commentators on this myth, which does not mean that the worldview expressed in this myth is any less influential in the conduct of American individual and collective life. Slotkin himself, saying that myths are not natural but are cultural inventions with identifiable names and dates, calls for the making of a new, alternative, American myth.

> The history we have been tracing suggests that the discourses of myth are not fixed and programmatic – not a "prisonhouse of language" from which conceptual and political escape is impossible. The mythic codes to which we subscribe derive their authority not only from their linguistic form, but from their historicity – which also renders them vulnerable to modification when they are tested against current experience. The poetics of myth entice the naive audience into false or dangerous identifications, as when the captain in Vietnam invited Michael Herr to play cowboys and Indians. But the rules of mythic expression are not fixed constraints, like those that govern games. They are more like the structures that define the forms of expressive art: the study and practice of such rules ultimately give the artist the freedom of the form, the power to transgress and transform its regulations. When we understand the history and structure of our mythic

PHOTOGRAPH: BARBARA JO REVELLE

> language, we acquire the power to innovate and invent in that language – to alter the *practice* of the myth and thereby (potentially) to initiate an adaptation of its basic *structure*. The language of myth allows us to speak in terms that are fully intelligible and rich in historical resonances for the people to whom we speak. Through the discourses of myth we can recover a true (or truer) understanding of our history. (Slotkin, 656)

Choragraphy is intended as a way to enact Slotkin's proposal, by providing the flash reason needed to conduct augmented prudence. What Slotkin means by "myth" is *an actual historical reality put into discourse and made to serve as a precedent guiding further action.* "Myth" has something in common with Benjamin's dialectical image, being a kind of Then-Now illumination, but fixed in the case of myth into a monument, mobilized as an emblem or impresa, an image-text, John Wayne (for example) in his star persona as Western hero, captioned by a phrase such as "a gun in the hands of the right man." While the Frontier is not the only myth by which America is imagined, by most accounts it is the defining one. "[Ronald] Reagan's appeal to the frontier myth was masterful but not unique. In the post-war period the political left as well as the right made use of what historian Warren Sussman called 'the official American ideology' – the belief that westering defined the nation's unique heritage and

PHOTOGRAPH: BARBARA JO REVELLE

that it amounted to the purest expression of American idealism. Accepting the Democratic nomination in Los Angeles in 1960, John F. Kennedy pointed to the 'new world here in the West' built by 'the pioneers of old' and coined the phrase that would define his administration when he called on Americans 'to be pioneers on a New Frontier'" (Hine and Faragher, 531).

Poetics of the Gunfighter

To be embraced by a public, a myth must be plausible, credible, grounded in the reality of history. However, once it exists as a representation, it acquires a certain autonomy, a capacity to generate further cultural forms. The Frontier, in other words, is a topos in rhetorical terms – a "place" storing commonplaces of information and logical operators for generating new instances of the discourse. The importance of Slotkin's account is his demonstration that the Frontier, condensed into a scenario, was (and still is) drawn upon equally by the culture industry of mass entertainment, and policy makers in government. Here is a fact of fundamental importance for choragraphy, whose purpose is to intervene in policy formation at the level of the myths and associated value habitus determining deliberative decisions. The case in point is twentieth-century American foreign policy, in which the developing or "third" world is treated as an extension of the Frontier.

> The policy-makers faced an imaginative task analogous to that of a screenwriting team which has been given a setting and a situation and is asked to turn them into a narrative that can be acted out. For the movie scenarists, the problem could be resolved by canvassing the conventions of a genre, in which mythic narratives are encoded as a set of standard models for heroic action. Imaginative work of this kind involves both the interpretation of "hard data" according to "objective standards" and the speculative canvassing and deployment of the full range of available interpretive procedures, including those that we have identified with the processes of myth-making. Mythic symbols encode paradigms or programs of real-world action, drawn from past experience or historical memory, which are projected as hypotheses about the outcomes of prospective action. Historical precedents are also encoded and recalled in the form of myths, and policy-makers canvass them for the same reasons that scenarists review the generic repertoire: to sample a range of interpretive models and alternative resolutions. (Slotkin, 408).

When it was suggested after the surprise attack of 9/11 that Hollywood screenwriters actively assist government security agencies in imagining attack scenarios, the proposal simply made explicit a collaboration already native to the popcycle. A mythography extracted from Slotkin's history includes the following operators.

1. Historical Crisis.

   A myth is created in response to a crisis. A community needs a guide for making political and ethical decisions. The crisis in this case was world-wide industrialization that shifted the economy from agriculture to industry (manufacturing). In these circumstances the "crowd" (immigrants, workers, freed slaves) began to lay claims to the promises of a democratic republic, putting pressure on American ideology of democracy and equality, thus threatening the hegemony of the White Anglo-Saxon Protestant consensus.

2. Historical Events.

   The myth expresses the past of the community as a story of "origins" anchoring and providing a measure for a consistent public identity. The defining experience by which America imagined itself was the colonization of the new world by European immigrants. The original "discovery" of the continent and the push westward included actual events, actions, places, characters: Natives (Indians), "wilderness," pioneers, mountain men, cavalry, ranches, cowboys, outlaws, gold, and the like. All these "happenings" are raw material that may be put into discourse in a multitude of ways.

3. Popular Culture.
Various media select certain incidents and characters and put them into narratives which have their own pre-existing forms and conventions. Newspapers set the agenda of discussion by framing events such as Custer's Last Stand as a parable reflecting on the ideological conflicts of the moment. The small heroic cavalry troop overwhelmed by a horde of savages emblematized the fate in store for WASP culture at the hands of the class race foreign "other." Dime novels glamorized the exploits of violent men on all sides of the law. Buffalo Bill Cody was perhaps the figure of transition from history into myth. Painters and photographers selected landscape views and fixed the setting, costumes, and all the props of the scene. The literary novel, specifically the Leatherstocking series by James Fennimore Cooper, fixed the archetypal story by combining the two most compelling tales into one: the captivity narrative (based on Mary Rowlandson among others), and the hunter-Indian fighter (Daniel Boone among others). Hawkeye is the "man who knows Indians" – a character who has the "trickster" power of operating on both sides of the cultural dichotomy (civilized/savage).
4. Theory.
At some point the narrative is made explicit in conceptual form, abstracted, named, put into an argument. Concepts tend to cover a range articulated as a polarity, so that the myth will have at least two theorists, one representing each pole of the concept.

Frederick Jackson Turner speaks for the populist pole of the Frontier myth. The crisis is expressed as the closing of the frontier. American exceptionalism is based on the special experience of open and free land which solved all dilemmas of quality versus equality. New frontiers – meaning new forms of economic bonanza – must be found to replace the loss of this free land. The populist values are "nostalgic" – agrarian, idealizing the Jeffersonian yeoman farmer. Turner couched his ideas in academic terms, using the language of social science, and counted on education to disseminate his vision.

Theodore Roosevelt speaks for the so-called progressive pole of the myth. For him the wilderness made Americans exceptional by testing them with the challenges of "the strenuous life." His is a Social Darwinist meritocracy, an elite of natural aristocrats, assumed to be WASP. Roosevelt embraces bourgeois managerial capitalism, but is nostalgic for the military virtues of the old aristocracy. The crisis of the closing frontier is the potential loss of WASP virility and with it the claim to superiority over the dangerous classes. Roosevelt couched his ideas in an autobiographical narrative that generalized his personal experience of ranch life into a model for the nation. He counted on political action to disseminate his vision.

The populist-progressive positions appeal to the same discourse, indicating an underlying consensus: the urban proletariat is the absolute enemy of the imagined community. The one option for dealing with industrialization that is foreclosed is socialism, siding with the proletariat.

5. Scenario.

Through myth the community's past becomes usable. The myth in turn is condensed into an emblem and mobilized through a scenario that is mapped analogically onto other circumstances to render them intelligible and to provide a guide for action. The Western genre is the clearest manifestation of the scenario.

"The geography of the Frontier represented in Western movies is that of a world divided by significant and signifying borders, usually marked by some strong visual sign: the palisade of the desert fort; a mountain pass or a river, especially one whose name is recognized as a boundary marker, like the Rio Grande; the white empty street of the town. Finally the hero must leave a good woman and a place of safety, enter that street, and confront the enemy who advances from the opposite end. Through persistent association, these border signs have come to symbolize a range of fundamental ideological differences. The most basic of these is that between the natural and the human or social realms, 'wilderness vs civilization.' This opposition is given depth and complexity by metaphors that liken it to social and ideological divisions: between White civilization and Redskin savagery; between a corrupt metropolitan 'east' and a rough but virtuous 'west'; between tyrannical old proprietors (big ranchers) and new, progressive entrepreneurs (small ranchers, homesteaders); between the engorged wealth of industrial monopolies (railroads) and the hard-earned property of the citizen (farmers); between the undisciplined rapacity of frontier criminals and the lawman's determination to establish order. The borderline may also be construed as the moral opposition between the violent culture of men and the Christian culture associated with women. It is nearly always understood as a border between an 'old world' which is known, oppressive, and limiting and a 'new world' which is rich in potential or mystery, liberating and full of opportunity" (Slotkin, 351).

"The action of the narrative requires that these borders be crossed by a hero (or group) whose character is so mixed that he (or they) can operate effectively on both sides of the line. Through this transgression

PHOTOGRAPH: BARBARA JO REVELLE

of borders, through combat with the dark elements on the other side, the heroes reveal the meaning of the frontier line (that is, the distinctions of value it symbolizes) even as they break it down. In the process they evoke the elements in themselves (or in their society) that correspond to the 'dark'; and by destroying the dark elements and colonizing the border, they purge darkness from themselves and from the world. Thus the core of the mythic narrative that traverses the mythic landscape is a tale of personal and social 'regeneration through violence'" (Slotkin, 352).

6. World-view

Although the mythic landscape provides a wide latitude for variation and innovation, it is constrained by an underlying world-view, existing as a state of mind, a mood or attunement, that is directly intuitively accessed by the emblem of the scenario. The emblem scene that emerged within the formation of the myth is in fact an invention with little or no historical basis. The scene is the showdown, performed by the emblematic gunfighter. "The image of the gunfighter as a professional of violence, for whom formalized killing was a calling and even an art, is the invention of movies like *the Gunfighter*, the reflection of Cold War-era ideas about professionalism and violence and not of the mores of the Old West" (384). The codification of this figure merging lawman and outlaw types coincided with the heyday of film noir. Film noir adapted hard-boiled fiction to cinema, and this fiction in turn was an updating of the Western, mapping the Frontier scenario onto the city, whose "night-town" red-light mean streets were the new wilderness. Although the heroic virtues and the border structure remained consistent, the noir mood was darker, less utopian, more pessimistic than the original Westerns.

PHOTOGRAPH: BARBARA JO REVELLE

"It may be more useful to think of the term 'gunfighter' as describing a style of action appropriate to a certain kind of world-view rather than a specific social role in a particular historical frame. The 'cults ' of the outlaw, the Indian, and the cavalry are constituted by the repetition of a particular kind of story, setting, historical reference, and ideological problem. The 'cult of the gunfighter' is constituted by the use of a particular character and style of action to resolve a wide range of conflicts in a nearly limitless variety of settings. The outward form of the gunfighter style emphasizes artistic professionalism in the use of weapons, but what justifies and directs the professionalism is a particular state of mind, a 'gunfighter' understanding of 'how the world works.' That understanding is essentially 'hard-boiled': the world is a hostile place, human motives are rarely good, and outcomes depend not on right but on the proper deployment of might. The fictive politics of gunfighter Westerns is worth attending to, because its symbols contribute to a public discourse on the role of power and force in American political and social life" (402).

This state of mind, this feeling about how the world works, Slotkin elaborates, is a "mood of entrapment." "One should not underestimate the appeal of the homely analogy between Ringo's [Gregory Peck in *The Gunfighter*] situation and that of the 1950s archetypal 'Man in the Gray Flannel Suit.' Ringo renders as heroic tragedy the malaise of the middle-class man grown weary of the 'rat race' through which he has prospered. But Ringo's situation is also a parody of Cold War, nuclear garrison-state psychology; he is at once the most powerful and the most vulnerable man in the world" (390).

7. Policy

There have been a number of "cowboy presidents" since Teddy Roosevelt initiated the role by applying the Frontier myth to policy formation both foreign and domestic. Slotkin emphasizes the development of American foreign policy during the Cold War, in the administrations of Eisenhower and Kennedy especially. He mentions Ronald Reagan's revival of the stance after the inversion of the scenario following the seeming refutation of the myth in Vietnam. His book is published at the time of the first Gulf war, and he mentions that Saddam Hussein is the perfect villain for the scenario. What is astonishing is the degree to which George W. Bush follows the scenario in the second Gulf War. Bush's cowboy image in the media, which at first seemed to be a parody, turned out to be reality, as enacted in his strict constructionist application of Teddy Roosevelt's progressivism. In the final days leading up to the American attack, news headlines declared that Saddam Hussein had "48 hours to leave town" or face a showdown. U.S. policy on Iraq demonstrated the Frontier worldview: a virile hero, capable of redemptive violence, confronts on behalf of civilization a savage other given to atrocities and incapable of reason and therefore an enemy who must be eliminated. At the same time, this violence is entrusted to professional killers capable of carrying out a surgical strike that confines destruction to the evil parties while sparing the innocent.

The "equality" respected in this vision is only that of "being equal to the situation." In the war on terror, the Middle East is the American Frontier, the Arabs (or Persians) are Indians, the U.S. military is the cavalry, the president is the "town tamer." Bush at the United Nations (which refused to endorse American policy in Iraq) no doubt felt like Will Kane (Gary Cooper) in *High Noon*, betrayed by the hypocritical citizens of the town he was trying to defend against the predatory gang of killers. The group of policy advisers based in the Pentagon's Office of Special Plans, headed by Paul Wolfowitz, Deputy Secretary of Defense under Donald Rumsfeld, was said to be deeply influenced by Leo Strauss, the conservative émigré scholar at the University of Chicago. "Strauss's influence on foreign policy decision making is usually discussed in terms of his tendency to view the world as a place where isolated liberal democracies live in constant danger from hostile elements abroad" (Hersh, 48).

Slotkin's assessment is that the Frontier scenario no longer describes the imagined community for several reasons, especially its reduction of everything to absolute antagonism, and its exclusionary elitism. The solution is not the (impossible) elimination of myth, but the revision and continuing invention of a living mythography.

> If a new mythology is to fulfill its cultural function, it will have to recognize and incorporate a new set of memories that more accurately reflect the material changes that have transformed American society, culture, and politics in the last forty years. The historical adventure of our national development will have to be reconceived to incorporate our experience of defeat and disappointment, our acquired sense of limitation, as well as the fabulous hopefulness that has perennially informed and energized our culture. Even in its liberal form, the traditional Myth of the Frontier was exclusionist in its premises, idealizing the White male adventurer as the hero of national history. A new myth will have to respond to the demographic transformation of the United States and speak to and for a polyglot nationality. Historical memory will have to be revised, not to invent an imaginary role for supposedly marginal minorities, but to register the fact that our history in the West and in the East, was shaped from the beginning by the meeting, conversation, and mutual adaptation of different cultures. (Slotkin, 655)

The point for now is that the propensity of Myami when Revelle and the FRE enter the scene is expressed in the prototype neo-noir mood of *Miami Vice*, heir and contemporary offspring of the Frontier scenario (the cowboy transforming into the hard-boiled detective). The policies governing the zone in general and affecting the Haitians in particular are those of the hard-boiled Frontier. The river as border must be mapped in order to prepare the virtual city for mythographic revision and invention.

CHAPTER 9

Mystory 3: History

In a personal mystory this third quarter would draw on the history of the egent's community. In a group mystory conducted as part of a consultation, the history is that of the zone, the problem site. During our table work the FRE began with the crisis of the image of Florida triggered by a series of tourist murders, mostly in the Miami area. Monitoring news reports from Miami produced a focus on stories related to Haitians. Revelle's site visit and attunement of the zone refined this selection further, leading ultimately to "Crossroads" as a categorial image (emblem). The final selection of the specific policy issue addressed in the consultation was determined by the chance encounter (*tuche*, *trouvaile*), the repeating signifier across the four quarters of the popcycle. "Crossroads" is a "decisive moment," extracted from the total documentation of the drift by the FRE acting as diviner, metonymically evoking this pattern, whose particular features will be analyzed in Part 3. For now the point is to review a sample of several of Revelle's video interviews with denizens of the river keying the attunement. The interviews are those with Haitians operating trading vessels caught in Operation Safety Net, the Caribbean Code enforced by the United

PHOTOGRAPH: BARBARA JO REVELLE

 | DOI:10.1163/9789004534643_010

States Coast Guard (see 1. 2). This scene of impoundment is the answer to Revelle's burning question, offered as an oracular scene, figuratively evoking the feeling of her own experience. The further point to make in the next section is to locate the default mood, the pre-existing commonplace determining the propensity of the zone. The Haitian traders find their readymade place in the topos of the *Chiffonier* (ragpicker), designated within art discourse as the prototype of a modern hero.

Documents; River Interviews

Aboard the 'God-Is-Able'

Fred: I no speak English.

Barbara: (Laughs) Yes you do! I heard you speaking it. I was talking to this guy named Lubin. Do you know him? Over on that boat over there – Simon Lubin. The one that's right outside El Capitan. You know him? He's from Haiti but he's had his boat stuck there for a year and three months.

F: Yeah I know him, yeah.

B: That's kind of a long time to be sittin' there. So, what do you do on this boat?

F: I'm the deckman.

B: The deckman? On the God-Is-Able? So what does the deckman do?

F: Do everything, do everything. There is paint, steering, do everything.

B: What's your name?

PHOTOGRAPH: BARBARA JO REVELLE

F: I am Fred.

B: How long have you been a deckman, Fred?

F: Oh, maybe three years, too much.

B: And you are from Haiti?

F: Yeah, I'm from Haiti.

B: You speak English pretty well.

F: Nooo.

B: Yeah, how'd you learn it so well? you speak Creole and French in Haiti, right?

F: Yeah Creole and French and English, I speak. sometime English.

B: How did you decide to go on boats? To work on boats?

F: I'm not gonna stay for too long. Going to Haiti, coming Miami. I wanna leave. I don't have no good conditions to stay. I don't have no good conditions to stay, working on the boat before. I wanna out. In Haiti, before the ship. I wanna live in my house. But no, nothing to do there.

B: Nothing in Haiti, right? it's hard times in Haiti? Hard to find a job?

F: Much problem down there. That's why I'm working here. Mucho problemo.

B: Do you have a family?

F: Yeah, I have my family. I have my wife and children. Yeah but it's very hard to make a living there – here is better.

(*Second man comes from the shadows inside the boat*)

B: Hi!

ELI: Hi!

B: What's your name?

E: Eli Loua

B: It's what?

E: El-e-low a

B: Where are you from?

E: I'm from Haiti.

B: How long have you been on this boat?

E: I was there before. I spent eight months here after I, I stay Haiti. I just come back again.

B: Aha! what's it like in Haiti these days, it's okay or bad?

E: In where?

B: In Haiti. It's getting pretty hard to live?

E: I like Haiti but sometime, you know, Haiti we got a lot of problem. You don't find nothing to do, to eat or nothing. Where to just come here to make some money after to go back to there.

F: Nothing is better in my country. All want get money by them say, all I want get money by them say.

E: They, they come harder, harder, harder. We don't got personal problem. We don't got that. The problem is, too much, too much black magic. Too much wrong thing, you know that the people there, we need god to change the thing, but people can't change anything Haiti. People can't change the thing Haiti.

B: Too much black magic? What do you mean, like Tonton Makouts kinda voodoo stuff?

E: Yeah, something stuff like that.

F: Ha ha ha

B: But, does it work?

E. Yeah, it does work. Yeah. That's the reason Haiti is like that, and I never think Haiti can change, if they stop some, doing those things, those bad things.

B: Too much black, black magic makes bad things happen?

(*Fred is cracking up laughing*)

B: You don't think so Fred, or what?

F: Aha ha ha

E: You think Haiti… I tell you what, they can kill you and treat you like you were someone else. They put you under the earth, after three days someone can see you work somewhere in the garden. Some place where, yeah.

F: Aha aha aha

B: Because of black magic or what?

E: Yeah… yeah black magic!

F: Aha aha aha.

E: No he know what I'm talking about; he know, he don't wanna say.

F: Nooo, I believe it but I never see that, I never see that. I don't believe it.

B: So, you're saying they could kill you, the spirits, and three days later you could be working around somewhere on the earth, because of black magic?

E: Yeah, yeah.

F: Ah aha aha ah ah aha

E: You know, let me tell you something. I see him and I talk with him. He dead twelve years. Twelve years! After twelve years they find him; after twelve years his family find him.

B: Where was he for twelve years?

E: Him? he work. He explain how when he sleep under coffin he hear everything but he can't talk. He hear everything. After, the same night they come, one come, his cousin kill him. He come look for him and they put him in the garden for seven years; after six years he left there that day. He eat. One woman pass near the garden send something that got salt, when they eat it they can SEE, they can… they can… they don't want to work for free no more, and after they kill their master, and after they work, they go look for a job. This man tell me. He find a woman. He make three children. After one there was no, he was dead. She left and after she go back, go leave straight from Haiti he find his family. I talked to him, like I talk to you. He explained me every... everything.

B: Because a lot of people have powers that make them, what? able to do this kind of stuff?

E: Oh yah yah yah they got powers to do bad thing. They can change the game um um saturn evil can do good thing only, only if you got a good job Haiti, if you don't serve god straight... they can do whatever they want to do. Some spirits they live long time from they come from when you know Haitians come from Africa and from... They basically live everywhere they live in Africa. In Africa, they live in America, everywhere, but in Haiti, but in Haiti they work more than every country else, you know?

B: That's really the main thing wrong with Haiti, is it?

E: Yah yah yah, yah the reason why I tell you, anyone, if every country put together money together. If and if every country in the world put together to change Haiti, they can't change it, they can't. Because Haiti got a power, that power only God can change it, can cut it out and change, can do something and I hope before Jesus Christ come back, Haiti will be changed. Haiti will be changed. I hope that.

F: The Haitian no like they country, this is the problem, a lot of Haitian have money, he no wanna help you, no wanna help Haitian, he collect money, he wanna go away go live on next place, comin in army spending money do everything you want to Haitian, go next next country just make money. In Haiti after you get money you go out, because Haitian no lucky country, this is the problem, no lucky country. I've seen lot of Haitians have money. You can do it everything down there. They lada lada lada factory, big job for Haitian can do it down there, nobody can, make money in come America and spending money. I need the money shut down and come back to Haiti, to make more money to try to collect more money. Haiti no lucky country, that's the problem.

B: Why do you think it's not a lucky country? Because of the black magic?

F: But there is something, you do it, making you see there is no lucky country at all there, something you do it before you can see it, no lucky country. You see, if you go see lucky country. U.S. lucky country, Nicaragua lucky country, You want coming back Haiti? Why coming down there? No how good they? No how good are they? Nothing down there. No lucky.

B: Didn't it used to be, you know, okay to live there, maybe thirty years ago?

E: Yah, after Jean-Claude Duvalier, Jean-Claude Duvalier in 1986 leave the country everything shut down. When Jean-Claude was there everything was good, yah–and I work. I go to school. I hope to, to become something, but now you go to school you don't, you don't find water to drink, to bathe, you don't find nothing, no job.

F: I don't like America, I don't like coming living here, but don't want to go to my country back there, you know, I have nothing to do down back there. Be taken long hour come and collect money.

E: Anything you can say, to live in Haiti you can live by faith, by faith in God you got faith in God you can live in Haiti, only that you, if you got faith in God you can live there because that the only way.

B: And can the spirits come over and follow you here to Miami or do they just stay over in Haiti?

F: The bad spirit? Yah yah they can come over here, yah 'cause they're everywhere, they're everywhere.

Coast Guard

On the River

COMMANDER: Hello, uh I'm Commander Bill Umberti. I'm the Executive Officer here at the Coast Guard Marine Safety Office in Miami. And presently, uh, before I had this job I was Chief in the Inspection Department, which was the person in charge of enforcing inspection laws for the Miami River here. Now, you've asked me to uh explain what a typical operation on the Miami River is. And I can show you that today by telling you how it's different from a normal way business is conducted on the Miami River. Really, what happens is a foreign ship will come in and she'll be empty, she won't have any cargo at all. And then she'll come in, she'll tie up at the dock and they'll send out handbills saying that this vessel's in port and that they'd like to solicit cargo for whatever port they're going to, usually Haiti. And so, what happens is people will come down and they'll bring their cargoes, their mattresses, buckets, bags of rice, bags of flour, old clothes, bicycles, whatever it may be, and they negotiate a price with the master and then the master then, that's how he makes his money to pay his crew and pay the expenses for the voyage. Then they go to Haiti. He sells them there. He pays off everybody and makes another trip back with no money to start this all over again.

B: And most boats that come in to either the river or into the Port of Miami come in empty?

Cmd: No. They do it differently. What's different is that a ship that comes into the Port of Miami comes in loaded with cargo, discharges cargo and picks up cargo. They don't have a what you would call a cash flow problem. So, in other words, if we stopped a vessel in the Port of Miami for doing something wrong and there had to be a fine to be paid, they could pay the fine and then move on. But if you stopped one of these vessels down here on the river before they get their cargo, then what happens is they can't pay the fine until you allow them to load and all that, so you have kind of a 'Catch 22' problem, you know, but fortunately that's not my problem.

(*new person comes in*)

B: Okay, and you are...?

Lt: Lieutenant Linda Sturgis, S-T-U-R-G-I-S. The Miami River is a complicated issue. It's not just a river, it's not just shipping. Uh, the people live on the river. Literally, their houses are up to the river, but there's industry that comes in and out of the river and they like the Coast Guard because we keep the river going; we keep things running. But then again they don't like the Coast Guard because we regulate 'em. And then with the agencies involved with the river, it's like one big Rubik's Cube. Because you have the Coast Guard, you have the people that do safety boardings, you have the part of the Coast Guard that does pollution prevention, and you have the part of the Coast Guard that does drug interdiction, you have customs, you have customs that looks for drugs, customs that gets the money, you have INS looking for immigration, you have the local sheriff's department, police department – so there's a multitude of different agencies.

Cmd: So when a boat comes into Miami a lot of organizations and agencies are interested in it. INS is looking to make sure there are no illegal immigrants on board, customs is making sure that the cargoes are declared okay. The Coast Guard goes on board to make sure it's meeting the international safety rules. And our search and rescue part of the Coast Guard, I shouldn't say our search and rescue part, our law enforcement people are going on board to make sure there is no illegal drug activity. So, we're all interested in making sure they meet the federal, the appropriate federal laws.

B: How does it work, actually?

Cmd: We hold 'em up at the anchorage and then we put a marine inspector on a boat, just like we're doin right now. And they'll drive out to the anchorage and board 'em out there at the anchorage. Under the law, all boats coming, all ships coming in to Miami have to give us twenty-four hour prior notice, okay. So, if we see someone out there who doesn't give us prior notice, right away, we would think something was up and we would send someone out.

B: The main thing that the Coast Guard gets involved is in the safety of the vessels coming in or seaworthiness...

Cmd: Seaworthiness, safety and enforcement. And drugs, and enforcement of laws and treaties, right? Many of the drugs coming in are cocaine and marijuana, generally speaking.

B: And is the Coast Guard involved, or are you the main enforcers of anti-drug laws?

Cmd: This office here, this group office and this station right here, yes. They have their targeting profile and if it fits a targeting profile then they go board it and tear the boat apart, ya know?

B: What's a targeting profile mean, exactly?

Cmd: Well, it has to do with, it has to do with, I mean, a lot of things, I mean you know, who would uh. It has to do with intelligence. Like we have agents all over the place, ya know. This boat left here with so much ya know, so many bags of whatever on board. Then, they'll send it to intelligence center, it'll go out to all of the ships, see if they spot it, that kind of stuff. Um, certain... the way it sits in the water, ya see what I'm saying. is it at a certain height, it tells you that it's probably carrying bags of marijuana ya know, all kinds of, I mean it's it's just a matter of being on a ship, I mean you can almost tell by lookin at one of these things if it fits the profile or not, so, like you get good at telling just by looking at the people. But anyway, that's, this command does that. Then, our command, the Marine Safety Office does the boardings for safety and making sure they're constructed correctly and that kind of stuff.

B: Is it dangerous?

Cmd: Yeah, our law enforcement people risk their lives every day when they go out on these boats here. So, so, they have a lot of times when they go on board: sometimes my boarding officers and marine inspectors go on board and a couple times uh the crew has been dead. They were shot at night for whatever reason, most likely a drug deal and all that, so we're very careful. When I send out the marine inspectors on the river I always send 'em out in twos, never by themselves. They always stay in close liaison with other people so we know what's going on.

Lt: I'm not afraid because it's hard sometimes when you first go on board a vessel for the crew to get some kind of trust from you. Some of the countries that we deal with are afraid of authority and think a person in a uniform is someone who's gonna hurt 'em, someone who's bad or someone who's gonna try to send 'em to jail. But what we do at the Marine Safety Office is we try to enforce international safety regulations, so when we board a vessel we're actually there to help the crew, and we're there to make sure that the waters stay clean. Ya know, we look at life preservers, life jackets, we look at fire pumps, we make sure that the engines are running properly. So, the crews normally aren't afraid of us when we come on, and we don't wear guns, and we go on and let them know we're there for their safety. And uh, so that's like a lot of the difference between us and some of the different agencies or even some of the missions that the Coast Guard does is you have to when you go on board give a presence that you're not there to hurt anybody. You're there to protect them, you're there to enforce safety and to promote safety.

B: I think this is so good of you to take me on the boat.

Cmd: Just tell your boyfriend that he owes us bigtime.

B: Okay. What is that big thing? That Casino Miami?

Cmd: It's a gambling boat. It goes out three miles, gambles. We got a law in FLorida that you're not allowed to gamble inside the three-mile limit. So, they go outside the three miles, and you go out there and gamble, drink and dance ya know.

Lt: See like that boat right there, that yellow one? Well, any small passenger boats that you carry with you more than six people for hire, like that guide boat – excursions ya now, sight seeing – our office also inspects those as well. Ya know like the ones that are bayside or the ones...

B: Inspect them for seaworthiness.

Lt: Yeah, right, yeah, for safety, for passenger safety.

B: So, I'm a... if I'm a Haitian boat, and I'm coming in here, it's pretty weird that I can come in in broad daylight, right past customs and right under the bridge and everything. Is that where they go through?

Lt: That's actually customs for the Port of Miami.

B: But they must know about the wooden boat – I mean the one that came through and all the Haitians got off and disappeared into the city...

Lt: Ya know, like they say, the best we can do is storm the troops in broad daylight.

Cmd: Ya know, they know now, I'll tell you that. Here's some better boats, see? Ya know, loaded properly, moored properly. The hull is in pretty good condition I mean, roughly, looking at it. And these are the new ideas, to have these kinds of boats in harbor here now. See, even this one's kind of beat but it's a little better than normal. See, now cargo ship safety code they don't even have to have a load line because they're below the convention guidelines, see. But under the Caribbean Cargo ship safety code they have to have the markings on the side for loading now. See, so what they would do is just load it as low as they thought they could load...

Lt: ... Since the Caribbean Cargo Safety Code was passed uh January, the number of bridge collisions we have had has decreased probably by about seventy-five percent less. We used to have at least two, three, four a week at least. Now we're down to maybe one or two a month.

B: What about the issue of pollution, are you involved in that at all?

Cmd: Yeah, our office does that.

B: So, if somebody sees somebody dumping oil or sewage they...

Cmd: That's us; yeah, you call us. Now, you've also noticed that... this is what was happening. We had a lot of midnight mystery spills, ya know what I'm saying? We'd get a lot of calls at night and you can't find the source and all that.

Since we've got, since a lot of these substandard boats have left, the amount of pollution on the river has gone down significantly too.

Lt: Oh, dramatically. The amount of pollution on the Miami River has decreased dramatically since the uh Caribbean Cargo Ship Safety Code because the uh, not only are we on board more but there are fewer vessels, and the vessels that are on board are uh taking better care of their equipment, uh we're enforcing re-sanitation device laws and things like that.

Cmd: The goal is to make sure that the vessels that come in and trade on the Miami River, or anywhere else in the Miami zone, meet all the applicable international safety and pollution standards; just like any other boat anywhere else in the United States has to be.

B: Now I've heard some stuff about dredging.

Cmd: Ok, that's a different issue. That involves the Army Corps of Engineers. That… on that booklet I gave you, that's addressed in there. But basically the problem is is that a lot of the spoils here are contaminated, so you can't dredge it and dump it offshore; you have to dispose of it.

B: What's spoils; what do you mean?

Cmd: The stuff you dredge up; the stuff you dredge. Yeah, ya know, so, so, so all the soil that's underneath here is contaminated pretty bad. And so you can't just uh dump it out in the ocean. So now you've gotta get rid of it and that costs money and makes things complicated. Mon Repos, is that on detention? I think so. You're detained here until you correct your deficiencies.

B: So he actually has to stay there and raise the money and stuff, to figure it out–how to fix his boat?

Cmd: That's right, yes.

B: That's pretty hard sometimes, for the people to get their hands on that much money, many thousands of dollars, right?

Cmd: That's his problem. Sylvania Express II. Hey, Linda, this one's on detention, Sylvania Express II?

Lt: It wasn't but it might be now.

B: What's "on detention" mean?

Cmd: Means she's on hold just like I told you for Mon Repos.

B: Ok, it's like they can't leave but they're not confiscated.

Cmd: Right. You just gotta correct your deficiencies and then you're able to load cargo and leave.

B: … like this right here, then these people go off into Miami and try to figure out how to get the money or…?

Cmd: Well, it depends. I mean if INS boarded 'em and said you're detained on board, then no – they are supposed to stay on the ship. Its illegal to go off, but that ya know, unless… Now see that? See all the postage stamps on that, See all that? That's unsafe.

B: That means what the…

Cmd: That's not the proper way to repair a hole in the boat is to just put a postage, what we call a postage stamp over it. This one, Hardness, doesn't look too bad.

B: Belize is a little more compliant than the ones going to Haiti, right?

Cmd: Not really. There's really no more Haitian flag boats left. All the wooden ones were Haitian flag. See what I'm saying. These guys are Haitian owners, they just don't register with Haiti see, cause Haiti really has no maritime administration, see.

B: Meanwhile, there's a lot of kinda fancy looking boats on the river.

Cmd: Oh, that's uh that's a party boat. American boat, you go out, you go out for a party and you buy ya know a ticket, get a tour of the Miami River – whatever they do.

Lt: You know the ones that you see with all the lights flashing and people drinking.

B: Do they ever sink or anything or are they safe?

Cmd: Oh, of course their safe. They've got a Coast Guard certificate of inspection. They're inspected by us.

B: Hey, I saw Titanic.

Cmd: I had to inspect ships in the wintertime up in the Great Lakes, ya know. I had my coat on and I'm still freezin to death.

Lt: We… Alaska Fishing Patrol, which is nothing but a big bathtub.

PHOTOGRAPH: BARBARA JO REVELLE

Cmd: Oh, yeah, yeah, and I had to go into the refrigeration holds and I had to inspect the fisheries patrols and stuff. It's freezin cold on those.

Lt: That's a vessel that we just recently inspected and gave it uh numerous, about thirty-five deficiencies, allowed to go back to, they asked to go back to Bimini to uh fix everything in the Bahamas. So we did allow it a one-time trip to uh the Bahamas to get uh fixed up. And then when they came back absolutely nothing was done. So now its on hold until everything is done. It's not going anywhere.

Cmd: Lots a times they pretend they don't understand, then you say well OK if you're detained it means you can't load cargo. "Oh, no load cargo." Ya see, then they understand that, ya know. But somehow mysteriously they can't understand when you tell them to fix the violations.

B: Well, this one shipping company was saying that they were feeling very bad cause they used to have sixty-five Haitian boats and now they have three, ya know, and they feel like they're being forced off the river and they think it's the developers that are doing it.

Cmd: No, it has nothing to do with the developer. In fact, that's the first time I've heard that argument that it was the developers, ya know. I've never heard that said before.

Lt: What company was that?

Cmd: I'm tryin to think who you talked to. Did you talk to Marchado?

Lt: Must have been Marchado

B; Marchado. Tell me about Marchado.

PHOTOGRAPH: BARBARA JO REVELLE

Lt: Um, we can't

Cmd: He was one of the – how can we say it nicely – yeah, ya know. He represented a lot of these substandard boats.

Cmd: So, naturally we had problems with him because they would operate irresponsibly and all that. He was their agent and we would hold him accountable and he didn't like that. Ya know, I mean that's the bottom line ya know so. And our answer back to him was that if you're not going to be accountable then we're not going to let you represent these guys. Ya see what I'm saying, you can't have it both, I mean if you let this boat in then we're holding you accountable for that, so you gotta take responsibility for this boat, so if you're not gonna take responsibility for this boat then we're not gonna let it in. Once we started doing that ya know, he realized it's very difficult ya know so.

Lt: Like any other uh, like any other company or like any other type of industry, ya know the shipping industry's extremely resilient. So when the Caribbean Cargo Ship Safety Code was coming in to play, a lot of the agents, representatives and owners had forethought and said ya know many of the vessels that I'm representing now aren't, aren't necessarily the vessels that are allowed in a year from now or six months from now or two months from now. So, some of the other agents, dock owners, representatives, uh owners of the vessels started looking around and started investing in vessels that are built to a more significant standard, vessels that would have…

Cmd: … that would meet the, meet the Caribbean Cargo Ship Safety Code.

B: And Marchado's not one of the ones that's really made the adjustment very well.

Cmd: Right, I mean you know I don't get involved in people's, how they conduct their business affairs. With us it's very simple: the ship either meets the requirements or it doesn't, it's either safe or not safe; whether you make money, lose money ya know, I want everyone to make money but that's not my primary concern ya know so.

Lt: So it's not really uh…. It has nothing to do with the Haitian boats, it has to do with the vessels that come up to a safety standard that's acceptable, cause uh most of the crews are still Haitian and the owners are still Haitian. So it's not really a Haitian thing, it's just vessels that are up to a safety code. Like if a cop stops you for an unsafe car, bald tires, something, and you say "but I have no money, what can I do"?

Cmd: Yeah do you think the cop says, "oh, o.k. sure, just go ahead then if you are too poor to fix it?" See? So it isn't our problem when we tell them stuff they have to fix–violations. Where they get the money from: to Fix the boats. Or not to fix them.

B: I was talkin to this one Haitian guy who was saying that he thought it was racism, ya know, that the only people who were hassled were the Haitians, things like that–that Cubans he says you guys never stop.

Cmd: That's not true, that's not true. Everybody, every ship that comes in here has to meet the same standards, I don't care what their flag is, so it makes no difference what their crew is, what their flag is. You have to meet the Caribbean Cargo Ship Safety uh rules. And ya either meet 'em or ya don't meet 'em, so uh…

Lt: We have vessels right now that are on hold on the Miami River and the Port that have uh US captains, some US crews, might be owned by a US guy, it has nothing to do with nationality, it has to do with is the vessel, does it meet the requirements of the safety code. We don't see nationality; we don't see color. We see ships, we see safety.

Cmd: Yeah safety.

Lt: I mean we're not out for any economic sanctions. We have no political agendas. Safety.

B: So if they can't meet standards its not your problem.

Cmd: Right. It's really not our problem.

The Oracle (Simon Lubin)

Haitian stuck on the river for one year and three months. Here is the figure Revelle and the FRE *recognized as the pivot, the turntable, the measure of the vortex. The recognition is retrospective, not immediate. An epiphany in choragraphy is designed, and the experience becomes possible within and through the design (the difference between Erlebnis, or immediate experience, and Erfahrung, an event undergone through representation).*

SIMON: My name is Simon Lubin.

BARBARA: And where are you from, Simon?

S: In Haiti.

B: And how long have you been working on the Miami River?

S: About two years to go.

B: Two years you've been making trips from Haiti and over?

S: No, I make a trip one month sometime because de Coast Guard give you all de job to do. Dat mean when you ready to make to trip by dem, dat mean you can't because dere's too much trouble in de river.

B: And so what, normally you just come over here from Haiti and you used tobe able to load up and take things back to Haiti and it would take how long?

S: Yes, just comin here you know, get to load up your cargo sometime, you know, to feed the people in Haiti because de family get hungry. Just you use yourself to help dem. But for America right now you can't because dey tied us to get out. Nobody goin nowhere right now so everybody bein tied.

B: How do they do that? How do they harass you?

S: Yah. Like dey give you like fifty thousand dollar job to do on your ship so when you do some you tell dem ok, you mean when I'm comin back to do dem. Dey say no and dat mean if you don't do it dey say don't come back anymore.

B: What kind of stuff do you usually take back to Haiti?

S: Like um we take ol' bicycles, ol' mattresses, rice and bean and sugar, something like dat you know, some ol' ting you know that for Haiti it can help poor people in Haiti. Like for right now, we can't do nothin now. All dis junk got to stay over dere cause de Coast Guard say don't come back over dere. Dey're tyin us up; dey're givin us hard time you know. I don't know why their reason.

B: Where are you from in Haiti?

S: I'm from Port au Prince, Haiti.

B: And so you've been doin – how did you get involved in being a shipper?

S: No, I was living here in America. I'm living here for thirty – eighteen years. And so I buy de ship while here to help with um my family in Haiti. So when I come in here I don't be in drug, I don't be in nothin. Any time my boat comin here dey don't find nothin to say. And den so dis is de problem. Dey just want to give you all de problem. Das it. Because dey think that we are all de Haitian, dey don't have no respect for us. This is the problem. That's it. And then dey don't give no Cuban, no other people dat's comin from other country, dey don't give dem de problem. Only Haitian people.

B: Why?

S: I don't know why. I don't know why. And den we come in here, workin hard, sometime de boat come in here – if you come in with twenty thousand dollars and pay de tax, de money to stay here, so I don't know de reason dey don't want us comin here to get something to take care of ourselves. I don't know why you see. Den I work here; for nineteen years I work in America. I'm come in America when I was fifteen years old, I'm comin here for right now. Dey don't get me in de drug, nothin. Just hard work to help my family. Now dey stop my business.

B: What is going on in Haiti?

S: In Haiti, like in American people say okay I'm gonna help Haiti. De government in Haiti, dey help de government of Haiti, but dey don't help de poor people in Haiti. Like American say ok government, I'm going to give you two hundred thousand dollars to do de job to help de poor people – no, dey keep dat money.

B: The government keeps the money?

PHOTOGRAPH: BARBARA JO REVELLE

S: You know dey keep dat money for theyself. But dey don't help de poor people. You have to work hard. Sometime de people in Haiti work all day for two dollars Haitian money. They can't even eat for two days or take care of dere child for two days. We have to do plant de banana, plant de coconut, plant de potato, you know, every year to get da food you see. That's how da people try to run to dis country everytime tryin to come to America. But if Haiti have a job to do we're not going to come in America. Because if you go over dere, if da people sick, no money to go to de doctor, no money to feed the child. From de school everywhere you know you have to pay school big money. Dat means nobody can get an education for your own self over dere – because no good government. But Haiti is a pretty country. Today big business over dere to de Haitian people get job, get food, but we stay home. But when you're hungry, when you work hard for dis money, you no have no food to eat, you can't do anything to eat. Dat mean nothin, nothin to do. You see, look at de carpet, mattress, ol' bicycle, whole thing; we just come in here to get de junk. We go over dere, okay, make a little money, help de poor people. But now we can't do de same thing anymore because American people say– not, not een American people – like de Coast Guard, Immigration. If they give you a job to do for like $60,000 to do de job on de boat, what de job? De paint, de boat very – de paint... When dey come in every day dey come and dey say do dat. You do it. Tomorrow dey come and dey say do dat again. You do it. But so what's – how about me, I'm not gonna get no dollar to pay de people, to take care of my muddah, to take care of myself, to take care of my family. Don't thinkin about the color. We're de same.

B: Right.

S: If it white, if it black, we same. Only one god is above us. See white American people just kick all – kick de Haitians' ass, but dey don't kick de Cubans, dey don't kick de other people, de Venezuela – only da Haitian people. Why? Like de government over dere – Clinton, Clinton do best to help de Haitian people, but Clinton don't know what goin on in America, in Miami right now. See, because he don't know, but...

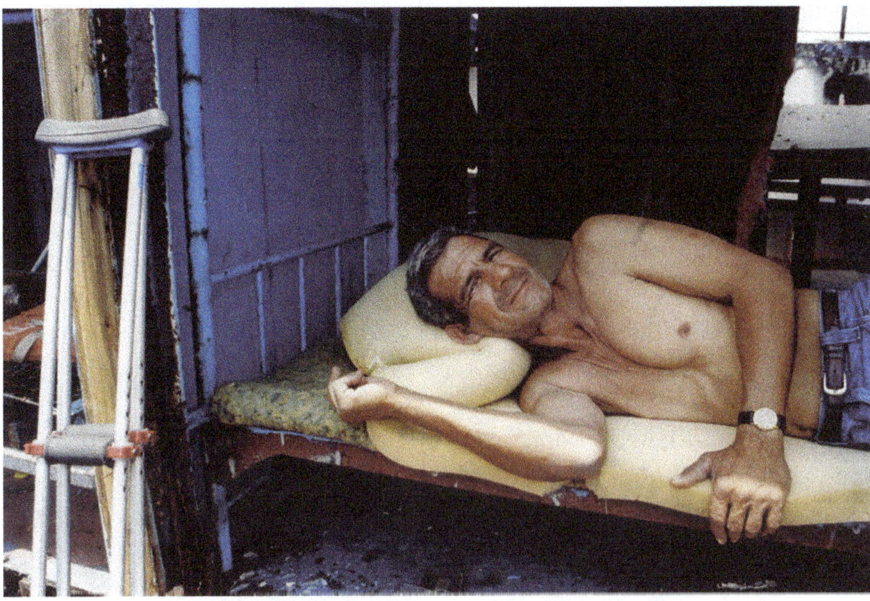

PHOTOGRAPH: BARBARA JO REVELLE

B: Know what I think, I think he's uh – there's this guy named Jorge Peres. He's a big developer. I read about him and they say that he's great friends with Clinton. And so Jorge Peres wants to develop the river, wants to put in all these condos and stuff like that, right...

S: Yah, yah. The working people, they have to get 'em off because um, they think that they pollute the river or this or that you know. The people don't want to see all the Haitian boats or something like that. It's pretty bad because okay, I'm a Haitian, They told me to get a Belize flag to put in my boat. Belize flag cost ten thousand dollars. Yah, Belize flag. That's what the Coast Guard say, you have to get a Belize flag before you get out. Okay, if you don't have the ten thousand dollars den stay docked over dere all year until you get de money. Then pay dock one hundred dollars per day – you pay de dock, But when you get de ten thousand dollars, you go buy de Belize flag. After de Belize flag dey come in with the – you have to get dere for ten, ten more thousand dollars. After, you have to get two life raft for about ten thousand dollars, you have to get two little boat again for ten thousand. After, you have to come in with de radio and put de boat in drydock for about ten thousand dollars. Where da money come from? Where's dat come from? If you don't do it – out from da business. Dat's right, dey put us out. Dey don't know how to tell you, okay get out of here. But dey put sometin in da middle. If you can't do it just get out. So, what you want me to do? So, God, dey supposed to know one day God is gonna come down. He's not comin to get me, to get you. He's comin to get everybody. Dat's it. For one year da Coast Guard charged me twenty-four thousand dollars for no reason. I don't have no money to pay it so I come in here. Pay one hundred dollars every day for dis river boat dock. So, when I can't pay, the dockman say okay, I'm gonna get your boat. Dey told me to do all, all dere job. I'm not drunk; I'm livin in America for eighteen years. I'm never goin to jail for de drug, I'm never goin jail for stealin from somebody. All dat I do, buy dis boat to help my family in Haiti. So in Haiti, I make some more money, I come in here to pay de tax, to pay my house. Now I can't pay nothin. Dey make me stand up over dere for one year and three months. People try to kill us. We want to go home and we can't.

B: But, you, so the only thing you know how to do right now is just stay here until he gets your boat?

S: Yah, if I stay here, if I don't pay, then he get my boat. Den I can't make some money to pay him, to pay everything. So dat's de time I stand up over dere because for da little piece boat I'm gonna stand up over dere if I can get some money to put dis boat out den to go help my family. Den i don't have it. I don't have de money and so…

B: This boat must be worth a lot of money.

S: Yah, lot of money, yah. When I buy dis boat, when I buy dis boat, dis boat cost me fifty-five thousand dollars. Yah, I was doin two job, two job. I start 1982 to 1995. I just keep dis money, put dis money in de bank. And so when my mother call me and say we're not comin here, I say you don't have no visa, somethin like that. I don't come in here with you all, you can't work, you can't do nothin. So, rest over dere, don't go away. I'm gonna take care of dat. Dat's what happened by da piece boat over dere. Both of the men get their money. Buy dis boat, put de junk over dere, put de flour for dem so to feed dem. For right now, one year, three months I been standin here. I can't even move.

B: What do you do all day while you're waiting?

S: Waiting, when I can't find no money to pay dem? Take care de boat. You know why de Haitian people don't complain? Because so many of dem don't know how to speak English, to let the American people know, de government know what's going on... Okay? Dey don't know how to speak to de government, to let de government know what's going on right here in America. Dat's it. And den Cuban, everybody mash up Haitian people because dey don't know de language.

B: Yeah.

S: Dis is de problem. Dis is de problem.

B: Why don't we let people come in here that want to come here? This is supposed to be this big, great, free country.

S: Because dey don't want dem comin in here because we're black. Dat's what dey think.

B: But there's lots of black people in Miami.

S: And so if its de Haitians, Haitians comin here, dey're not comin to steal, to do somethin – they're comin here to get de job, to work. We pay taxes more den some people comin in America. But we don't have no respect like that. Some Haitian get a green card, dey leave dere green card here, dey go back to Haiti. Why they go to Haiti? Dey cry because dey see dere family leave for dead, for hungry. I think Americans is hard to look at Haitian people, because when American people get independence hundred years, hundred years to go American people get independence. De Haitian people give American people some blood to fight for American people to get this country. But for right now, anybody can come in dis country, but de Haitian people, no.

B: Why is the country of Haiti fallen on such hard times? Why has it gotten so bad there; the government?

S: Okay, the Haiti – if you got a goat, you sell de goat to help you save; if you got a pig, you sell de pig to help you save. Every year de government send somebody to kill all de animals. Dat means if I'm poor, I got two goat, dey kill 'em, if I got two cow dey kill 'em so I can't find no money.

B: Why, why? How can they do that?

S: Because I don't know why dey do dat, dey just want to do dat.

B: They take them and they eat them or something like that?

S: No, dey kill em, just dead. When dey're finished killing em day say "if you want de meat, get de meat." Dey just kill em for no reason. See, so every month I go dere but I can't rest dere for long. Because too much somethin you don't like. Dat's what we – de poor people is not de, is not de problem because de big problem in Haiti is that they want to come in here. Even me, even you, if we don't have no food to eat, what you gonna do? You're not gonna rest to die. A lot of people die in de ocean, get de boat tryin to come in America. How many Haitians dey kill at de island over dere, dey shoot dem, shoot dere boat and de boat sink and so a lot of people was died. About every year about, maybe about ten thousand Haitian people die every year. Why? To come in America.

B: Why?

S: Hungry. So you got two child, you got four child. So dey want to go to school, you can't put dem in de school, you can't feed dem. In de morning you can't even give dem food. In de night if you find little food by twelve o'clock to feed them – dat can't be, see.

B: Yah, and it's very, very bad that we don't let the boats go back to Haiti and take the stuff that you take, because the stuff that you take is…

S: If they don't let us come inside Miami, if dey don't want us come inside Miami, dey say okay, we're gonna clean da river, dey don't want de Haitians in here because dey're gonna clean de river. Okay. Dey say no, no go back, don't come in, go. Okay, so how many, look in Miami now, got de big ship, de big ship comin Miami for dere little country. Dey bring dere junk, dey bring dat, but de American people never say nothin. But if de Haitian people come in here, dey find one plaintain in de boat to eat, dey give you problem for dat. So why do Americans call dem da freedom country? It's not freedom country. Dey're not freedom country. America is not freedom country for Haitian people. Because why? American people, we like American people so much because American people, we think we can get da freedom from American people. But now it's not. Everything dey say dey're gonna do for da Haitian people and Haiti. When do you think they're gonna do something? It's not true. Send the money to Haiti? We don't get no job, no nothin. The government keep da money to make good da house, good livin. So, make de poor people still de poor people, rich is still rich. Dis is de problem, man.

B: Why can't you just go? I mean, they'll stop you?

S: Yah, they'll stop you and we can't go. They have to give you papers to go. They have to…

B: It's only two days away, right?

S: Three days. Yah, so one year, three months I been here. The Coast Guard, the Coast Guard is the main people, not the government. The government don't even know what goin on right down in Miami. The government don't know what goin on in Miami. Dey put, dey put somethin dey have to put in de news, look like dey do somethin, dey don't do nothin. But dey don't put de real thing in de news to let da government know what goin on right here. Take care of da Haitian people very badly in America. Very very least. Because why? We leave our country to come in down here. We don't supposed to get somethin for ourself. Okay, go back, go back, like dogs. God don't like somethin like that.

B: Very bad.

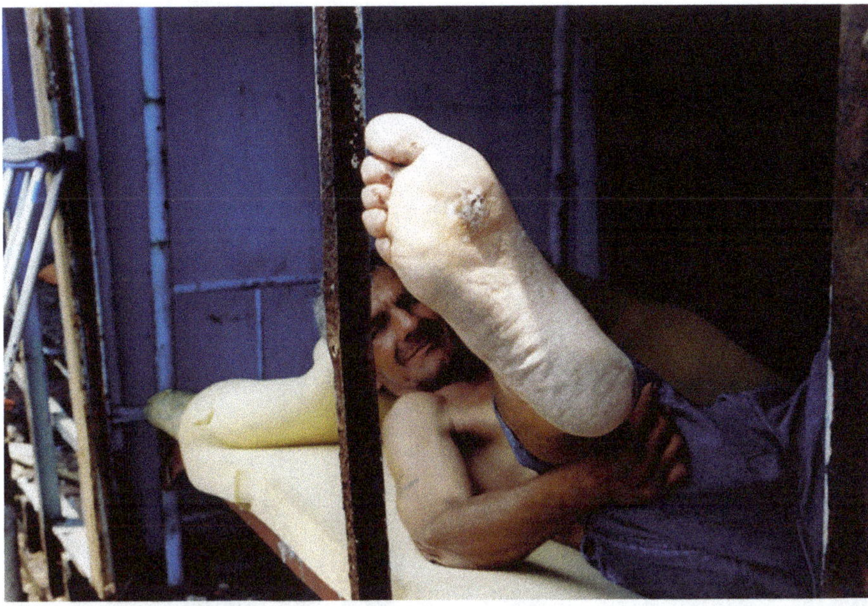

PHOTOGRAPH: BARBARA JO REVELLE

S: Look at all de mattress in the street everywhere. Nobody now to taken dem because dey stop us. We can't come here to take de dirty mattress to sell over dere to make some money to take care of our family – you know to do something for ourself. Dey stop us. Dis is de problem. We can buy the new bicycle, like the little child can sell you the old bicycle, so you buy it for de child, for the family, and so when you go over dere, dey're happy. You know, dey tink dey're livin good because you take care of dem. But for right now, I can't anymore because God has stopped us. Sometime dey come in here, or what dey do over dere, dey call us dogs, dey, you know, de animals. Look at de dock over dere – still I tell you – hundred dollars day. You see, when American people go to Haiti, even the white American people go to Haiti, there's a nice respect. We work wit dem, we show dem everything, they happy. We walk wit dem everywhere dey want, you know, make dem feel dey're comin in a black people country, they don't have no problem wit person in Haiti. But why we come in here dey take care of us like garbage? So that's what you have to you know – be patient. When God put you out – but now you know nobody helped me to do, to tell me "okay, Lubin, you been sit up over dere so long, try to get out." More dey help me dey tie me. One year, three months.

Simon Lubin, Haitian trader, standing on the deck of his boat in the Florida sunshine, the boat piled with used bicycles, mattresses, plastic buckets, in response to the egent's questions, to explain his motives, says: "Hungry. Hungry." We hear in his calm, objective tone an instance of what Spinoza called conatus: the principle that every creature strives to persevere in its own being. Appetite is metaphysical.

PHOTOGRAPH: BARBARA JO REVELLE

CHAPTER 10

Mystory 4: Theory

The Topos of Wisdom

An image category is a chora. A choral category correlates the collection of *topoi* circulating in a given geographical and cultural region, setting the default moods that guide the organization of personal and collective life (determing the common sense guiding decision). When the Haitian traders and immigrants arrived in Miami they found a place prepared in advance for them, within the Frontier world-view. The Frontier myth continues to organize American experience, both in the Entertainment mode of neo-noir such as *Miami Vice*, and the community policy formation exemplified by the Caribbean Code. In this scenario the Haitians are other, part of the *crowd*, the dangerous classes of workers, immigrants, and freed slaves. In the default mood structuring the disciplinary discourses of the FRE, however, the Haitian traders have a different pre-assigned place. They are a type of the modern hero. The ethics of cultural studies is as predisposed to side with the Haitians as the Frontier scenario is to exclude them. The point to make in this section is that the Career quarter of the popcycle, registering the specialized knowledge of the consultant, has a propensity determining its default mood just as much as do the other quarters of the popcycle. It is not that the consultants have an objective stance overseeing and evaluating the Family, Entertainment, and History quarters, but that each of the quarters has its own default moods, its own topics. The work of attunement is needed to correlate all four quarters into a chora. An image category supports "divination" through the correlations to be found between the microcosm and macrocosm. In our post-Enlightenment era, however, these correlations emerge not from the anagogic (cosmological) but from the symptomatic (singular) dimension of a local-global situation. Moreover, the correlations are not pre-determined, but are constructed in each attunement.

Ataraxy

The media narratives called attention to a situation, associated with the state of the State of Florida in general and with the city of Miami in particular. The imperative was that we should tour the site of the problem to see for ourselves, to witness and give testimony on behalf of the public. This tour of the ground would in turn provide the basis for the design of an Internet site, a virtual Miami, leading ultimately to flash deliberation: public prudence conducted

 | DOI:10.1163/9789004534643_011

at light speed. One goal of the Internet tour of Miami/Myami is to open this image to the problematic of the critique of realism that informed the revolution in representation within the arts and sciences of the twentieth century. The policy context for the crisis of the image of Florida is an opportunity for citizens to update their cognitive and affective mapping skills to include the lessons of modernism and postmodernism.

The initial idea was that the theoria would go to the Miami River to attest to the situation of attraction – the convergence of tourists and immigrants. We discovered, however, that in so doing we were participating in a genre that is just as formulaic as *Miami Vice*, even to the point of its implicit title: "Miami Virtue." The scene of a shipwreck with spectator (as in the nightscreams episode) is a topos, as Hans Blumenberg explained, a commonplace of the sages who used it since ancient times as a moralizing anecdote. The discourse of the sages of the ancient theoria that we had adopted as a relay for the new consultancy expressed the state of mind known as "ataraxy," a kind of serene indifference to the accidents and tribulations of the external world. Blumenberg wrote a history of the shipwreck paradigm or "model open to multiple possible actualizations. By tracing the history of these actualizations, Blumenberg discerned changes in the way human beings have imagined their relation to the 'Lebenswelt,' or life-world. He noted that one of the most pervasive actualizations of the life-as-sea-voyage metaphor includes a spectator who observes the distress of those at sea from the safety of dry land. For Blumenberg, this spectator embodies theory (the Greek word 'theoria' derives from 'theoros,' spectator)" (Rendall, in Blumenberg, 1997: 2).

Blumenberg traced the history of the metaphor comparing life to a sea disaster from its origins in Hellenistic philosophy (from the Stoics for example who emphasized that the point of the scene was not pleasure in the misfortunes of others – "Schadenfreude" – but the feeling of one's own security), through the Enlightenment reading of shipwreck as a necessary risk associated with acting upon a curiosity about the world; to the modern dissolution of the spectator's position, noting that one has no choice but to embark and the voyage ends necessarily in the shipwreck of death. A common theme running through all the Myami scenarios (consulting, advertising, noir, wisdom) was that of a "safety crisis." The function of the metaphor was not to think about sea-faring as such but about happiness. The practice of conceptual metaphor demonstrated by George Lakoff and Mark Johnson as definitive of the Western philosophical tradition (the generation of philosophical theories by means of mapping folk theories and everyday life experience onto abstract questions) is manifested also in this topos. "Since ancient times, what theory was supposed to do was not to make life possible but to make it happy" (Blumenberg, 1983: 232).

The common concern of wisdom in politics and ethics is whether, when, and in what way to act in the world. Such decisions depend upon what the agents assume is possible, to distinguish what is necessary from what may be changed, along with some sense of measure: a concept of the good life (well-being).

> What we should look for may be a conceptual framework which we can apply to particular cases, which articulates the reciprocal relations of an agent's concerns and his perception of how things objectively are in the world; and a schema of description which relates the complex ideal the agent tries in the process of living his life to make real to the form that the world impresses, both by way of opportunity and by way of limitation, upon that ideal. In coming to form this situational appreciation, however, we cannot rely on maxims or precepts. In no case will there be a rule to which a man can simply appeal to tell him what to do. The man may have no other recourse but to invent the answer to the problem. (David Wiggins, cited in Kekes, 1995: 27)

The understanding of "how things objectively are" is heavily influenced by image. The decision of the classical sages was that "life" was a "shipwreck," so there was little point in attempting to "get anywhere." "Happiness" as a goal does not itself have any content, but refers to a state of mind whose specific nature remains to be determined situationally. The feeling experienced in observing the shipwreck sets the agent on the path of self-knowledge. The EmerAgency ultimately consults on the "pursuit of happiness."

The ideal that emerged among the different classical schools of thought, reflecting their understanding of what was possible in their life-world, was "ataraxia." Ataraxy is to the wisdom genre what nihilism is to noir. "All agreed in affirming that happiness is to be found in 'ataraxy,' that is, in tranquillity of spirit. Pyrrho sought tranquillity of spirit in complete renunciation, in complete indifference, and in a state of insensibility; Zeno sought it in 'apatheia,' in imperturbability, that is, in the suppression of all passions of the soul. Epicurus finally sought it in 'aponia,' in the suppression of physical pain, and in 'ataraxy,' in the elimination of all disturbances of the soul" (Reale, 1985: 13). While the classical sages turned away from the shipwrecks of the external world to pursue the inner life (that portion of reality they deemed to be within their control – their own desires, which they suppressed as much as possible), ataraxy persisted as the ideal even in the modern fusion of the positions of victim and spectator of shipwreck (being witness of one's own life as a disaster from which there is no escape). Wisdom directed one not to "solving the problem of shipwrecks" but to reflect on one's own life situation.

Schopenhauer makes the sailor in the midst of storms a stoic. Nothing need be said about the ship's voyage and goal, because it has become wholly the vehicle of surviving and rising above life. "Just as a sailor sits in a boat trusting to his frail barque in a stormy sea that is unbounded in every direction, rising and falling with the howling mountainous waves, so, in the midst of a world of sorrow, the individual man sits quietly, supported by, and trusting to, the 'principium individuationis,' or the way in which the individual knows things as appearances.... Just as Lucretius's spectator no longer needs a sailor in peril at sea, because he projects his own past or future distress into the image of the raging sea, so the sailor in his little boat no longer needs a spectator on the shore, because he himself has become a world spectator" (Blumenberg, 1997: 65).

The success of the movie *Titanic* proved that the shipwreck as the basis for reflection on life continues to be relevant to the spirit of our time. The sinking of the *Titanic*, like the explosion of the *Hindenberg*, gave a premonition of Virilio's "general accident." This film itself included a *carpe diem* story as a placeholder for the viewers' own existential attitudes about the good life in a world in which "such things are possible." The outpouring of historical materials associated with the blockbuster remake reminded us that this event provided the allegorical basis for a wide range of political and philosophical positions, as the blurb from Steven Biel's *A cultural History of the Titanic Disaster* makes clear: "Everyone found moral and cultural lessons in the sinking of the

PHOTOGRAPH: BARBARA JO REVELLE

Titanic – suffragists and their opponents; radicals, reformers, and capitalists; critics of technology and modern life; racists and xenophobes and champions of racial and ethnic equality; editorial writers and folk singers, preachers and poets" (Biel). The wreck was not so much the target of explanations, but the source; it did not have one meaning but was associated with the most troubling conflicts and problems of the day, exposing the anxieties emerging in response to modernity.

The disaster became a "myth" by means of which "American culture in effect thought out loud about itself" (8). As a myth it was used to make sense of the most contested issues of the day – matters of race, women, labor. "What emerged in the days after the disaster was a myth – not because it was false necessarily but because it located a disturbing event within routine structures of understanding. 'Myth,' as Richard Slotkin has written, 'is invoked as a means of deriving usable values from history, and of putting those values beyond the reach of critical demystification. Its primary appeal is to ritualized emotions, established beliefs, habitual associations, memory, nostalgia'" (24). For example, at the time of the sinking, the heroes of the stories were the first-cabin, upper-class men (part of the WASP hegemony). A motive justifying James Cameron's remake was to change the hero to correspond with changes in ideology – interpellation – (the hero is from steerage this time). Myth, that is, provides the "logic" of entertainment narratives. Simon Lubin and the impounding of Haitian trading vessels in general, and by extension any disaster scene, may also function as myth in this sense: as parables applied by egents to understand their own situations.

Before setting foot on the banks of the Miami River, then, the FRE was already in a scene, a schema – a topos – and we could not pretend that our inquiry would be any more objective or free from "perspective" than was that of the empirical consultant, the journalist, the booster, the TV script writer, or the policy maker. Chora as category makes it possible to think all these topoi together, holistically. The EmerAgency (as fifth estate) proposes the possibility of citizens appropriating from journalism the function of first responders in the ongoing collective practice of mythography. The continuing relevance of the shipwreck topos to contemporary philosophy may be seen in Walter Benjamin's attitude to his situation. "In April 1931 he had described himself as 'a shipwrecked person adrift on the wreck, having climbed to the top of the mast which is already torn apart. But he has the chance from there to give a signal for his rescue'" (Buck-Morss, 1989: 37). The nightscreams event or the Rose-Marie Express, or the Deliverance Express, whose distress we watched from the safety of our academic shore, showed us nonetheless, if not objectively then metaphorically, allegorically, how things are, our own human condition, whose features induce a certain state of mind, depending upon what

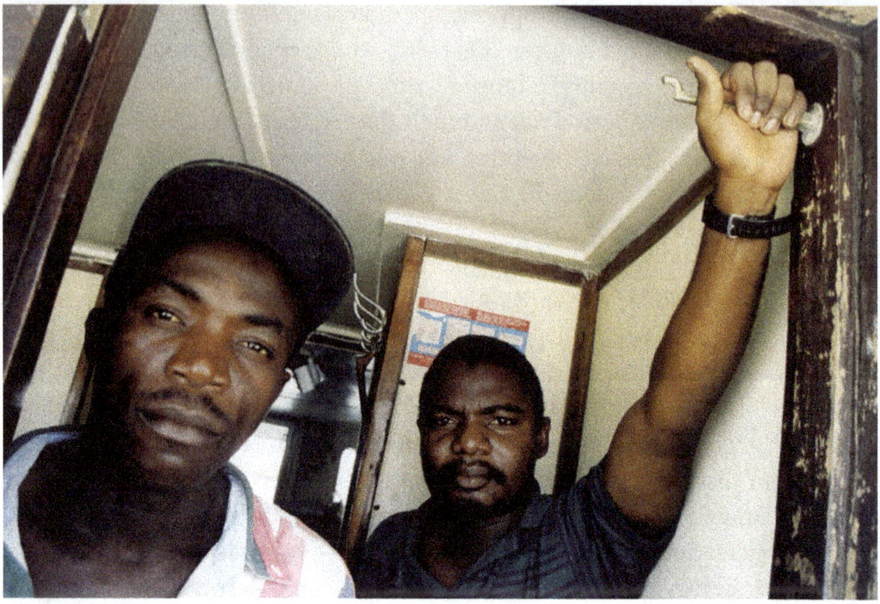

PHOTOGRAPH: BARBARA JO REVELLE

we bring to the scene. The default moods are those established by consulting (progress), boosterism (nostalgia), noir (nihilism), and wisdom (ataraxy). In America these topoi all are modalized by the Frontier myth (confrontation). Choragraphy opens these defaults to revision.

The Quilting Point

Much of the research conducted prior to the Miami tour concerned making an inventory of assumptions conditioning our reception of the site, in order to map the relationship between Miami and Myami. Apparatus theory suggests that the means of the electrate practice (testimonial) in which we planned to compose the consultation already exist or are emergent within entertainment discourse. One relay (an example that does not provide a model but a direction for further work) for the testimonial is the convention of testimony in advertising – the endorsement of products by celebrities, or characters representing "everyperson" (ethos). Almost any advertisement could serve to demonstrates the device by which entertainment discourse (including news) transforms problems into feelings of well-being (makes catastrophe pleasurable). Advertising does not avoid the conflicts of life. On the contrary, many ads initially raise at least momentarily profound existential questions concerning identity, power, sexuality, and all the other categories of ideology. The ads also foreground the problem-solution structure (conflict, product, resolution). "The problem was to be dealt with from a perspective closely related to initial pain and rediscovered happiness," to take just one example. "The campaign was made up of a series of four testimonies, thirty seconds each, that were, in principle, supposed to incite the spectator to make use of the phone network" (Haineault and Roy, 1993: 94). "'I was furious with him, he called me, he explained, now everything's fine,' declares a young housewife, smiling next to her telephone." The ads are typical in moving from a difficulty to "a secure sense of well-being" – "an electric bearer of happiness."

Advertisers have the right idea, and deserve credit for accomplishing on a massive scale (as only the commodity form could) the conversion of modern people to an aesthetic orientation to the world. The new axis added to human faculties by Kant in the historical moment of the invention of aesthetics is that of pleasure-pain, to supplement the existing axes of right-wrong (institutionalized in religion, representing the oral apparatus historically), and true-false (institutionalized in school, representing the literate apparatus). This third capacity is institutionalized in capitalist entertainment, and is specifically suited to the nature of image metaphysics. To adapt to electracy, the society must make the viscera intelligent. Advertising, however, as the house version of epiphany, uses the device on behalf of ersatz well-being. The tactic is a kind of "bait and switch" – the ad raises a life-world problem, but then sutures the gap thus opened with a "homily" (an ideological commonplace of the given society) (119). United Colors of Benetton carries the device to the extreme, testing its limits, by simply juxtaposing its logo with real or staged images of the most severe public problems (AIDS, war, diaspora, birth defects, and the like), a technique characterized as "logo-shock." Indeed, one of their ads invokes the shipwreck topos precisely in our context: the green and white Benetton logo (the same colors as the Rose Marie Express) is inscribed in an appropriated news photo of a Coast Guard ship conducting a rescue at sea of a boat overloaded with Haitians. "Why does the advertiser, in order to seduce us, resort to a fundamental, eye-opening gash, a gash from which the advertiser then steers us by reassuring us that it is really only a scratch? The process is surely doubly perverse: the fundamental problem is evoked in a properly accessible real situation and then, right on the spot, a needlessly sparked flame is extinguished. Well beyond the consumption such maneuvers incite is the corrosive and constantly banalizing usury of an authentic desire that, from one message to another, repeats itself insidiously" (106). Although in one respect contemporary consumer culture seems to embrace an attitude towards happiness that is at the opposite pole from that of the classical sages – the former hyperbolizes desire while the latter extinguished desire – both seek the same state of mind: the well-being of invulnerability (ataraxy). "Safety" is the ideal.

As the Benetton advertisers know, the dominant or hegemonic discourse in control of Myami already has a place for the Haitians swarming up the sides of the cutter (and, by innuendo, swarming onto our shores and into our public institutions, overwhelming national values). Commercial testimonial circulates in the order of faith or belief and not of science, and as such its purpose is to address a different dimension of reality than is considered by empirical consulting. As Slavoj Zizek explained, regarding a psychoanalytic theory of society, belief or "ideology" involves a mode of thought that is both prior and external to the one who thinks. The unconscious order of the psyche is "extimate" (to use Lacan's neologism) to the subject, at once outside and inside, having the topological structure of a Klein bottle or Moebius strip. Thus the experience of epiphany or "anagnorisis" proper to tragedy is accounted for in this context not as a divine or natural revelation but as an apprehension of extimacy, in which the Symbolic order, the Other of law and institution formation of all kinds, is inside me. The "auratic" effect of the object returning my gaze is an expression of Symbolic identification ("identification with the very place from where we are being observed" (Zizek, 1989: 105) – that is, the position from which our identity is constructed – as distinct from the secondary imaginary identifications with celebrities or fictional heroes representing an ideal self. The "unconscious" describes a distributed subject. This is the thought of event rendered accessible to personal experience within chora. The image category maps this transversal bordering. The edge has shifted dimensions in electracy.

As if the metaphysics of Aristotle had been transferred from "episteme" to "doxa," the Identity of collectivities is made possible by a fantasy that operates not as illusion but on the side of reality: "a basic Lacanian proposition, contrary to the usual thesis that a belief is something interior and knowledge something exterior (in the sense that it can be verified through an external procedure). Rather, it is belief which is radically exterior, embodied in the practical, effective procedure of people.... The most intimate beliefs, even the most intimate emotions such as compassion, crying, sorrow, laughter, can be transferred, delegated to others without losing their sincerity. Lacan speaks of the role of the Chorus in classical tragedy: we, the spectators, came to the theater worried, full of everyday problems, unable to adjust without reserve to the problems of the play, that is to feel the required fears and compassions – but no problem, there is the Chorus, who feels the sorrow and the compassion instead of us – or, more precisely, we feel the required emotions through the medium of the Chorus" (34). A description of testimonial in structuralist terms, as a transformation of tragedy, might say that in testimonial everyone is in the Chorus.

Everything that Zizek says about the place of the Jew in ideology applies in our case to the Haitian. The Jew, or "the jews" (to use Lyotard's term for all scapegoated others), are fantasy objects whose function is to support and make consistent the identity of a collective, serving as what Lacan called the "quilting points" or "upholstery buttons" holding together an ideological field ordering reality.

> The French term *point de capiton* is variously translated in English editions of Lacan's work as "quilting point" or "anchoring point." To avoid the confusion resulting from this variety of translations, the term has been left in the original French. It literally designates an upholstery button, the analogy being that just as upholstery buttons are places where "the mattress-maker's needle has worked hard to prevent a shapeless mass of stuffing from moving too freely about," so the *points de capiton* are points at which "signified and signifier are knotted together." Lacan introduces the term in his 1955–6 seminar on the psychoses to account for the fact that despite the continual slippage of the signified under the signifier, there are nevertheless in the normal (neurotic) subject certain fundamental "attachment points" between the signified and the signifier where this slippage is temporarily halted. A certain minimum number of these points are "necessary for a person to be called normal," The *point de capiton* is thus the point in the signifying chain at which the signifier stops the otherwise endless movement of the signification and produces the necessary illusion of a fixed meaning. (Evans, 149)

PHOTOGRAPH: BARBARA JO REVELLE

In the prototype case of the Jew as upholstery button for the Nazi, "at first, 'Jew' appears as a signifier connoting a cluster of supposedly 'effective' properties (intriguing spirit, greedy for gain), but this is not yet anti-Semitism proper. To achieve that we must invert the relation and say: they are like that (greedy, intriguing…) because they are Jews.… The appearance of tautology is false: 'Jew' in 'because they are Jews' does not connote a series of effective properties, it refers again to that unattainable X, to what is 'in Jew more than Jew' and what Nazism tried so desperately to seize, measure, change into a positive property enabling us to identify Jews in an objective-scientific way" (96–7). No system or field is complete or completable, totalizable, not even mathematics (as we know after Gödel), and certainly not any society or state. The ideological function of fantasy is to cover over this hole, to plug the gap, mask the inconsistencies and failures of coherence (the Being-One, self-identical). "The 'Jew' is the means, for Fascism, of taking into account, or representing it own impossibility" (127). Advertising is propaganda, of course, and the tactic of bait and switch was carried to its logical (extreme) conclusion by the Nazis: a real problem was raised (the troubles of Germany after World War I) and the homiletic solution proposed was "final."

The Haitians are the jews of the Miami Frontier. They show us the place of Autrui, *other. They are the* autre *in Miautre. All the materials needed to form a pattern have been registered. The ideogram, constellation, may flash into coherence at any moment. You may already get it, the constellation that gathers a virtual city. Chora enables an image metaphysics.*

PHOTOGRAPH: BARBARA JO REVELLE

The Collective Sinthome

The paradox of chora as a category is that it gathers order from a singular position. In the style of "horseless carriage" it may be called during this transition a "strange concept" or a "conceptual image." "Symptom" is sometimes written in the archaic version, *sinthome*, to distinguish the specialized function of a repeating signifier in Lacan's theory. *Sinthome* names the trait, the signifier without signified, whose insistence transforms nonsense into a plane of consistency (Lacan's analogy is with the way style informs an artist's *oeuvre* with a unique identity). The dimension ontologized through these aesthetic means is "manner," the embodied experience of signification. Guattari's name for this non-sense trait, this vinculum, is "existential refrain," thus calling attention to its musical quality. "Like Bakhtin, I would say that the refrain is not based on elements of form, material or ordinary signification, but on the detachment of an existential 'motif' (or leitmotiv) which installs itself like an 'attractor' within a sensible and significational chaos. The different components conserve their heterogeneity, but are nevertheless captured by a refrain which couples them to the existential Territory of my self" (Guattari, 1995: 17). An implication for electrate identity is that a unique refrain, a singularity, may be the clasp that holds together a collectivity (that the nation, so to speak, "hangs by a thread").

A preliminary question for us was: how are we interpellated by the Benetton shipwreck? What do we recognize in the Haitian disaster framed within advertising as a commodity? What did the makers of this ad assume in advance about us? It is not a question of our conscious sympathies, since it is as easy for intellectuals to identify with the haitians as it is for minorities to take up the stereotyped positions reserved for them in the hegemonic order. Rather, it is a question of locating that extimate point that knows before we do what to think (the default mood). Every ideological order has its jews to embody the antagonism that delimits the totalization of any system. The haitians are a good candidate for this role in contemporary America. For Zizek the limit on democracy is the existence of incompatible fantasies. Jacques-Alain Miller added that the reason for hatred of one group by another has to do with the difference of "jouissance" (the surplus of pleasure, like surplus value): "Racism is founded on what one imagines about the Other's 'jouissance'; it is hatred of the particular way, of the Other's own way, of experiencing 'jouissance.' We may well think that racism exists because our [Haitian] neighbor is too noisy when he has parties. However, what is really at stake is that he takes his 'jouissance' in a way different from ours" (Miller, 1994: 79).

To be come electrate, the multitude must discover and manage the third axis of human faculties now institutionalized in Entertainment – the axis of pleasure-pain (jouissance). Electracy supplements literacy and orality, displaces but does not replace them. The feature that serves as the attractor for placing haitians in the fantasy position of jews in American ideology is the fascination with Voodoo. "Tales of zombies, 'loups-garous' (sometimes identified as werewolves and sometimes as female vampires), snakes, and red-eyed 'bakalou baka' or 'pinga maza' (an evil spirit, translated literally as 'prenez garde,' or beware), though peripheral to the more basic beliefs and practices that articulate the multiple and ambiguous versions of Haitian socioreligious life, have always appealed to foreigners" (Dayan, 1997: 14). Or, as Nora Zeale Hurston put it in her study of Haiti: "Zombie – the word which never fails to interest tourists" (Paravisini-Gebert, 1997: 37). Zombies interest not only tourists but theorists as well. The Enlightenment tradition of rational knowledge is founded upon the differentiation between literate and oral civilizations (between science and religion). Emerging in electracy is the third kind – chora, receptacle of generation for remix, for accomplishing mash-ups pulverizing the old antagonism between knowledge and belief.

European literate societies in their contact with the colonial other established a clear hierarchy subordinating all manner of magic or anything to do with non-Western religion to the rationality of science. The great insult of rationalism to its other is summarized in the metaphor of the commodity fetish, whose purpose was to dramatize the extent to which freedom eluded modern people in their reified alienated lives. Kroker's *Possessed Individual* surveys contemporary French theory to show the extent to which postmodern identity is conceived of by means of the metaphor of possession – one of the fundamental experiences of Afro-Caribbean religions in general, and especially of Vodou in particular. More specifically, the zombie served the tradition of critique as the most extreme, the strongest image possible, of the inhumanity of living in a totally commodified civilization. The insight of the metaphor is the phenomenon of agency separated, reified and displaced into the outside, giving rise to the primary experience of electrate identity: the uncanny. A postmodern subject, as David Hawkes explains in his summary of Adorno's *Negative Dialectics*,

> may construct his or her "personality" out of an amalgam of consumer goods, fashion statements, "lifestyle" choices, identification with celebrities – in other words, out of objective phenomena which are administered to them by the culture industry. As Adorno shrewdly sees, this is a shabby ideological ruse: "to a great extent the subject came to be an ideology, a screen for society's objective functional context and a palliative for the subject's suffering under society." There are thus two subjects: a "real" subject, which is repressed by exchange-value, and an "objective"

> subject, which is not "real," and which is the product of exchange-value. This might not matter, except for the fact that the relationship between these two subjects is antagonistic: we are faced with "the subject as the subjects's foe." The "objective" subject is like a zombie in a horror movie, preying upon and destroying the "real" subject. (Hawkes, 1996: 184)

Adorno concluded, "we are the dead." According to Zizek, the fundamental fantasy of our moment has to do with zombies: "if there is a phenomenon that fully deserves to be called the 'fundamental fantasy of contemporary mass culture,' it is this fantasy of the return of the living dead: the fantasy of a person who does not want to stay dead but returns again and again to pose a threat to the living" (Zizek, 1991: 22). The archetype is George Romero's *The Night of the Living Dead,* in which the undead are portrayed "as sufferers, pursuing their victims with an awkward persistence, colored by a kind of infinite sadness." Hawkes noted that George Orwell used this metaphor, "connecting the motif of the unsuspecting zombie to totalitarian politics" (Hawkes, 1996: 185). In our entertainment discourse haitians personify otherness in this negative sense used to mark our borders, "the not-me we use to embody the sense of catastrophe" (Robins, 70). Accurate as far as it goes, this ideological account of identity shift in electracy, relative to the conditions of spectacle or dromosphere, remains caught within the assumptions of literate metaphysics (critical abstraction). The denigrative application of "zombie" to describe contemporary subject formation is revalued as "avatar" in electracy: my image being separates and circulates in the spectacle.

The signifier by which haitians are "recognized" metonymically is the zombie. This fantasy threat is contained by "myth," as in the James Bond thriller, *Live and Let Die,* in which the Voodoo ritual is shown to be a fake and a cover for the machinations of the villain, and thus the problem is resolved, dissipated through the narrative. In our context, however, the repulsion felt for the zombie is continuous with its attraction: it embodies the defense of projection. Within Haitian belief itself, of course, the zombie is the figure of the very worst, the polar opposite of possession (visitation of a god). Within the punk aesthetic of popular culture, the zombie stereotype is a hit motif on the party scene. As we set out to witness the Haitian shipwreck, we already had located the upholstery tack in our psychogeography – the zombie. The zombie figures in Deleuze's film theory, to evoke the "automaton" in the image apparatus separated from the organic eye. "The 'model' is a modern psychological automaton, because it is defined in relation to the speech act, and no longer, as before, by motor action (Bresson was constantly thinking about automatism). Similarly Rohmer's puppet characters, Robbe-Grillet's hypnotized ones, and

Resnais' zombies are defined in terms of speech or information, not of energy or motivity" (Deleuze, 1989: 266). For Deleuze the zombie affords an insight into contemporary memory.

> In the modern world, where the suprasensible "has lost its effective force in history" (Heidegger), order has been undermined, "incompossible" eventualities intrude and confabulate our own, and not even the dead can be counted on to remain that way (in the modern era, what is memory if not resurrection?). This explains Deleuze's consistent recourse to Resnais, whom he considers the cine-philosopher par excellence because, like all philosophers – those who "have passed through a death," who are perched "between two deaths" (past and future) – Resnais creates images that have returned from the dead, whether from the Holocaust (*Night and Fog*), the atomic blast (*Hiroshima, mon amour*), war (*Muriel*), or some other unspeakable trauma (*Last Year at Marienbad*). The image is an aftermath of the trauma, the shattering of the present into so many shards that, like flaneurs, we labor to collect, each of us consigned to tracing different, incommensurable lines of this distended world-memory. (Flaxman, 42)

The very familiarity of this image, our "recognition" of this uncanny entity (especially the realization that this recognition is a specific experience that

PHOTOGRAPH: BARBARA JO REVELLE

we had to learn to notice), constituted an explicit part of the consultation. A function of the categorial image made operational online is to facilitate the composition of this world-memory.

The virtual city is home to zombies. Electracy supports a collective identity without zombies (without quilting points). Or rather, choragraphy transduces "zombie" into "avatar," making the distributed subject capable of possession, that is, of undergoing the w/hole. Simon Lubin collaborates with Lacan.

Ethics

The first priority of Revelle's attunement of the Miami River was to replace the sage's default mood of ataraxy with the Stimmung encountered on site, locating the border, the affective map connecting individuals with the collective order of the zone. The mystory allegory reveals the different default moods contributing to the choral w/hole. John Caputo clarified what it means for our consultation to be a matter not of "law" but of "justice" (of the irreducible conflict between law and justice). The jews, the other (haitians), the indigenes of the Miami River, are not partners in a dialogue or conversation, within this view. Rather, outside of any principle or rule, any idiom (phronesis, prudence,

PHOTOGRAPH: BARBARA JO REVELLE

good judgment), there is some event, a specific case, that touches me (the egent), that gives me a feeling of obligation: something shocks me. “Disasters do not produce a result. That is what is meant by a disaster. An obligation is a matter of being bound to a disaster. That prevents you from looking for a result. It makes the search for a result an obscenity” (Caputo, 1993: 30). Caputo’s point may be compared with Claude Lanzmann’s position that guided the making of *Shoah* (which consists mostly of the testimonies of Holocaust survivors): “It is enough to formulate the question in simplistic terms – Why have the Jews been killed? – for the question to reveal right away its obscenity. There is an absolute obscenity in the very project of understanding. Not to understand was my iron law during all the eleven years of the production of *Shoah*. I clung to this refusal of understanding as the only possible ethical and at the same time the only possible operative attitude. This blindness was for me the vital condition of creation. Blindness has to be understood here as the purest mode of looking, of the gaze, the only way to not turn away from a reality which is literally blinding” (Lanzmann, 1995: 204). In short, a testimonial does not compete with an empirical consultancy since the two pose entirely different question, from different paradigms (different apparati).

Caputo’s starting point is a feeling that the other has a claim on me. Moreover, the other that holds such a claim is not the best but the worst, the lowest, as measured by a eudaemonistic ethics of “the Good,” “excellence,” “self-sufficiency,” and this is Caputo’s revision of Heidegger’s Hellenism, an ethics of the “great” that forgets the “lowly” (Caputo, 1993: 35). Caputo’s rationale is the fact that Heidegger forgot about the Jews. Caputo calls attention to the syncretism that formed our civilization, which is not just Greek (or Greco-Teutonic as Heidegger would have liked), but jewgreek. Caputo proposed as an alternative to an ethics of virtue or rules, a poetics of obligation. “What if someone were to propose a poetics of the outcast, of everyone an-archical: lepers, prostitutes, tax collectors, Samaritans, prodigals; the mad, homosexuals, blacks, Jews, the HIV-positive? ‘Les juifs’ all. Could there be a jewgreek poetics of everything that is not great, of the disenfranchised, the remnants, the leftovers? Is that not, in any case, something poetry has always been?” (159).

We share this much with Caputo – the Haitian disaster (nightscreams) included us, obligated us. It also revealed to us the peculiar conditions of electracy that requires citizens to deliberate directly with “disaster,” or rather, with the general accident. Our experiment is not to remain within the default mood of the sage (ataraxy) any more than it is to leave intact the Frontier myth itself, the nostalgia of tourism, or noir nihilism. Our consultation modifies Caputo’s guidelines in this respect: the haitians are neither jew nor greek, but *creole*. We hoped to open up this opposition between “the good” and “the lowly” received from the tradition. The haitians would have to teach us the nature of this alternative, and Caputo's recommendation that we be guided by a poetics was just right, since there is no "ethics," no phronesis, no schema available, no myth for this post-witnessing. Nightscreams is an "event" in Lyotard's terms. "Events are often disasters in search of an idiom; the idea behind Marxism, e.g., is to find an idiom for the suffering produced by capitalism. The differend is an event of suffering deprived of an idiom. Events are characterized by their intensity, singularity, disruptiveness, multiplicity, heterogeneity, irreducibility, and openendedness" (270). Yet Marxism goes too far (for us), in taking the extreme position "that it is not possible to solve any particular question without solving them all" (Zizek, 1989: 3). Marxism has its own version of a final solution: revolution. Virilio suggests that conditions in the dromosphere have surpassed that possibility: the horizon of expectations of civilization passed from the condition of revolution (including the industrial revolution) to that of total war, and from war now to the general accident. Our commitment in response is not to move too quickly, neither into solutions, nor even into narratives, but to make a map adequate to the dromosphere.

EmerAgency consultants abandon problem/solution frames of reference, in order to gain access to a general economy of impossibility. Why do we resort to myth and religion in an age of science? Michel Serres explained it is because there still is pain.

The Modern Hero

We are thinking through a zone, framing a disaster as parable. As Simon Lubin explained in his interview, he and his peers came to Miami to carry away the refuse, the unwanted, broken, and second-hand cast-off items of the region for resale or recycling in Haiti. We recognized him, and greeted him as incarnation, as avatar, of a position native to our own disciplinary anagogy. Here are a few documents to indicate the status of the ragpicker (*Chiffonnier*) as one of the heroes of modernism from Baudelaire to Derrida.

> There is a third figure of defiant destitution in the metropolis [after the industrial worker and the prostitute] who takes on a heroic visage in the Paris writings: the rag-picker. "When the new industrial processes had given refuse a certain value, rag-pickers appeared in the cities in large numbers. They worked for middlemen and constituted a sort of cottage industry located in the streets. The ragpicker fascinated his epoch. The eyes of the first investigators of pauperism were fixed on him with the mute question as to where the limit of human misery lay" (Benjamin). For

PHOTOGRAPH: BARBARA JO REVELLE

Benjamin, the ragpicker "is the most provocative figure of human poverty. He is a *Lumpenproletariat* in a double sense: he is dressed in and preoccupied with rags [*Lumpen*]." The ragpicker is a social outcast who derives his or her precarious subsistence precisely from that which is cast out. He or she is, as Wohlfarth notes, both the "bottom of the barrel" and the "scraper of it." The ragpicker, as a despised figure of suffering in the city, is the counterpart of the urban prostitute. Both are demeaned by their socio-economic circumstances and by a degrading activity. But the heroism of the ragpicker, unlike that of the prostitute, does not stem from martyrdom.

Benjamin's account of the labor of the chiffonnier is derived from Baudelaire: "Here we have a man whose task it is to gather the day's rubbish in the capital. Everything that the big city has thrown out, everything it lost, everything it despised, everything it broke, he catalogues, he collects. He examines the archives of debauchery, the stockpile of waste. He sorts through, making an intelligent selection; like a miser gathering his treasure, he brings together the rubbish that, chewed over by the divinity of Industry, becomes useful and pleasurable artefacts." The ragpicker assembles the urban detritus; he or she is a collector of what no one else desires. The ragpicker rescues from complete destruction the broken, the obsolete and the despised, refunctioning them and making them useful once more. Frisby notes: "it is the rag-picker who makes use of the scraps, the fragments, since he is aware of their history. The fragments can be used again, they can be reassembled in a context that renders their mosaic intelligible." The ragpicker inhabits and recycles the ruins of modernity. He or she is an urban "archaeologist" who unearths the old-fashioned commodities that in turn reveal the truth about new ones: namely, that they are the same old rubbish.

Wilson writes of the ragpickers: "of all the bizarre kinds of work that the growing urban scene produced at this time, none was more symbolic than theirs." The ragpicker is a heroic figure for Benjamin because he or she is a model for redemptive practice. Missac points out that for Benjamin, "the activity of the rag-picker represents a form of allegory for the work of the historian." The clearest evidence of such a view is Benjamin's description of Kracauer: "A ragpicker early in the dawn, who with his stick spikes the snatches of speech and scraps of conversation in order to throw them into his cart, sullenly and obstinately, a little tipsy, but not without now and then scornfully letting one or other of these discarded cotton rags – 'humanity,' 'inwardness,' 'depth' – flutter in the morning breeze. A ragpicker, early – in the dawn of the day of the revolution.' The historical materialist, like the rag-picker, is concerned with the salvation of objects and people from the oblivion of forgetting, with collection and recollection. The ragpicker not only embodies the deprivations of modernity, but

PHOTOGRAPH: BARBARA JO REVELLE

> fundamentally resists the fragmentation of experience in the urban complex and the growth of modern amnesia. The rag-picker is the quintessential figure who 'excavates and remembers.'" (Gilloch, 165–166)

The ragpicker is not only an individual and a social phenomenon, but an allegory as well in the worldview of modernist artists and intellectuals. The sorting and selecting activity of this work emblematizes equally well the categorcal and classifying operations of chora. A topos not only stores the commonplaces of a type or figure, but also includes a set of logical operators, a poetics, for generating new instances of the kind. The same Kracauer to whom Benjamin alluded, in his theory of cinema, compared the camera to a ragpicker, and associated the scavenger's work with that redemption of physical reality underway in the Western tradition in the era of photography. Kracauer agrees with Benjamin's characterization of the camera as supporting an optical unconscious, bringing into representation a range of things and actions that otherwise escape attention or fall out of memory.

> The refuse. Many objects remain unnoticed simply because it never occurs to us to look their way. Most people turn their backs on garbage cans, the dirt underfoot, the waste they leave behind. Films have no such inhibitions; on the contrary, what we ordinarily prefer to ignore proves attractive tot hem precisely because of this common neglect. Ruttmann's Berlin includes a wealth of sewer grates, gutters, and streets littered with rubbish; and Cavalcanti in his Rien Que les Heures is hardly less garbage-minded. To be sure, shots in this vein may be required by the action, but intrigues inspired by a sense of the medium are often so devised that they offer the camera ample opportunity to satisfy its inborn curiosity and function as a rag-picker; think of the old silent comedies – e.g. Chaplin's *A Dog's Life* – or pictures which involve crime, war, or misery. Since sights of refuse are particularly impressive after spectacles extolling the joy of living, film makers have repeatedly capitalized on the contrast between glamorous festivities and their dreary aftermath. You see a banquet on the screen and then, when everybody has gone, you are made to linger for a moment and stare at the crumpled tablecloth, the half-emptied glasses, and the unappetizing dishes. The classical American gangster films indulged in this effect. (Kracauer, 54)

Drucilla Cornell extends the allegorical reach of the ragpicker into the metaphysical, finding the *chiffonnier* to embody the qualities expressed in Jacques

Derrida's critique of Hegel's totalizing philosophical system. Against Hegel's *aufhebung*, the dialectic that cancels out and appropriates into itself everything external to it, the *chiffonnier* evokes a non-violent relation to the Other, to the outside external of a logic or metaphysics. Derrida is the philosopher who practices this ethics that Cornell calls "the philosophy of the limit," and that respects the difference and singularity of what exceeds conceptual (literate) ontology. As in the previous allegories, the ethical ragpicker's gathering and sorting activities concern a categorial "remainder," the wasted or unwanted status of the leftovers falling outside a system of value. The rags by now refer to people, language, cultures, histories, civilizations.

> There is an alternative reading that locates the drive behind deconstruction in an ethical desire… In the name of the elusive residuum left over once the relentless machinery of the Hegelian dialectic has finished its work. The subtle phrasing of Derrida's opening question [in *Glas*] acknowledges that we cannot separate the question of what remains of Hegel from the question of the remains of Hegelianism. What of the rest that has been pushed out of the system? To ask the question is already a kind of tribute to the forgotten Other, whose remains have been scattered. *Glas* attempts the only salvation of the rest that remains possible through the work of mourning itself, but the project is still one of "salvation" and not just disruption. Indeed, for Derrida, it is only through the work of mourning that we can remember the remains because there has never been, nor can there be, a gathering of the rest that makes fully present what has been shut out: For what has been shut out is literally not there for us. Even so, the work of mourning the remains demands the mimetic persistence to scrape through the debris left over from Hegel's system at the same time that we recognize that, "the rest, the remain(s), is unsayable." *Glas* does rather than says. Derrida may well be out best salvage man, our ultimate *Chiffonnier*. (Cornell, 1992: 63)

Since subject formation and the experience of identity itself are constructed within a language apparatus, the allegory extends inevitably to all of us. To be a subject (an identity experience) is to be a rag-picker of memory. "Subjectivity is not constituted in the present, nor does the subject exist as a presence in and for itself. Instead the subject recollects himself in the act of remembrance of the Other in himself; an Other, however, that is beyond his memory, since she remains other. In spite of the limit of memory, the remembrance of things past

PHOTOGRAPH: BARBARA JO REVELLE

is the story of the subject, the only one he can tell. For Derrida, the subject only becomes a self in and through the possibility of mourning" (76).

The ragpicker cleans up after the showdown. What you refuse is refuse. You cannot throw away your own thrownness, certainly not when identity is a zone, a field, a relationship, and not an isolated bit.

On the Frontier of Language

The allegory traces a path between the concrete material and the abstract conceptual dimensions of institutions and discourses. This other (matrix/mother) in us turns out to be language itself. To attend to the refuse, the dimension of the remainder in language, is to notice, as Heidegger said, that we do not speak language; it speaks us. *Die Sprache spricht*. The metaphysics/category formation of literacy – conceptual thinking – produced the world-view of science, of instrumental reason, calculative thinking, utilitarian pragmatism. The notion of the "remainder" in linguistic theory refers to those features of language and writing excluded by the powerful abstractions of conceptual, analytical reason. Jean-Jacques Lecercle inventoried all these excluded features, basically everything associated with dreamwork (the unconscious) and poetic play, gathered under the heading of a linguistic "rag-bag." "Describing the remainder is like gathering a rag-bag. And if, according to Judith Milner, formulating a rule of grammar is

PHOTOGRAPH: BARBARA JO REVELLE

a negative process, describing the remainder is the negation of this negation – a matter of reversing the practices of science and dealing, in the manner of Courbet, with the exceptional, the irrelevant, the agrammatical" (Lecercle, 55).

Significantly for us, Lecercle refers to the border of the classification system that distinguishes between language and its remainder as a "frontier." "This is my aim: to describe the other side of language. The grammarian and the linguist but also the philosopher of language, choose their objects carefully. The first simplifies in order to formulate clear rules, the second separates the relevant from the irrelevant phenomena, the third is intent on saving rational communication and therefore on ridding natural language of the various impurities that obscure it. These are the frontiers that the remainder subverts and that I wish to cross. I shall dwell in the exceptional and the agrammatical, in the irrelevant and the excluded aspects of language, in the ambiguities and impurities of natural idioms" (52). Electracy does not reject literacy, but rather accepts the metaphysical choice it made to ontologize the semantic slope of language. The task of electracy now is to ontologize the other slope, whatever it might be called, the dimension favored by poets, the musical or semiotic dimension that lends itself to manner and style, to music.

Why venture into the *wilderness* of language? The political and ethical reasons concern the project of the EmerAgency, to design in Miautre a virtual civic sphere in which the netizens participate in the creation of their imagined

PHOTOGRAPH: BARBARA JO REVELLE

community at the level of mythography, and of value formation itself. Habitus goes deep, and expresses itself in manner. To use the language of Althusser's ideological state apparatus that Lecercle insists is still useful, Choragraphy maps the interpellation, the hailing of subjects, operating in a zone. The first experience of being spoken by language is not the aesthetic ecstasies of a poet playing musically with the rhythms of speech, but the ventriloquism by which one repeats the commonplaces found readymade in the topics of the popcycle discourses. Alluding to Deleuze and Guattari's assertion that the "slogan" (imperative) is the fundamental mood of discourse, Lecercle calls attention to the collective nature of language. "In Deleuze and Guattari, the subject is only a mouthpiece for a collective source. The word 'arrangement' stresses that the source is not a collective subject, but a non-subjective organization of discourse. The institution speaks, and its speech is not the sum of the discourses of its members. Even if someone invents the words, by the time they become a slogan they have lost their subjective character.... In other words, they assert that all speech is indirect speech. What I believe to be my own words is only a reported utterance" (44).

This view of all speech as slogan and citation is related to Althusser's account of the indoctrinating function of institutions and their discourses. "There is a theory of the constitution of subjects by language. 'Each subject [Flahaut claims] accedes to selfhood from and within a system of places which exceeds him by far; there is no word that is not addressed from a place and does not ascribe his correlative place to the hearer.' ... The place that parents unconsciously assign to the unborn child, for instance by giving it the name of a deceased sibling or a revered grandparent, will partly determine both its identity and its destiny. The same is true of the place the social formation assigns to its members within the relations of production. 'Social and interpersonal relationships make up a "game" which nobody is allowed to opt out of, and in which relations of force are endowed with a violence which constantly overrules the conventions which attempt to contain it'" (249).

Chora, as a category supported by digital technology, makes it possible to perceive and think with this vortex of default moods through which the collective forces of history address my body in a situation. Those Creoles collecting the used leftovers of the city unfold into the linguistic theory of the remainder by means of which creative change becomes accessible to reason and communication. Yet this insight comes with a caveat: the Creoles are the heroes of the discourse that speaks us, the intellectuals (Shipwreck with Academic observers). Our theory has a default setting that guides our consultation, in the same way that the Frontier myth guides the foreign policy (by default) of the administration of George W. Bush (*le cowboy* as he is called in France). The settings of a chora may not be changed, however, until the category system is internalized by the society. Miautre renders a map of this correlational "frontier." Academics are the chamber of commerce for creoles. We are willing to renegotiate our own archetypes, when all archetypes are on the table of well-being.

Choragraphy goes beyond critique to heuretics; beyond analysis of ideology to the default moods through which interpellation creates its identity borders. The purpose of Miautre is not only to map the chora by gathering the topoi and correlating their atmospheres into an ideogram. The goal is to learn choral poetics: to become mythographers of a virtual city against the Internet Accident.

PART 3

Miautre – Oracle

Those who anciently wished to exemplify illustrious virtue to the whole world, first ordered well their own states. Wishing to order well their states, they first regulated their families. Wishing to regulate their families, they first cultivated their own characters. Wishing to cultivate their characters, they first rectified their hearts. Wishing to rectify their hearts, they first made their thoughts sincere. Wishing to make their thoughts sincere, they first extended their knowledge to the utmost. This extending of their knowledge to the utmost lay in the investigation of things.

THE GREAT LEARNING

CHAPTER 11

Reading

Mood

Tour the site of the problem. To select the image answer to the burning question, sort the photographic and textual record into a mystory structure. Note any signifier that repeats in each of the four quarters of the mystory.

The premise of an electrate consultancy, addressing the impossibility of a condition, is that the egent is a categorial operator. The operating frame of the consultation is that the psychogeographic drift is conducted as an act of divination – choramancy. This stance makes evident the fundamental feature of the image category: that it is Stimmung that allows circumstances to matter to a person. Neuroscience supports this ontology, in demonstrating that brains store and receive information not like library card catalogues, but transversally by means of shared feeling tones. Divination as a reading strategy exploits this human quality to simulate (educate) ethical responsibility. "Agency" in individual and collective identity is caught up in a play of modalities, as part of the shift from the metaphysics of literacy to electracy and a different order of categorization based on the image. In electracy there is a short-circuit or pun switch crossing between the "mood" of linguistics and the "mood" of state of mind. One does not write the disaster in a literate way, with definitions and concepts (the way of homogeneous science, the expert consultants). "State of mind" is the dimension of mental life foregrounded in electracy. Jean-Francois Lyotard, in his study of Kant's Critique of Judgment, and especially his "Analytic of the Sublime," defined the aspect of mentality that is enhanced in choragraphy.

> The possibility of universal communication, required of knowledge, must also be required of the "mental state" accompanying knowledge. For all thought occupied with knowing an object is at the same time affected by its act of knowing and thus finds itself in a certain "state": for example "conviction," which is a delight due to knowledge. By reversing the order, one could even say that this subjective "condition," this "Stimmung," i.e., the state in which thought finds itself when it knows, is also a "subjective condition of… knowing," a condition this time in the sense of a condition of possibility. If there were not a disposition of thought about which it was immediately

 | DOI:10.1163/9789004534643_012

> informed and was favorable to its act of knowing, the latter would not exist for thought and thought would not know. (Lyotard, 1994: 100)

The questions that Lyotard pursues in his close reading of Kant are: "How can feelings orientate a critique. Why should the latter have any need for them?" (9). The answer that he evolves over an extensive reading is that the "state of mind," a fundamentally aesthetic nuance, constitutes the reflective dimension of judgment upon the object of thought. In literacy, the apparatus split apart these dimensions of thought in the institutional practices, manipulating the object of thought in science, and the state of mind in art. The outcome has been a hypertrophy of the object of knowledge and the atrophy of state of mind (to put an extremely complex history into a simplistic binary). In electracy a state of mind is as writable as the object of thought (testimonial is an experiment in developing such a practice).

PHOTOGRAPH: BARBARA JO REVELLE

Rodolphe Gasche describes how Heidegger used his etymological poetics to show the relation of the two fundamental powers of language. To define a thing (*bestimmen*) is made possible by attunement (*Stimmung*) of state of mind through mood. "The three essential determinations of *Stimmung* – Dasein's thrownness or facticity, the disclosure of its being-in-the-world as a whole, and the fact that something can 'matter' to it – constitute existentially Dasein's openness to the world. In *Stimmung*, in the attunement of a state-of-mind,

Dasein, which experiences itself always already factically (knowingly or not), is shown to be capable of being 'affected' by the world and of directing itself toward things in a world that in every case has already been disclosed to it. Dasein's being-attuned in a state-of-mind is the existential a priori of all possible linkage, connecting, or relationship" (Gasche, 1999: 116). Our argument is that literacy ontologize the linguistic potential of *bestimmen*, and electracy does the same for *Stimmung*.

A testimonial does not define (analyze) the policy problem but discovers its mood. To bring mood into active deliberation on policy formation sets a place at the political table for well-being. Through attunement the policy problem *matters*, and matters first of all because through mood I discover that it has already taken place in me. Mood in linguistics concerns the attitude speakers take towards their statements (indicative, interrogative, conditional, imperative, subjunctive). To invent the practices of electracy, and enabled by the digital prosthesis of memory, we may put some pressure on our language resources to support an additional modality of voice: the virtual. The mood (linguistic/existential) of a consultation is virtual, in the way Deleuze and Guattari use the term to describe an abstract machine such as Kafka's novels. "It consists in prolonging, in accelerating, a whole movement that already is traversing the social field. It operates in a virtuality that is already real without yet being actual (the diabolical powers of the future that for the moment are only brushing up against the door" (Deleuze and Guattari, 1986: 48). We recognize in this

PHOTOGRAPH: BARBARA JO REVELLE

vocabulary Aristotle's updating of Plato's chora, identifying the receptacle in a one-world (immanent) metaphysics as "potentiality" (*Dunamis*). The egent, like Kafka, is a "bachelor machine," possessed by and articulating the collective multiplicities passing through "me" that signal a potential beyond the actual default given. "When a statement is produced by a bachelor or an artistic singularity it occurs necessarily as a function of a national, political, and social community, even if the objective conditions of this community are not yet given to the moment except in literary enunciation" (83–84). Here is an anticipation of the categorial egent.

Kafka the Oracle

The EmerAgency opens public policy problems to citizens educated in the logic of bachelor machines. In a consultation the egent writes not the hierarchy of law (or concept) but the field (ecology) of justice, following the metonymic connectivity, contiguity, the intensities of desire and drive indexed in machinic assemblages. The vocabulary alerts us to the metaphysical implications of the example. In Kafka's stories the people or animals are indexes of Europe's future: "'The Metamorphosis' forms a complex assemblage in which the index-elements are Gregor-animal and the musical sister; in which the index-objects are the food, the sound, the photo, and the apple; and in which the index configurations are the familial triangle and the bureaucratic triangle.

PHOTOGRAPH: BARBARA JO REVELLE

The bent head that straightens up and the sound that latches onto the voice and derails it also function as indexes of this sort in the majority of the stories. There is thus a machinic index each time a machine is being built and is beginning to function, even though one doesn't know how the disparate parts that make it up and make it work actually function" (47). Each consultation introduces some set of machinic indexes into the Internet prosthesis, where they may be added to the virtual map forming in the collaborative mapping of the field (the gaze). The mattresses form a machinic index of the Miami River.

It is easy to become impatient with the philosophers who are attempting to imagine the new paradigm, to bootstrap themselves from the paradigm of literacy (site of their expertise) into the coming spectacle without being deterred by the force of the dromosphere. The argument in translation is: electrate metaphysics as category formation (the skill-set of choragraphy) will have been invented by the modernist avant-garde. Paris is the Athens of electracy. Delueze and Guattari point to Kafka and declare: there, that is how to design an image category!

Miami Ad-Vice

How does "Crossroads" record the attunement of the Miami/Myami River? The attunement is figured in the detail that produced the epiphany, that triggered something "in me" (in us), the querent, egent, consultant. This detail forms the dialectical image that links the river scene (external, Now) with a past (Then) scene of memory, forgotten, repressed, or stored in archives. The goal of choragraphy as an electrate practice is to learn how to support in a digital network – how to enable, enhance, and extend as a procedure of reason – the flash of spirit. Choramancy (applied choragraphy) attempts to be for the experience of insight what electricity is to lightning. As the theoria recollected the developed photographs and viewed the tapes accumulated over the five-week tour, we noticed a mattress that insisted, repeated, performed in one location after another: among the homeless in Jose Marti Park; in the ruins of the demolished apartments, along with other detritus from the displaced Guatemalan family; in the warehouses and on the decks of the ships trading with Haiti; in the bedroom of the Inn where Revelle stayed, and where Ron visited her.

The mattress is a natural feature of the tourist discourse, which gives pride of place to the bed. The two most memorialized destinations are battlefields and bedrooms, according to Diller and Scofidio's *Back to the Front: Tourisms of War*. For Georges Bataille, the bedroom is to heterology what the laboratory is to empirical science. "Happy action is the 'sister of dreams,' on the very bed where the secret of life is revealed to knowledge. And knowledge is the ecstatic discovery of human destiny, in this guarded space [the bedroom] where science – as much as art or practical action – has lost the possibility of giving a fragmentary meaning to existence" (Bataille, 1985: 229). The bed and bedroom are a place and position – a commonplace, a topos – in the mode of wisdom as well. "The bedroom I will talk about," Mary Carruthers states, explaining that the standard posture associated with meditation was that of lying prostrate and weeping, "is what both the Romans and the medieval monks called 'thalamus' or 'cubiculum'. The monks associated it most often with the bridal chamber of the Song of Songs. And while all the sexual associations of fertility and fruitfulness resonate in this bedroom mystery, its goal is cognitive creation, and its matrix is the secret places of one's own mind, the matters secreted away in the inventory of memory, stored and recalled, collated and gathered up, by the 'mystery' or craft of mnemotechnical invention" (Carruthers, 1998: 171).

The profundity of psychoanalysis, its direct relevance as an initial account of logic in electracy, is that it approaches human beings as sexualized bodies, just as philosophy in its origins approach humans as rational beings. In retrospect it all looks inevitable, as a mystery always does when read back to front. Our account is a rationalized simplification, for the purposes of establishing a prototype. The first recognition occurred when Revelle remembered the game she played (forbidden games) with her "white trash" friends in the neighborhood across the lake that her father contemptuously dubbed "Skeeversville" (back of beyond) – the game known as "Perfume in the Black Hole." This connection produced an experience of the "uncanny" (and opened a passage towards "satori"). That the mattress or any object could conjoin so many discourses is a possibility prepared in advance within allegory. "The sign 'lion' functions in Chretien de Troyes' YVAIN as a remarkable shifter between several discourses: that of an allegorical moral system, that of a natural or 'physical' system, and that of a new and purely social system that we may call heraldic or totemic" (Vance, 1987: 105). We selected "Crossroads" because it contained the switch or shifter that became the strange attractor of our consultation. We may review this document.

Skeeversville

> Linda McDonald's house – with the basement where her Daddy did upholstering when he wasn't drunk, and where the kids from Skeeversville played sexual and scary games in the dark amidst the mattresses, old chairs, and moldering upholstering material. Two brothers from Skeeversville, Ricky and Bobby Eason – were often in on those games. There was one we called Perfume in the Black Hole. I can't quite remember the perfume part, but I think we divided into two groups, boys and girls, and went to

> parts of the basement where we couldn't see the other group. The girls passed around a perfume bottle, and put some on in turn, and the idea was that where you put the perfume was related to the "secret you were required to remember and tell to the rest of the group." We went around the circle, and when a player couldn't think of any more secrets to tell, she 'lost' – and the punishment was that she had to crawl into a mildewed and spider-infested space created by a folded-over broken-open mattress. This was the "black hole." Then someone from the boys side of the basement, chosen through some more manly ritual we never got to witness but were told had to do with their "privates," would crawl inside and try to guess the losing girl's identity by feeling her body. Nancy Moyer and I talking earnestly and often about whether we could be pregnant because of these games. (Revelle, "Testimonial")

The Other Mattress

The epiphany gathering the constellation was initiated by the uncanny (contingently necessary) repetition of mattresses across the different levels of the popcycle, which transformed the mattress from a signifier into a sinthome.

Allegory (Quartenary)

> The *Literal Mattress* exists within the history of Haiti, the trade in used goods between Haiti and Miami that caused the stacks, piles, heaps of mattresses to be gathered at the docks of the Miami River. This detail is a metonym that opens onto a field (the object becomes a field) constituted by the history of the Caribbean. The drift returned three mattresses, marking three different policy problems (homeless, gentrification, trading vessels). The Diviner (the FRE) proposed these three scenarios to the querent. Revelle selected the one she recognized as her own.
>
> The *Moral Mattress* of the Family discourse appeared in Revelle's memory of the forbidden childhood game, Perfume in the Black Hole. The field opened by the object at this level is that of Revelle's relationship with her father, engaging then with the same dynamic that interested Lacan in Joyce's life: the way that Joyce decided to use writing to prop up the failure of the Name-of-the-Father in his "family romance."
>
> The *Anagogical Mattress* appears in the figure used by Lacan to explain the formation of an ideological field around a quilting point or upholstery button. "The French term 'point de capiton' literally designates an upholstery button, the analogy being that just as upholstery buttons are

places where 'the mattress-maker's needle has worked hard to prevent a shapeless mass of stuffing from moving too freely about,' so the 'points de capiton' are points at which 'signified and signifier are knotted together.... The 'point de capiton' is thus the point in the signifying chain at which 'the signifier stops the otherwise endless movement of the signification' and produces the necessary illusion of a fixed meaning" (Evans, 1996: 149). Lacan's theory of the human subject (representing metonymically the group of French poststructualists) provides our cosmology, that we apply in turn to problems in the civic sphere.

The uncanny correspondence of these levels was confirmed as a recognition by the further fact that Linda McDonald's father (denizens of Skeeversville) was an upholsterer by trade, which accounted for the accumulation of mattresses in that residence (the Black Hole game was played in the McDonald's basement). This entanglement of threads or narratives produces the experience of a *dense moment*. Belatedly, through Revelle, the mattresses led theoria to an experience of eureka, if not yet satori. These patterns are not sent, but only received, being potential or vitual until actualized by (a) being.

Jealousy

What about the *Allegorical Mattress* proper; how does it appear in the Entertainment discourse? It is evoked in principle by the neo-noir status of *Miami*

PHOTOGRAPH: BARBARA JO REVELLE

Vice. The pattern already in place suggests a "bedroom drama," the seductive role of the femme fatale in the noir genre. Yet in practice (significantly, in our context) this aspect of the female position was erased from the *Miami Vice* formula in practice: there was no such position in the TV program; the women characters remained within the gender roles permitted by the noir forebearer, the Western. This absence of one of the generic features has the perverse effect of calling attention all the more to this border war underway in all the ideological identity categories. In the Family discourse the cause of Revelle's relationship problems was precisely Ron's intense jealousy, which is a defining quality of a noir male protagonist.

> The pleasure and fascination of the fantasy of the duplicitous woman in *film noir* are, no doubt, as varied – or limited – as the different forms the fantasy takes. In certain films the fascination seems close to the compulsively repeated pattern Freud describes in "Special Type of Object-Choice Made By Men" (1910), that of falling in love with a woman who is another ma's property, that is, his wife or mistress, but who is sexually promiscuous, giving rise to the lover's suspicions and jealousy. This jealousy is not, however, directed at the "lawful" possessor of the woman, but at all the other men she associates with (seen, for example, in Johnny's behavior in *Gilda*). Moreover, the man seeks to "rescue" the woman he loves from moral decline, or from poverty, or from a vicious lover or husband. Freud connects this type of object-choice to the man's oedipal desires, so that the duplicitous woman is a mother surrogate. (Cowie, 124)

The goal of the consultation being to translate the River zone into a mnemonic scene, the question arises of how to interpret the allegory, both as an answer to Revelle's burning question, and as an attunement of the zone for the group subject (the virtual city). The point to establish for now is to notice the correlations emerging across the registers of the popcycle. As Greimas noted in his semiotic study of the passion of jealousy, whole cultures may be jealous. The issue is "exclusivity."

> Exclusivity is based on a *having-not-to-be*, whether it be cognitive or logical. Two irreconcilable propositions are declared exclusive. In the epistemic, what is recognized as impossible is "excluded." In the juridical, a privilege or a right that is reserved for one person or group is "excluded." The withdrawal of a given integral unit from the partitive totality is regulated by a *having-not-to-be*, on two levels. On the one hand, it is the relation of the collective subject with the selected object of value that

PHOTOGRAPH: BARBARA JO REVELLE

> *should-not-be*. On the other, it is the relation between the single subject and the collectivity that also *should-not-be*. Exclusivity prepares the way for rivalry. In attachment itself, the collectivity brings itself into the equation negatively, as an actantial presence with which the subject establishes, through presupposition, polemical relations. Against the backdrop of this breaking of the consensus, this rejection of the partitive totality, we see a rival emerging. In this sense, the rival is nothing other than a concretization (actorialization) of this presence that is at the same time rejected and postulated by exclusivity. (Greimas, 1993: 134)

This scenario of conflict and antagonism organized around exclusivity is precisely what Slotkin insisted limited and rendered obsolete the Frontier myth. The WASP elite was jealous of its privileges and power, and took umbrage at the claims made by the crowd. Thus a certain aspectualization of value correlates the popcycle: Ron is made jealous in his relation with Barbara; the noir protagonist is made jealous by the femme fatale; the WASP elite is made jealous by the crowd. What this alignment brings out is that this state of mind is expressed from one perspective (the male protagonists). The consultants' proposal is to open up the actantial position, to include the perspective of Revelle, the female, the crowd. What are the policy implications of this revision to the national narrative? It is already happening socially, but the national myth remains to be revised. In the dimension of mood, we are still jealous. All parties to the scene share this atmosphere that Slotkin described as "entrapment," to name the mood of the *Noir* Frontier.

It must be admitted: the FRE *is jealous, being excluded from the circulation of knowledge in the community. In choragraphy citizens deliberate in parables, through identification augmented in the prosthesis to categorial power. I may not understand an issue conceptually, but I am jealous (resentful, miserly, fearful, the gamut of passions).*

Sublime Policy

Select a specific public policy issue as the target of the consultation.

A difference between an inventional agency such as the EmerAgency and a conventional agency is that the nature of the object addressed is different. In the order of belief, objects are categorized and made intelligible by means of stereotypes, clichés, the archetypes of opinion. In the order of knowledge objects are categorized and made intelligible by means of concepts, by identifying and enumerating attributes and properties of the things in question, according to the basic assumption of literate metaphysics (science). The electrate object, an object of mood, in contrast, is neither schema nor "substantial" but exist solely as relations, as pure difference without properties or attributes. Certainly the empirical object, the one covered by concepts with attributes, is present to belief, but some extra quality "X" enters into the object that is more than it, marking its participation in a fantasy. The FRE "mattress" was in a block with Haitians, "haitians" and "spider women," intelligible not through its attributes but through a relationship. Moreover, this meaning is not intended or communicated, but emerges in the field of relationships specific to a given person (the querent). A fantasy object may not even be perceived apart from the disturbance of desire associated with it. This experience is related to "fetish," of which the commodity is the prototype, according to Marx, so in principle the operations of desire and drive are familiar enough in everyday life. A theory important in our methodological hybrid – critical divination – was (poststructural) psychoanalysis, if not for its clinical claims, then for its status as a logic that is at once the end of literacy and the beginning of electracy. Fetish logic admits contradition, in order to hold together simultaneously two incommensurable thoughts. Its formula is, "I know, but all the same…." Choral inference goes beyond contradiction, whether excluded or accepted, by means of holistic images.

The Mattress shows how an image becomes "categorial." At the micro level of form and medium of photographic recording, a third meaning is to digital recording what the pattern of words relating to a concept was to alphabetic recording (such as the variations on *dike* – justice – noticed by the Classical Greeks who abstracted the first concept out of the oral epics). That a "bed" may demonstrate a metaphysics we know from Plato's discussion at the heart of his great dialogue on justice, the *Republic*. There are three beds in his "Realism": the real bed is the Form created by god; the actual bed is the copy made from this Form (idea) by a craftsman; an image of the actual bed appearing in a work of art (mimesis) is furthest removed from the real bed. Greek metaphysics does not account for the fantasy bed triggered by the photographed mattress in "Crossroads." In the new realism of the virtual city, this fantasy bed takes priority.

Objectity

This mattress with the X in-it-more-than-it is a different sort of object, a quasi-object, which Lacan called the "object *a*" (small a, "autre" or "other" in French – the object @). As Jacques-Alain Miller explained, object *a* falls outside the range of ontology, which "tells a priori what can be said about objects," based on the categories, a sorting into kinds based on what features or properties objects may share.

> But when we speak of object "a," we speak of another objectivity – let's say of another "objectity," an objectity that is not summoned opposite the subject of representation.… However, this new objectity is such that one cannot avoid experiencing it. It is an objet articulated not to the subject but to its division, to a subject that does not represent to itself the objects of the world but that is itself represented. For this reason, we cannot say that the structure of this object is identical to that of the "enonce" [statement]. There is no specificity of the object in the Other, where nonetheless, the object "a" does not dissolve. It escapes categories because it does not have the same structure as the statement. By using the medieval reference reactualized by Yankelevich, we can say that here it is a matter of a "quod," in the sense of difference between "quodity" and "quidity." We could also say that it is a question of the difference between existence and essence, or something that there is, but the essence of which one cannot define in the Other. One can say that it is – that is, "quodity" – but one cannot say what it is. There we have a kind of paradox of the "quod": something exists but without "quid." (Miller, 1994: 82–83)

"What" is a matter of substance (literate category). "That" is a manner of accidents (electrate category).

Hole-Ism (holEism)

The Mattress looked back at us; delivered its belated blow, the punctum, producing an effect best described as "uncanny" (something unfamiliar in that we have never seen it before, is nonetheless recognized as familiar). As Freud said, in the uncanny something that should have remained secret has come to light. Attempting to name the feeling specific to the era and the industrial city, Freud found in a lexicon this term, the uncanny (Unheimlich, the not-at-home). The term was attributed in the lexicon to Schelling, who introduced it to translate the Greek word "Nemesis" (the event of retribution). In the experience of the return of the repressed, my own alienated agency comes back as revenant, like the ghost of Hamlet's father, to hold me accountable. *Problems B Us*. Here we have the name for the experience giving rise to electrate subject formation, supplemental to oral "spirit" and literate "self." The artists were registering the same phenomenon in the city. Breton spoke of the uncanny as something "troubling." TROUble. This word includes the "trou" of "hole" that Lacan played on to link the event of the "lucky find" ("la trouvaille") to the "hole" in the real marked by the object @ (Cohen, 147). The real, in Lacan's pun, is "troumatic" (Foster, 1996: 136). The challenge to electracy is that the new object is traumatic, an event without subject. The challenge of chora is to receive Nemesis as belonging-to-me, that is, to recognize it as limit or measure. In the creole dimension of the method such effects are explained in terms of magic. This effect happens in the real. The satori Barthes spoke of, however, will not occur without further collaborative work. For now we have the fact that the theoria recognized the Mattress.

PHOTOGRAPH: BARBARA JO REVELLE

This "thing" is "das Ding" in theory – the object of desire (the object-cause-of-desire): Mattress. "The pleasure principle is the law which maintains the subject at a certain distance from the Thing, making the subject circle round it without ever attaining it" (Evans, 1996: 205). The Lacanian graph showing the triangular path of the drive around the "hole" or void of jouissance could be imposed on the river as a map for a psychogeography. The Thing or object @ in terms of categorization has the effect of a "strange attractor," being some part of the scene that becomes, temporarily, a point of order and coherence of the whole (hole). The visceral was acknowledged and assigned a place in the previous apparati. In electracy it finally is given equal status, if not granted priority through (mere) inversion. Taboo and Sin must be transvalued in order for the public sphere to access what escapes the dragnet of problem-solution.

Traumatic kernels/voids are *that* around which the Symbolic order circulates, graphable as strange attractors. The holes found in the Tequesta Indian site excavated by the developers at the mouth of the Miami River uncannily refer to the importance of the hole (which is holism with a difference – addressing the silent "we" of "whole": w/hol/e). "Crossroads" grasps the river zone not as a whole, but as a hole: hence the neologism "holEism" to name the peculiar way of gathering. "One learns about the whole from the hole." And we should add, "one learns about the block from the block. Psychoanalysis uses topology to show the hole of the subject. Freud identified the uncanny in his own experience, when, lost in a foreign city, his wanderings brought him back three times to the same neighborhood, a red-light district. His returns were noticed and misunderstood. He was brought into the orbit of a hub.

What is pleasure-pain, such that it had to be framed as virtue-vice for two millennia? Zizek identified the sinking of *Titanic* as exemplary of the traumatic Thing, which accounts for the fascination of the wreck/wreckage. The egent reads this example as a template: "your disaster here." If concepts define "things," choras render the Thing. "The symptom is conceived as a real kernel of enjoyment, which persists as a surplus and returns through all attempts to domesticate it, to gentrify it (if we may be permitted to use this term adapted to designate strategies to domesticate the slums as 'symptoms' of our cities), to dissolve it by means of explication, of putting-into-words its meaning" (Zizek, 1989: 69). The zone as place and as category resisted the gentrification or reduction of the extra X that disturbed the policy makers and seemed to be the matter of what was the matter. The Thing for us shifted a series of Haitian boats, from nightscreams and the Rose-Marie Express that initiated our sense of obligation, to the traders God-Is-Able and M/V Deliverance Express that produced the feeling of recognition; from the smuggling of aliens to the market in used goods. Used. Goods. Egents listen with the third ear and hear "commodity sign."

PHOTOGRAPH: BARBARA JO REVELLE

There is a measure already in the hole: money. It is better than Nothing. Zizek's analogy suggests that no amount of empirical solutions can "fix" the zone, since the zone is a relationship, is extimate. The "cause" of zone problems, that is, is "non-local." The switch point, the swivel, the twist in the moebius band, the "lip" of the Klein Bottle that may appear is the Mattress (in this reading, attuning this situation). Here was the zone's preliminary response to the querent. The zone says you are already fixed: in a fix. The fix is in.

Look for the sublime (impossible) dimension of a public policy issue.

The warped mattress focused attention on the Haitian trading vessels affected by the Caribbean Code safety policy. The consultation looks for the sublime dimension of any public policy problem (such as the policies of the Caribbean Code or the Miami River Planning Commission). The querent treats the problem data as a parable in order to notice the dynamic topological warp governing the movement of the zone through time: the propensity of das Ding. The point of entry is that order of "feeling" associated with the family of "recognition." "The logic at work in the experience of the dynamical sublime is therefore as follows: true, I may be powerless in face of the raging forces of nature, a tiny particle of dust thrown around by wind and sea, yet all this fury of nature pales in comparison with the absolute pressure exerted on me by the superego, which humiliates me and compels me to act contrary to my fundamental interests!" (Zizek, 1994: 51).

The scene of "Crossroads" is promoted in our report to this same status (a scene become iconic of a theory – a picture theory). One purpose of the testimonial is to register this experience of the sublime, to make it legible as a force – to map the outline of the ethical obligation felt within by means of the disaster investigated outside. In electrate metaphysics, the accident is essential. The Miami River in me. The measure of my attunement, how things are with me (in us) is to be inferred from the TROUbles of the river. *Weltinnenraum.* Objective correlative. Dialectical image. Transference. Prudence.

The attunement happens only through an act of recognition in which the consultant takes ownership for the external scene that reveals in a flash that she has "been there before." It is the effect of the middle voice, of reflexive agency. The theoria report takes the form of a metaphor, or metonym (is figural). The poetic method of research – the tuning – depends upon such events. It is not that the correlations are already there, as in pre-modern epistemology. The attunement brings them into alignment and legibility. The rewriting of the sublime formula has to be modified to accommodate the traumatic nature of the mediated disaster. The river does not "express" something I already felt, but reveals a correspondence between subject and object (micro- and macrocosm). "True, I may be powerless to do anything about the condition of these Haitians, yet their difficulties figure and reveal to me the pressure of the obligation that otherwise I am unable to feel."

The haitians are Nemesis. They implicate me in an ecology to say that the world is thus because I am that. Make a parable joining the personal domain of the burning question with the policy problem of the zone, in order to hear what the disaster knows.

The Mattress opened the field of Haitian shipping and led to the story of the impounded traders, caught in the "safety net" of the Caribbean Code. The Coast Guard boat is a rhetorical vehicle patrolling the borders of identity, expressing a default public mood, a public that had grown tired of rescuing substandard entities. The connection came through Revelle's tuning, seeming all too obvious in retrospect. Prompted by the Mattress, she recognized (in an illumination) the relevance of Ron's biography to the inquiry – *that at one time he served in the Coast Guard.* This coincidence constituted a "lucky find." The immanent real neither conceals nor reveals, but intimates. Revelle recognized in the full sense of the "sublime object" that her story with Ron, or the feeling that she got from being an actor in this story, found its voice, its mood, in the situation of the haitians. A mapping among the levels (Literal-Moral-Allegory) turned up a block (in Lyotard's sense) of correspondences, including her identification as a fifty-something woman with the old ships no longer considered to be economically viable (The "Barbara-Jo Express"). In answer to her burning question ("what is my love relationship?") the zone showed her *an impounded boat*. The mystory functions as a map to link this scene with other images, to dialectize or put the subject in motion through the popcycle. As in any divination, the action that follows from the answer is open to interpretation by the querent. The key point for category formation is that the classifying feature is not "in" the image. Rather, as in the mnemonic practice of meditation, the image is a cue, a reminder whose "auratic" power is due to the network of associations that it calls forth.

Impoundment

The consultation produced a proportional analogy: Revelle's relationship with Ron is like the relationship of the haitians with the Coast Guard. Here is the attunement. *How do things stand with me?* I am (we are) *impounded*. We are discovering the correlations operating in contemporary cosmology (microcosm to macrocosm). At the anagogical level it was easy to see that the impounded circumstances of the haitians connected with the psychoanalytic impasse that marks the interruption of the unconscious in the smooth running of daily business. At the level of experience, "repression" happens as "impasse." The problem or TROUble isolated by the policy decision (the Safety operation) was the visible dimension of trauma that extends into several other dimensions that resist both representation and repair. Specifically, Lacan's maxim that "there is no sexual relation" (however good the sex), suggests that the impasse is "irreparable." As Agamben said about the irreparable condition of life, the world (the catastrophe) is the same whether one's state of mind is sad or happy. Uncannily, the image emerging out of the theoria resonated with the images of spleen inherited from Paris. We had come upon a scene, a screen, of repression, characterized in the theory as an experience of hopeless deadlock. A pattern arose (the point is worth repeating) linking Miami with Baudelaire's Paris, Joyce's Dublin, Eliot's London: *paralyzed cities*. Are we still so modern? It is not difficult to understand why "ataraxy" is the default attitude of wisdom in nearly every tradition in the world. It is not adequate or justifiable in a dromosphere.

Moments of "objective chance" that give rise to the uncanny acquire their power because they expose not just a repressed desire, but an irreducible drive. The egent is also the hole, and so must function as measure, which is what sinthome is for. The impasse of the subject's condition is evoked in the very use of knot topologies in Lacan's theory. We are bound and tied up in knots. The "subject" is not a "thing" but a trajectory, a path, a propensity, movement along a chain in which one link, or one signifier, is missing (expelled to the outside). Such is the inner path we follow with our geographical movements. An apparatus aligns physical, mental, and institutional formations, willy-nilly, which doesn't mean that metaphysics happens by itself. During transference, a patient in analysis runs up against a "master signifier," which "covers the subject's lack." This signifier is a placeholder of the "letter" (the sinthome is the set of such letters). The name of Revelle's childhood game recurs in a commentator's description of the symptom/sinthome. "These metonymic ruins mark the place of an absence the way extraordinarily strong blasts of energy radiate out from the spot where a black hole is swallowing up matter and energy from the universe. In the case of symptoms, including those produced by the transference-neurosis, the black hole is that of the unique being of the subject" (Chaitin, 199).

The disaster in me.

PHOTOGRAPH: BARBARA JO REVELLE

Whatever

The Mattress punctum provides the categorial mark of *whatever* in "Crossroads." Giorgio Agamben theorized the quality that Frow found shared in Haiku poetics and tourist snapshots – the condition in which things become signs of themselves. The fundamental fact of the post-alphabetic recording technology is that now not just the sounds of speech, but any sound and any sight may be stored and retrieved, and thus are available for "writing." Agamben's *The Coming Community* takes up the question where Barthes' *Camera Lucida* left off in its efforts to pinpoint the quality of signification native to the new media. To theorize this quality Agamben revived a category from medieval philosophy that became obsolete with the fall into irrelevance of most of Scholasticism. The category is "whatever"–a translation of "quodlibet." He remotivated the concept in order to identify the peculiar position of the "I" or first person subject in language. This position is transcendental and is not experienced, and yet it is fundamental to how an immanent world manages death. In our context Whatever is the quality by which an obtuse meaning is generalized without resorting to conceptualization. The egent splits between body and image to accomplish choragraphy. In electracy the transcendental "I" comes into appearance as *avatar*.

Agamben's procedure is to approach Whatever asymptotically by gathering a constellation of terms and notions from which the meaning of Whatever may be inferred. "Quodlibet" as a musical form, in any case, is a kind of bricolage that evokes (from phenomenology) a figure that explains how the w/hole is grasped in mood – the way "melody" is recognized in a sequence of notes. One of the most straightforward entries is "example." "Neither particular

PHOTOGRAPH: BARBARA JO REVELLE

nor universal, the example is a singular object that presents itself as such, that shows its singularity" (Agamben, 1993: 10). Another instance is expressed in the heresy of Amalric of Bena, who was burned at the stake for relating the Apostle's "God is all in all" to "the Platonic doctrine of the 'chora'": "God is in every thing as the place in which every thing is, or rather as the determination and the 'topia' of every entity. The transcendent, therefore, is not a supreme entity above all things; rather, the pure transcendent is the taking-place of every thing. God or the good or the place does not take place, but is the taking-place of the entities, their innermost exteriority. The being-worm of the worm, the being-stone of the stone, is divine. That the world is, that something can appear and have a face, that there is exteriority and non-latency as the determination and the limit of every thing: this is the good" (15).

Our digital technology now records not just the words of a language, but the embodied world, putting into an image the gestures and sounds of these bodies – animal, vegetable, mineral: a manner of being. The present institutionaliization of this machine is capitalist, which brings electracy into existence in the mode of commodification. Yet there is nothing inherently capitalistic about electracy, anymore than there was anything inherently religious about linear perspective, even if the latter evolved in the service of Church propaganda. Like Barthes noticing the obtuse in a photograph, Agamben noticed Whatever, which is specifically a photographic effect: "The task of the portrait is grasping a unicity, but to grasp a whateverness one needs a photographic lens" (49). The equipment makes possible "special effects": "Each dancer was filmed separately and later the single pieces were brought together over a single sound track. But that facile trick, that calculated asymmetry of the movement of long legs sheathed in the same inexpensive commodity, that slight disjunction between the gestures, wafted over the audience a promise of happiness unequivocally related to the human body" (47). In this ad the best and the worst coexist. Commodification mediates between the concept and chora. "Thus the glorious body of advertising has become the mask behind which the fragile, slight human body continues its precarious existence, and the geometrical splendor of the 'girls' covers over the long lines of the naked, anonymous bodies led to their death in the 'Lagers' (camps), or the thousands of corpses mangled in the daily slaughter on the highways. To appropriate the historic transformations of human nature that capitalism wants to limit to the spectacle, to link together image and body in a space where they can no longer be separated, and thus to forge the whatever body, whose "physis" is resemblance – this is the good that humanity must learn how to wrest from commodities in their decline" (50).

This promise of happiness must be wrested from the commodity, with the proviso that some other measure is put in its place. Here is what baffles the planners in the mode of sufficient reason: objectity. *Clear off, clean up all those mattresses! But the Mattress Thing may not be expunged from the group subject (the unconscious). The sinthome (in its singular well-being) replaces money as world measure.*

CHAPTER 12

Invention

What may be learned from a used mattress? The attunement consultation categorizes Miami/Myami as (still) a modern city, The Dark City topos of Blake's London, Baudelaire's Paris, Joyce's Dublin, stuck in the default moods of entrapment, paralysis, alienation – impounded. The advice, based on the capacities of electracy as an apparatus, is to design a new mood, or, as the Situationists said: construct a situation. The policy issue of the Caribbean Code, as the anchoring sample, manifests the inertia of the obsolete Frontier myth. We have already seen the poetics of this myth, making available the mythography that produced it. The point of intervention for the EmerAgency is to compose an alternative myth, or at least to propose the possibility for such a composition. The immediate task is to review the inventions already in place, the defaults that are available and currently guiding the scenario of (in our case) an American state-of-mind. Decisions made in Real Time improvise on default schemas. An intimation from the Real about how Real Time decision works is the scene of George W. Bush, participating in a reading lesson at Booker Elementary School, when he was informed of the attacks on the Twin Towers, September 11, 2001. Michael Moore used this footage to criticize Bush, but the lesson of the scene (Bush reading along with "The Pet Goat") in our context is that he did not defer his decision. The decision (the event of reception that Heidegger named *Ereignis*) was made already, during his childhood, as it is for all of us who say "I," who are capable of occuping the transcendental register of a civilization.

The Cowboy

The prototype of iconic inventions for our purposes is the "cowboy," invented by the ensemble of Theodore Roosevelt, Frederick Remington, and Owen Wister.

> Simultaneous with Turner's promotion of the frontier thesis, three prominent easterners were doing their part to bring the myth of the frontier to popular attention: politician Theodore Roosevelt, artist Frederic Remington, and writer Owen Wister. Born into prominent families in the era of the Civil War, educated at Harvard or Yale, each of these men went west seeking personal regeneration at a critical point in their early twenties. Historian G. Edward White persuasively argues that these sojourns

 | DOI:10.1163/9789004534643_013

> convinced each man that only by coming to grips with the experience of westering – with the myth of the frontier – could Americans preserve important aspects of their culture being swept away by the rush of industrialization. Most important, they sought to encourage a rugged version of American manhood. Their heroes were all "men with the bark on." (Hine and Faragher, 495)

Roosevelt asked Remington to illustrate his book, *Ranch Life and the Hunting Trail.* Roosevelt and Wister were classmates at Harvard. "They saw their class as the trustee of traditional American values imperiled by the struggle between the proletariat below and the plutocracy above. After meeting Remington, Owen Wister wrote that he was an 'excellent American' who shared the belief 'that this continent does not hold a nation any longer, but is merely a strip of land on which a crowd is struggling for money'" (Slotkin, 95). The ideological stance of these WASP paragons regarding their Social Darwinist views on race and class were worked out in Wister's *Virginian* in terms of gender relations (setting a pattern for the genre).

> Wister uses their relationship to articulate a gendered allegory of politics in which the Virginian's virile "realism" is opposed by Molly's inconsistent mixture of genteel class snobbery and philanthropic-sentimental "egalitarianism" – which Wister takes to be the ideology of an emasculated and intellectually exhausted American upper class. She initially sees herself as superior to the Virginian by birth and culture. Hence, as a would-be lover he is supposed to look but not touch (or hope to marry), and as an intellectual companion to be her pupil rather than her teacher. But her sense of class superiority and sexual reserve are 'unmanned' by the cowboy's overwhelming masculine sexual appeal. The demands of Darwinian nature, which urge her to breed with the most virile male of her race, conflict with the values of class. Wister of course values both and his game is therefore to show that class privilege is as "natural" a phenomenon as sexuality; that the difference in the gifts and powers belonging to the senses is a paradigm for distinctions between different orders of men; and that biological urge that drives Molly to subject herself to the virile mastery of the Virginian is therefore her best guide to both thought and action. (177)

The crisis of industrialization and the rise of the modern city populated by the dangerous classes were a threat to the virility of the "Teutonic" WASP male. During his presidency Roosevelt formed the Immigration Commission to reform the policy of unrestricted immigration. On the Commission was John

R. Commons, whose report on "Races and Immigration in America" reflected the ideas found in Roosevelt's *Winning of the West.* His views summarize the virtues valued during the so-called Progressive era, and codified in the Frontier myth. "The racial gifts of Americans are those of organization, executive ability, and above all 'manliness, which the Romans called virility, and which at bottom is dignified self-respect, self-control, and that self-assertion and jealousy of encroachment which marks those who, knowing their rights, dare maintain them'" (190).

Salome (The Fatal Woman)

About the same time that the cowboy icon was being invented, the "fatal woman" who would ultimately represent the greatest challenge to the tradition of virility found a new representation in the figure of Salome.

PHOTOGRAPH: BARBARA JO REVELLE

The femme fatale emerges as a central figure in the nineteenth century, in the texts of writers such as Theophile Gautier and Charles Baudelaire and painters such as Gustave Moreau and Dante Gabriel Rossetti. If, as Christine Buci-Glucksmann points out, the archaeology of modernity is "haunted by the feminine," the femme fatale is one of its most persistent

> incarnations. She is associated with the styles of Decadence, Symbolism, and Art Nouveau as well as with the attention to decoration and excessive detail linked to a persistent and popular Orientalism (in the constant return, for instance, to the figures of Salome and Cleopatra). Her appearance marks the confluence of modernity, urbanization, Freudian psychoanalysis and new technologies of production and reproduction (photography, the cinema) born of the Industrial Revolution. The femme fatale is a clear indication of the extent of the fears and anxieties prompted by shifts in the understanding of sexual differences in the late nineteenth century. As Buci-Glucksmann argues, this is the moment when the male seems to lose access to the body, which the woman then comes to overrepresent. (Doane, 1–2)

As Doane points out, the fatal woman is still an invention of men, even if they are European bohemians. The cowboy and the fatal woman begin their careers independently, but eventually meet and interact in the composite genre of film noir. Meanwhile, one of the biggest cultural events of 1876, the year of Custer's massacre, was an art exhibition featuring representations of Salome. Gustave Moreau's painting, "Salome Dancing before Herod," drew more than five hundred thousand visitors at the Salon of 1876 (Buci-Glucksmann, 146). The relevant point in our context is that "Salome," like the cowboy, is an iconic invention, whose existence is connected with the names and dates of her creation.

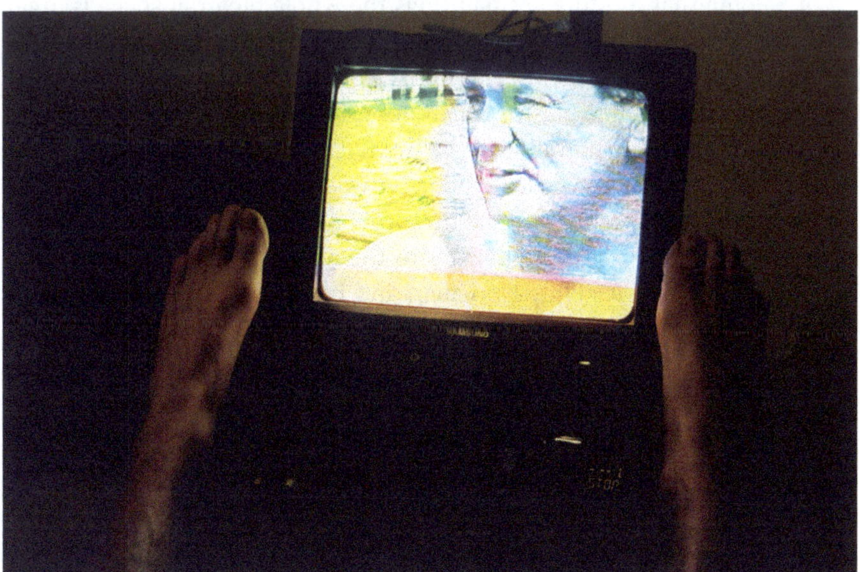

PHOTOGRAPH: BARBARA JO REVELLE

> Salome is among the most exemplary of these fin-de-siecle myths, the one in which the issues and forms of the baroque paradigm offer themselves for interpretation. Its unprecedented fascination is evident from the names of Flaubert, Mallarme, Huysmanns, Wilde, Moreau, Beardsley, Klimt, Strauss – just to mention a few of those who gave it representation in the literature, painting and music of the second half of the nineteenth century and the early part of the twentieth. Already a legendary history steeped in the Gospels had given rise to an exceptionally rich and charged iconography: in the sixteenth century alone, the figure of Salome bearing John the Baptist's severed head on a silver platter appears with matchless allegorical power in, among many others, the work of Jacob Cornelisz, Josef Heinz, Bartolommeo Veneto, Bernardino Luini and Titian. But her decisive importance comes with Oscar Wilde's play (1890), Gustave Moreau's paintings, and Richard Strauss's opera (1905) – which together "invented" Salome. (145)

As was the case with the Frontier myth, in which the cowboy crystallized into an emblematic scenario of the showdown, the fatal woman myth found its emblematic scene of the dance of the seven veils exchanged for the head of a saint.

> The gestural – which includes dance – is so much the figurative element of Salome that in Moreau's paintings the whole pictorial space is organized by her outstretched hand, resolute, frightened, vengeful, clutching a lotus flower that is the metaphor of sexuality. It is the gesture of desire become plastic force, desire which designates its own "part object": the auratic, radiant, bleeding head of the dead Jokanaan. The gesture is a kind of impossible pictorial hyphen between two bodies: Salome's body of seeming appearance and the mystical body, the apparition, of John.... In taking as the object of her desire the absolute Other of sainthood, Salome presents a highly Christian, and hardly Jewish, relationship to the body. For the Christian religion simultaneously requires that there should be a body (the body of Christ crucified, the body of the Church, the body of the sacramental host) and that there should not be a body – or that it should be cast off or doomed. On the one hand, the prohibition expressed in saintliness robs desire of any body; on the other hand, the flesh is guilty in itself. As Michel de Certeau writes: "The 'corpus mysticum' – the *Apparition* – always presents itself as the quest for a body." Indeed, he defines mysticism as "the production of a body," "a cinematography of the body" which refers back to the origin of Christianity in

> "the loss or disappearance of a body." This exacerbated relationship to the present/absent body, which is a feature of all baroque, also characterizes Salome, with one additional nuance. Salome rejects Christian "mourning work": instead, the affirmation of a woman's desire, as burning as it is virginal, supposes and "devours" the body of the Other." (146–147)

In every case, it is a question of the body as a value. The time-image as filmed category is organized by a gesture. The lesson for choragraphy: mythologies are invented.

"Acephale"

The terms of this allegory indicate some of what is at stake in the confrontation between the gunfighter and Salome when eventually they meet on the mean streets of *film noir* nighttown. There are several ways for the consultation to intervene in this situation. One purpose of Miautre as choragraphy is to map the defaults determining the propensity of the zone, in order to discover alternate routes, trajectories, lines of flight, as well as to raise awareness of the defaults triggered in decision. How to unstick the zone, overcome the paralysis of impoundment? Yes "aporia," but of what sort? The impoundment is generalized, not confined to the literal level of the allegory, but is the atmosphere of a scenario guiding collective judgment. The method is to shift attention away from the default gestures, to consider the margins of the scene, to locate opportunities for displacement. It is still not a matter of inventing anything new, but only of exploiting the augmented memory of Miautre to notice further possibilities within the allegory. We need to know more about virility, since that is the focus of the crisis in values, keeping in mind that all the policies in force in the zone manifest a commitment to a regime of values.

The point of departure for this displacement, in search of a line of flight capable of unfreezing the vortex, is the gesture found in the fatal woman's scenario Salome with the decapitated head. This decapitation, from the point of view of electracy, indicates a direction, figuratively at least, as an allusion to the operations of the image apparatus. The camera, that is, is associated with decapitation, taken as a metonym for all those terrible or catastrophic phenomena that overwhelm consciousness of the observer, but for which the camera has an affinity (thus establishing its status as optical unconscious). "The medium has always shown a predilection for events of this type. There is practically no newsreel that would not indulge in the ravages of an inundation, a hurricane, an airplane crash, or whatever catastrophe happens to be at hand.

The same applies to feature films. One of the first film strips ever made was 'The Execution of Mary Queen of Scots' (1895); the executioner cuts off her head and then holds it in his uplifted hand so that no spectator can possibly avoid looking at the frightful exhibit" (Kracauer, 1960: 57).

The camera itself functions as a kind of "guillotine" of time, "The ability to withdraw the appearance of the moment and preserve it from time's deliquescence looms large in the history of our fascination with the camera. Barthes suggests: 'looking at a photograph I inevitably include in my scrutiny the thought of that instant however brief, in which a real thing happened to be motionless in front of the eye.' It seems to matter little that this 'decisive moment,' which Henri Cartier-Bresson fashioned into a complete photo-ontology, is mythical; a techno-magical revelation performed by the guillotine action of the camera's shutter, which can suspend a running man over a pool of water forever awaiting the imprint of his foot on its mirror" (McQuire, 107). This technical guillotine is displayed formally in the close-up shot, the "talking head" now ubiquitous on television. Early cinematographers were reluctant to cut up the body of actors, instead using the camera merely as a way to record theater. Once editing was discovered, however, the close-up revealed a new dimension of experience. "Any huge close-up reveals new and unsuspected formations of matter; skin textures are reminiscent of aerial photographs, eyes turn into lakes or volcanic craters. Such images blow up our environment in a double sense: they enlarge it literally; and in doing so, they blast the prison of conventional reality, opening up expanses which we have explored at best in dreams before" (Kracauer, 48). The point for electracy is that photography makes possible and even necessary a different category system, a different metaphysics, than the abstract conceptual categories of literacy.

> Alexander Rodchenko, one of the first photographers to appreciate the implications of the camera for the notion of a static and eternal truth, asserted: "With photography we can refute the idea that one set [of characteristics] exists for a given subject." Unlike the projects of Galton or Bertillon, Rodchenko's advocacy of portraits assembled from divers "photo-moments" conforms less to statistical attempts to totalize the subject than to a profound questioning of the very notion of identity. Extending this line of thought, I would argue that the issues the camera raises for the social relations of meaning – perspectivism, seriality, the mobility of signs and the proliferation of centers of representation – go beyond simply disappointing certainty of meaning according to a regulated polyvalency which would leave unity of meaning untouched as the ultimate goal. In displacing the center of gravity which previously regulated concepts such

> as intentionality, context and totality, the camera entrains a profound epistemological shift in which the meaning of meaning has itself been irreversibly transformed. (McQuire, 47)

The infinity of accidents that threatened the borders of the concept become writable and thinkable holistically in the photographic image, capable of recording an infinite serial nuance and variation.

Allegorical Headings

The decapitation of the mystic John instigated by Salome's desire allegorically evokes a paradigm shift in the apparatus, indicating a reorganization of the sensorium in relation to knowledge. The old hierarchy established in literate civilization by Plato's philosophy and Judeo-Christian theology – head, heart, viscera – is inverted in electracy. The body, that is, gains new respect as the seat of wisdom. The power of reason acquired by the image in digital equipment includes institutional practices and identity experience. Virility itself becomes acephalic, a version of the decapitation scenario available in the theoretical register of the allegory. Georges Bataille agreed with the inventors of the cowboy that the problem of the modern world centered on the question of virility.

PHOTOGRAPH: BARBARA JO REVELLE

His answer to the crisis perhaps represents the anti-cowboy in his speculations on an acephalic virility, as in the following description of Andre Masson's drawing, "Acephale." "Man has escaped from his head just as the condemned man has escaped from his prison. He has found beyond himself not God, who is the prohibition against crime, but a being who is unaware of prohibition. Beyond what I am, I meet a being who makes me laugh because he is headless; this fills me with dread because he is made of innocence and crime; he holds a steel weapon in his left hand, flames like those of a Sacred Heart in his right. He reunites in the same eruption Birth and Death. He is not a man. He is not a god either. He is not me but he is more than me; his stomach is the labyrinth in which he has lost himself, loses me with him, and in which I discover myself as him, in other words as a monster" (Bataille, 1985: 181),

Bataille introduces a new icon, a new myth, attempting to act upon Nietzsche's call for the revaluation of all values. "The acephalic man mythologically expresses sovereignty committed to destruction and the death of God, and in this the identification with the headless man merges and melds with the identification with the superhuman, which IS entirely 'the death of God'" (199). Acephalic (headless) man – meaning a humanity that renounces all idealist systems of value in science, religion, politics, and personal life – relates to time as an object of ecstasy. "Ecstatic time can only find itself in the vision of things that puerile chance causes brusquely to appear: cadavers, nudity, explosions, spilled blood, abysses, sunbursts, and thunder" (200). Allan Stoekl explains that Bataille's escape from the "head" remains allegorical but is not merely an inversion of Western metaphysics.

PHOTOGRAPH: BARBARA JO REVELLE

> Bataille is not simply privileging a new object (excrement, flies, ruptured eyes, the rotten sun) over the old one (the head, the king, spirit, mind, vision, the sun of reason). If, as Angus Fletcher has pointed out in *Allegory*, the medieval allegorical imagination posits a fundamental congruence between hierarchy in the body and the guaranteed, stable meaning of allegory (in the body, the highest element is the head; in society, the king; and in the universe, God), then we must conclude that a theory that simply substituted one hierarchy for another (a hierarchy that favors the high replaced by one that favors the low) would only inaugurate a new metaphysics and a new stabilized allegorical system of meaning. Filth would replace God.
>
> But Bataille's approach is not that simple. Fully conversant as a medievalist with the theoretical implications of allegory, Bataille precisely recognizes that the fall of the elevated and noble threatens the coherent theory of allegory itself. This is not to imply that allegory is simply done away with in Bataille, but rather that what Bataille works out is a kind of headless allegory, in which the process of signification and reference associated with allegory continues, but leads to the terminal subversion of the pseudostable references that had made allegory and its hierarchies seem possible.... The filth does not "replace" God; there is no new system of values, no new hierarchy. (Stoekl, in Bataille, 1985: xiii)

This shift in status of the body in the hierarchy of values is relevant to chora, but Bataille's mythography finally goes too far, is too "original" or avant-garde for the purposes of the EmerAgency. It is included in the default database, however, to confirm that "virility" is this issue. His experiments with a "mythical anthropology," for example, introducing a "lawless" phantasm into science, is unconcerned with plausibility, as in his description of the "Pineal Eye." "The eye, at the summit of the skull, opening on the incandescent sun in order to contemplate it in a sinister solitude, is not a product of the understanding but is instead an immediate existence; it opens and blinds itself like a conflagration, or like a fear that eats the being, or more exactly, the head. and thus it plays the role of a fire in a house" (82).

Detecting Transitions

We are following the track of a certain mythography – the inventions that converged to form the current version of the Frontier myth – the scenario organizing policy and entertainment narratives in America. The final cultural invention to account for is the detective protagonist. All the types involved in the scenario are part of the response to the crisis of industrialization that gave rise to the modern city. The detective is no exception, obviously, and was just one of the ways the culture attempted to cope with the shock, the lack of intelligibility, of the city. A relevant point in our context – the problem of the virtual city – is that the visibility and intelligibility of the city assumed by Virilio was an effect of specific practices and constructions.

> By the nineteenth century, the city had become a new problem: a problem of visibility. Too big to encompass in a single glance, populated by illegible strangers rather than familiar neighbors, the city threatened to exceed the capacity to envisage, represent or – key point – govern it. To get around or behind this opacity required a new way of relating to the city: investigation. Hence all those administrators, reformers and journalists braving the nightmare depths of the city to shed light on its terrible secrets....
>
> The cities of nineteenth-century fictional narrative share this will to visibility. Pedagogic as much as representational, they teach ways of seeing. Walter Benjamin noted the historical link between new techniques for fixing identity, and so creating individual records as a feature of urban surveillance, and the enigmatic city of the detective story. "The detective story came into being when this most decisive of all conquests of a person's incognito had been accomplished. Since then the end of efforts to capture a man in his speech and actions has not been in sight." What was essential in such stories was less the crime than the urban staging: "the pursuer, the crowd, and an unknown man who arranges his walk through London in such a way that he always remains in the middle of the crowd."
>
> The pedagogic reaction is then to wrestle it into a narrative. Read the clues, however trifling, the novels teach us; make the links between them, however obscure. There are still subterranean networks and secret social connections at work which make the city ultimately systematic – and so narratable, and so thinkable.... Holmes prefigures a popular modernism in which T.S. Eliot's "unreal city" is recycled as the blank urban space of twentieth-century detection, even if that space is filled out with distinctly pre-modern forms of subjectivity. Here, in all its boozily sentimental machismo, is Raymond Chandler on the private eye as hero: "Down these mean streets a man must go who is not himself mean, who is neither tarnished nor afraid. The detective in this kind of story must be such a man. He is the hero, he is everything. He must be a complete man and a common man and yet an unusual man. He must be, to use a rather weathered phrase, a man of honor, by instinct, by inevitability, without the thought of it, and certainly without saying" (Chandler)." (Donald, 2000: 57)

The detective hero is the vehicle for a transition from literacy to electracy. Walter Ong observed that the detective is the quintessential protagonist of literate narrative, just as the epic hero (Achilles) was for narrative in the oral apparatus. Semiotic analyses such as Roland Barthes's *S/Z* demonstrated with his five codes that all literate narratives are structured around the hermeneutic solution of a series of enigmas, and hence, in essence, are detective stories. This literate worldview of analytical investigation culminates in the epistemology of cognitive psychology, in which thought as such is framed in terms of problems-solutions: to think is to be a detective. In the shift from puzzle-solving ratiocinative detectives of the Sherlock Holmes type to the Hard-boiled private eye the epistemological virtue at least begins to move away from the abstract objective logical observer toward an immanent participatory immersion in embodied experience.

At the same time, whether performing in the wilderness or the city, the (American) hero remains consistent at the level of ideology. The merger of the Western and detective heroes in the hard-boiled genre reflected a crisis in the value of competitive individualism celebrated in the Frontier scenario. "One can no longer gain nobility and mobility by pitting oneself against a hostile wilderness. As a result, the focus is shifted to the conflict between persons and this struggle is often brutal, cruel and stealthily dishonest. Success is problematical; one cannot win a loyal wife, wealth is suspect, and the participants are tarnished by the combat. However, even at its bleakest, the hard-boiled detective formula does not utterly abandon the individualistic ethic. The same ideological assumptions are defended by it as were supported by the Western; only the focus of the thematic argument has shifted" (Hamilton, 29). As James Naremore observed, Joseph Shaw, the editor of *Black Mask*, the magazine that first published what came to be known as hard-boiled fiction, "sounded like a disciple of Teddy Roosevelt: "The ideal consumer of *Black Mask*," he wrote, "is vigorous-minded, hard, in a square man's hardness; hating unfairness, trickery, injustice, cowardly underhandedness; standing for a square deal and a fair show in little or big things" (Naremore, 50). One of these readers, it is worth noting, was Ludwig Wittgenstein.

Franco-American Syncretism

The adaptation of hard-boiled narratives to film throughout the 1940s and early 1950s, and their reception in France in the post-war years, resulted in the syncretic invention known as *film noir*.

> *Film noir* then comes to occupy a bizarre position [in France]: it can be the rallying cry of those who do not want to be caught up in the Communist

> Party and the moralism with which it was associated after the war, as it can be the love-object of those who want to hate the United States but love its cinema. By affirming that *film noir* is a harsh critique of American capitalism, by explaining the importance of European directors and cinematographers, by emphasizing the membership of Dashiell Hammett, presented as the father of the hard-boiled novel and thus of the genre, in the American Communist Party, film critics (that is, those cinephiles to whom writing is necessary) gave themselves the means to justify a love that was forbidden – whether it was the war (the impossibility of seeing the films), the Communist Party (whether one was a member or an opponent of it), or the supporters of a morality of hardship (often the same), if not all three, that forbade it. Finally, the invention of a new genre, at a few years' distance from the new wave, is one way among others for a generation coming onto the labor market, forgetful or ignorant of pre-war production, to force recognition of its own signature. *Film noir* thus finds itself to be literally (but also in all sense of the term) a critical object: invented by French criticism, it allows one to love the United States while criticizing it, or more exactly to criticize it in order to be able to love it. (Vernet, 6)

The ideal noir hero admired by the French was the opposite of John Wayne, Naremore points out, with Humphrey Bogart being the perfect embodiment of the new type – "psychologically he is passive, masochistic, morbidly curious; physically he is 'often mature, almost old, not very handsome'" (Naremore, 20)

> The discourse on American film noir was initiated by two generations of Parisian intellectuals, most of whom declared the form extinct soon after they invented it. Many of the films they discussed had been directed by European emigres (mainly Germans), who used tough, Hemingwayesque dialogue and American production values to bestow a kind of glamour upon the dark emotional moods favored by Continental artists of the postwar decade. For the French especially, an American star like Bogart epitomized these moods. Bogart's persona was tough, introspective, emotionally repressed, and fond of whiskey and cigarettes; within certain limits, he suggested a liberal intellectual, and he was sometimes cast in the role of a writer or director. Hence the Bogart thriller became a mirror in which European cineastes could see their own faces. (27)

The men who formed the Situationist International where among those French intellectuals who responded so favorably to the mood of the American films. Part of this French reception included, however, an inversion of the

PHOTOGRAPH: BARBARA JO REVELLE

American attitude or judgment of the Dark City. That part of the city –nighttown – disapproved of by the official bourgeois establishment, associated with crime, vice, poverty, and the dangerous classes – is precisely the city with which the Situationists romantically identified. This changing attitude to the Other was initiated in the hard-boiled story with the private eye operating between the opposing factions on the ideological frontier, thus merging the features of outlaw and detective. "In the myth/ideological landscape of the industrial metropolis he thus represents a 'third force,' identified neither with the propertied and managerial classes nor with the 'dangerous classes' and radicalized labor organizations (socialists, anarchists). In Chandler's novels, and in the best work of his successors, the hero maintains this position of precarious and delicate moral/social balance" (Slotkin, 219).

Naked City

The icon of the Situationists, their most famous "map," refers by its title to one of the most influential of the American films, setting the tone for the documentary-noir style. The Situationist "Naked City" in the mystory Theory discourse aligns with "Miami Vice" in the Entertainment register of Miautre as chora.

> Situationist maps accordingly declared an intimacy with the city alien to the average street map. The narration to the opening sequence of the 1948

PHOTOGRAPH: BARBARA JO REVELLE

> film *The Naked City*, an aerial view of lower Manhattan tracking northward, must have summed up the mood, the penetrating realism, and the sense of humanity that Debord and Jorn sought in maps: "As you can see, we're flying over an island, a city, a particular city, and this is the story of a number of its people, and the story, also, of the city itself. It was not photographed in a studio. Quite the contrary… the actors played out their roles on the streets, in the apartment houses… This is the city as it is, hot summer pavements, the children at play, the buildings in their naked stone, the people without makeup." Rather than float above the city as some sort of omnipotent, instantaneous, disembodied, all-possessing eye, situationist cartography admitted that its overview of the city was reconstructed in the imagination, piecing together an experience of space that was actually terrestrial, fragmented, subjective, temporal, and cultural. (Sadler, 82)

Directed by Jules Dassin and produced by Mark Hellinger, *The Naked City* was included in the body of work on which Raymond Borde and Etienne Chaumeton (writing in 1955) based their *Panorama du Film Noir Americain*, the book that defined *noir* style. Borde and Chaumeton considered the films as what in our context might be called an attunement specific to the era of the 1950s. "It exists in response to a certain mood at large in this particular time and place. Accordingly one who seeks the rot of this 'style' must think in terms of an affected and possibly ephemeral reaction to a moment in history. This is what links productions as diverse as The Shanghai Gesture and The Asphalt Jungle" (Borde and Chaumeton, 19).

> In *The Window*, director Tetzlaff and screenwriter Dinelli, whose pacing is expert, powerfully recreate the claustral atmosphere and apprehensive mood of wartime film noir, and of Woolrich's text, using actual New York locations as expressionistically as if they were pieces of scenery on a sound stage. The iconography of the city in early film noir was a brilliant minimalist response to the restrictions of wartime production. Hirsch describes these studio-bound cityscapes as "simplified and semi-abstract, deliberately lacking the fullness and density of real life," but "pregnant with meaning – an evocative background for dramas of entrapment." After the war, when camera crews were able to move outside the sound stages and venture onto the streets of New York and other cities, in the approved post-war manner, the controlled studio images of urban desolation and abandonment were soon supplanted by the more open documentary style that incorporated the random bustle of pedestrians and vehicular traffic. Jules Dassin's *The Naked City* (1948), the most famous example of pseudo-documentary post-war film noir, epitomizes this historical moment. Tetzlaff was one of the relatively few directors in he late 1940s who used these new conditions of production to intensify, rather than dilute, the thematics of the abandoned city that pervaded wartime film noir. Blake Lucas writes, "Such a world represents the inverse of the American dream of freedom." (Reid and Walker, 85)

The Situationists locate the line of flight moving from literacy to electracy, in their reading of a film such as *The Naked City*, to which they respond not as fans or cinephiles but more in the style of surrealist active viewers. "The surrealists were 'dreaming' cathected details from the cinematic mise-en-scene, but not just any detail caught their eye. They were profoundly attracted to the cinema of the 'social fantastic,' to stories of doomed erotic love, and to thrillers with Sadeian titles. Among their particular favorites were movies about gangsterism and murder, in part because such pictures depicted violent, antisocial behavior, and in part because they bestowed an aura of the marvelous upon urban decor. As Aragon wrote in 1918, American crime films 'speak of daily life and manage to raise to a dramatic level a bank note or which our attention is riveted, a table with a revolver on it, a bottle that on occasion becomes a weapon, a handkerchief that reveals a crime, a typewriter that's the horizon of a desk'" (Naremore, 18).

The European intellectuals, that is, were learning how to see films in a filmic way, ignoring the plot and related literate motivations of narrative and focus entirely on the image and its atmosphere. We are tracing here a development that repeats for the image category the evolution that produced the literate category. Plato is credited with inventing the first concept (justice) by finally stripping away all remnants of dramatic action found in epic and tragic dramatizations of instances of justice and injustice, to focus exclusively on the abstract idea of "justice" as such. He posed the ontological question that marks the beginnings of philosophy: what IS "justice"? In our time the emergence of the categorial image may be observed in the stripping away of conceptual as well as narrative framings to arrive at the mood of the image as such.

Forensic Aesthetic

The camera, as Kracauer noted, has an affinity with details, especially the most commonplace and neglected detritus and refuse of the physical environment. In the apparatus of literacy the attention to detail was motivated by hermeneutic investigation. We are moving in electracy towards a relation to the detail that no longer involves the state of mind of analytical, hermeneutic investigation of problem-solution. The commentators we have been citing made it clear that the noir mood was more profoundly inherent in the experience of the city than the inventors of film noir perhaps recognized. The crime scene featured in thriller narratives was not to be taken only literally, but as an expression of bourgeois fear, apprehension, anxiety, regarding their hold on power. This mood gathers the American and European histories into a set in our map of the Miami chora. The Situationist inventors of psychogeography named the map of the Parisian vortex "The Naked City." Dassin and Hellinger's *The Naked City* paid homage in turn to Weegee's photo book by the same name (1945).

> In 1938 [Weegee] acquired a police radio in his car: "At 12 o'clock I'd get in the car, ride around and if anything happened on the police radio, I was right there. Sometimes, I was right in the same block where the thing happened. I would get there before the cops did." The results were dramatic, fulfilling the logic of photography's obsession with death and evidence by capturing murder victims at the scene of the crime. Weegee used an intense flash that flattened the scene, saturated the blacks and often provided what he called a "Rembrandt light," evoking the rich chiaroscuro of the Dutch painter. Even then the power of the moment never seemed enough for Weegee and he always sought out an ironic twist. In

> his 1940 picture, On the Spot, the title seems self-evident as your attention is grabbed by a body in the foreground, lying on his back with blood running stereotypically into the gutter. The flash obliterated all details of the man's face or clothing, or the faces of the four policemen in the middleground but illuminated the bar sign above – The Spot Bar and Grill.
>
> This cultivated hard-boiled personality came straight out of the film noir with which it was contemporary. Film noir set aside the established realistic conventions of Hollywood cinema in favor of dramatic lighting with deep shadows, unusual camera angles and a jaundiced view of humanity.... Weegee borrowed its visual style for his work to such an extent that it can be called photo noir. He modeled himself on the detective character Nick Carter and was later to work as an actor in a number of Hollywood movies. His book Naked City was adapted into a 1948 Oscar-winning movie by Mark Hellinger, directed by Jules Dassin, that became a film noir classic in its own right. (Mirzoeff, 80)

Scene of the Crime

Weegee was known as the "American Atget." Atget photographed the streets of Paris in a way that made the whole city evoke the aura of a crime scene. "The surrealists came across Atget's work by chance. He happened to live on the same street as Man Ray. In 1928 Albert Valentin wrote an article about Atget's photographs for *Varietes*, noting that 'seen at closer range, these dead ends on the city outskirts registered by his lens constitute the natural theater of violent crime, of melodrama.' Valentin wrote this after seeing the collection of Atget photographs that Berenice Abbott, a friend of Man Ray's, had purchased after the photographer's death in 1927. Presumably this was the article that Benjamin had in mind when he first wrote that 'not for nothing were pictures by Atget compared to those of the scene of a crime' in a review of Abbott's 1930 monograph on Atget. What struck both Valentin and Benjamin was the sensation of the uncanny (Valentin's word) given by Atget's attachment to 'extremely strange sites where seemingly nothing aroused any interest,' which drew us nonetheless into a 'beyond' that existed 'in the margins, in the filigree, in the mind'" (Wollen, 30). The difference between Weegee and Atget, of course, is that the places Atget photographed were empty, devoid of people let along corpses. The mood of crime was somewhere between the street and the onlooker.

Benjamin, for his part, considered the bourgeois apartment as the ideal setting of the detective story, as a place that demanded to be the scene of a crime. He explains why an entirely empty location, whether street or apartment, has the feeling of a crime scene.

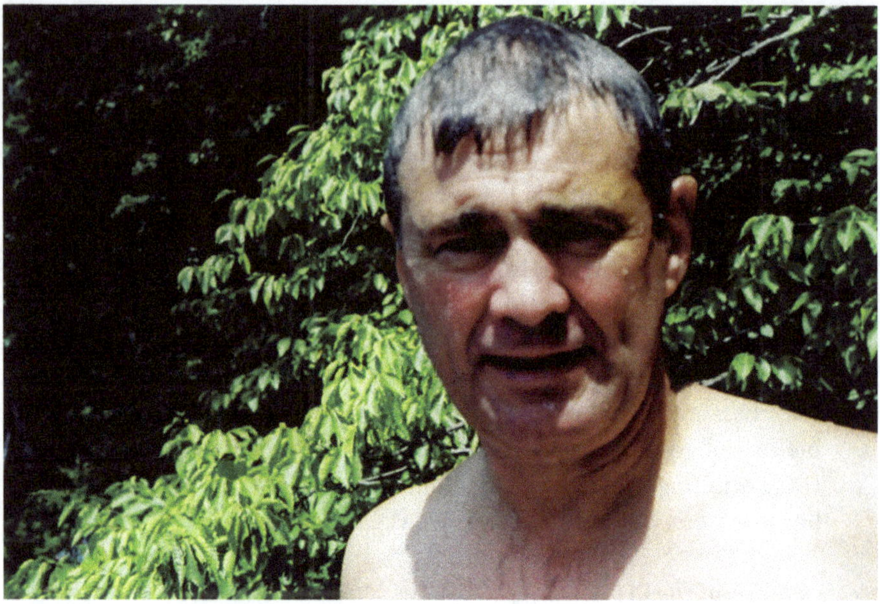

PHOTOGRAPH: BARBARA JO REVELLE

The bourgeois interior of the late nineteenth century is, on the one hand, the quintessential site of the dying, yet, on the other, the setting from which the dying are unceremoniously removed. It is the scene of the murder and the appropriate home for the corpse, but the body itself is absent. This paradox is intelligible as follows. While the interior is for Benjamin the domain of decay and demise, any association with the natural history of the human body, is gradual disintegration and eventual death is denied. The interior becomes "ageless," the sense of "bourgeois security that emanated" from the middle-class home stemming from "timelessness," from the denial of transience. The space of death, the murder, simultaneously becomes that of immortality, of permanence.

Benjamin's analysis of the interior highlights the bourgeois fear and detestation of the physical and corporeal the erotic is hidden in the "closets and crannies" of the middle-class apartment, while the dying are removed and confined tot he hospital.... Death is cast out of the bourgeois interior, and sexuality is consigned to its miserable recesses. The vitality of the living and the integrity of the dead are simultaneously denied in this space. As a result, it becomes lifeless, sterile and inhuman. Although the site of the 'murder scene,' the interior is not so much a space of death as itself "dead space." It is a site of imprisonment with its own incarcerated subject: the child. (Gilloch, 82)

These commentaries place the Miami mattress and the attunement of impoundment in the context of intellectual history (the default mood). The Haitian boats had to be excluded from the Miami River for the reasons that Benjamin itemized: their alleged decrepitude threatened the bourgeois worldview – its denial of time (ruin). Benjamin found the ultimate manifestation of this bourgeois denial of death in the Egyptian obelisk placed at the center of the Place de la Concorde in Paris. The very movement of this monument from its original site in Egypt to France contradicted the intention of the French revolutionaries. "The meaning of the monument changes in the course of its own history. Monuments to victory are inevitably transformed in time into those of defeat. As an object falls into a state of ruination, the pretensions which accompanied its construction crumble, and its truth is unfolded" (73). Georges Bataille connected a similar insight with the role of the Guillotine in that history (the site of the execution of the King of France).

PHOTOGRAPH: BARBARA JO REVELLE

> A moving perspective, represented by the shadows and traces of the successive generations of numberless dead, extend from the banks of the Nile to those of the Seine, from the angles of the pyramids to those of the monolith erected before the Gabriel palace. The long span that stretches from the Ancient Empire of Egypt to the bourgeois monarchy of the Orleans – which raised the obelisk on the Place de la Concorde "to the applause of the

> immense crowd" – was necessary for man to set the most stable limits on the deleterious movement of time. The mocking universe was slowly given over to the severe eternity of its almighty Father, guarantor of profound stability. The slow and obscure movements of history took place here at the heart and not at the periphery of being, and they represent the long and inexpiable struggle of God against time, the combat of 'established sovereignty' against the destructive and creative madness of things. Thus history endlessly repeats the immutable stone's response to the heraclitean world of rivers and flames. (Bataille, 1985: 216)

The obelisk replaces the guillotine at the center of the *Place de la Concorde*, at the center of the city of Paris, generalizable, Bataille says, to the public square (the civic sphere) of every modern city. "The *Place de la Concorde* is the space where the death of God must be announced and shouted precisely because the obelisk is its calmest negation. As far as the eye can see, a moving and empty human dust gravitates around it. But nothing answers so accurately the apparently disordered aspirations of this crowd as the measured and tranquil spaces commanded by its geometric simplicity" (215). This vortical movement around a central void, an auratic place where "something happened," is the feeling of the forensic aesthetic, calling attention to the crisis of the body in our time.

> Of equal weight is the forensic's emphasis on physical traces and the way it calls attention to the body's ghostly presence in an image-driven society. In contrast to much of the corporeal art from the last decade, which often tries to directly represent and "restore" the body, this vein of art functions like the chalk outline placed around the corpse at a crime scene. The body is envisioned neither as an innocent repository of nature nor as an existential symbol of isolation, but as an artifact that leaves traces and in turn is a surface for recording them. In the sublime chaos of Pollock's splattered paint, the forensic aesthetic finds a buried body that has been dissolved and dispersed. At the same time the forensic approach reminds us that our corporeal experience, our feelings of pain and pleasure, remains forever outside the domain of symbols. Indeed, this absence is the structural point, the black hole, around which our symbolic systems orbit. (Rugoff, 104)

Revelle's game – Perfume in the Black Hole – links to the Place de la Concorde through this forensic feeling, manifested formally in Miautre through the VRML nodes with their 360 degree pans at regular intervals along the river. That *CSI* (Crime Scene Investigation) is today the most popular series on television is evidence of the collective internalization of this noir worldview. *CSI-Miami* must be added to *Miami Vice* in the Entertainment register of our allegory. The forensic aesthetic as the last stage of literate reason prepared the "thing" for mutation into a support for the image category. "The document is the site of meaning and analysis," Peter Wollen observes, commenting on an art exhibition devoted to the forensic aesthetic. "How, then, can it be stripped of significance and become blank, frozen outside history? In the first instance this is achieved by suppressing the image of actual human beings" (Wollen, in Rugoff, 28). The last feature to strip away from the object is the literate detective frame of mind itself, the common sense in which life is a game of "Clue."

The Mattress Thing

The mattresses circulating through the river zone are transitional objects in more ways than one. With the sublime mattress, in its objectity, the default melancholy of the forensic aesthetic enframing the "thing" in literate epistemology is exposed and begins to dissipate, along with all the other default moods gathered together in the chora. The object supported digitally is capable of a different metaphysics, as the operator of an image category. Zizek alludes to how the sinthome in an artist's *oeuvre* functions in this categorial way when he uses the term "interface" to name such objects. The sinthome works as interface when a master signifier (Father, King, Nation, God, all those positions associated with the head and monumentalized in the obelisk) fails, when the "suture" holding together the ideological belief construct of a society slips, so that it is no longer able to control representation "inside" its borders of that which is external to it (the other). Postmodernism continued the Nietzschean "death of God" to which Bataille alluded – the transvaluation of values – by announcing specifically the irrelevance of the Meta-narratives of modern society. The EmerAgency consultation takes up the question of what these stances mean in practice. Miautre, the prototype, demonstrates the possibility of removing the quilting point of "haitians" – the other suturing "America" as system – and to replace its identity work with a distributed connectivity of interlinked sinthomes.

The cinematic effect of the suture is created by controlling the point of view shot, as in the conventional use of shot/reverse shot, so that the look of the camera is always aligned with and motivated by the look of a subject. In this way the object may never return the look, and the field of the gaze is suppressed. But when the suture fails, that is, when this device of foreclosure is rejected, the event of "interface" emerges. "The uncanny poetic effect of these shots resides in the fact that it appears as if the subject somehow enters his/her own picture – as Lacan put it, not only is the picture in my eye, but I am also in the picture. What, then, happens when the exchange of subjective and objective shots fails to produce the suturing effect? Here enters the function of interface.... Kieslowski [director of *Decalogue*] was the great master of making the spectator perceive this dimension of interface in an ordinary scene – a part of drab reality all of a sudden starts to function as the 'door of perception,' the screen through which another, purely fantasmatic dimension becomes perceptible. What distinguishes Kieslowski is that, in his films, these magic moments of interface are not staged by means of standard gothic elements (apparitions in the fog, magic mirrors), but as part of an ordinary, everyday reality" (Zizek, 2001, 39).

The relevance for our consultation concerns the attunement of the zone as "impounded." The mattress as sinthome or object @, experienced as impasse, marks a site of repression. The feeling of paralysis in psychoanalytic terms is precisely the mood of repression, symptom of the operations of the unconscious. The formal feature in cinematography that Zizek describes as serving to register within a work that which otherwise is irreducibly excluded from it is a virtue of photography as optical unconscious. These operations and effects in electracy pass from being sites of "failure" to positive logical operations. What does it mean to take up Bataille on his challenge to live "headless"? "The human being arrives at the threshold: there he must throw himself headlong into that which has no

PHOTOGRAPH: BARBARA JO REVELLE

foundation and no head" (Bataille, 1985: 222). In ideological terms this means holding together the collective order of identity without a suturing master signifier. The collective does not require a scapegoat – jews, haitians, or any other – to be abjected, since the digital apparatus brings into coherence the heterogeneous accumulation of singular sinthomes as a constellation, an ideogram, a pattern sustainable and intelligible through the interface of the EmerAgency.

The Mattress, in short, liberated from the investigative hermeneutics of the forensic aesthetic, functions as what Deleuze called (after Leibniz) a *vinculum*: "a special hook, a sort of vinculum on which he can hang and sort through his different things" (Deleuze, 1993: 109). "It is a strange linkage, a bracket, a yoke, a knot, a complex relation that comprises variable terms and one constant term" (110). Miautre as a choral category is held together by a used mattress that functions structurally as vinculum. There is no meta-narrative for the set of stories juxtaposed in our allegory. The policies of negotiation that implore partisans to read the other's narrative are literate, not electrate. The w/hole is not structured as a narrative, but the default forensic structure leads us to the pivot, the switch, objectified in the mudered corpse (the desiring body).

The egent seeks the identity of the corpse. "It is yours."
The egent asks who is the victim? "Well-being."

CHAPTER 13

Jazz

There is another cultural, iconic invention to record, appearing along with the cowboy and Salome around the time of the closing of the Frontier. This invention too is associated with the crisis of the city, but from a different point of view. It answers back to the WASP elitists with the voice of the crowd, and provides a relay for a new civic practice. Our point of access to the public sphere is not directly at the point of a specific policy. Or rather, a given policy, such as the Caribbean Safety Code, is treated as an effect of a fundamental value circulating through a zone, embodied in and expressed through a myth, such as the myth of the Frontier. We have traced this myth and its scenario through its evolution up to our own moment, beginning with the Western wilderness setting, moving into the hard-boiled city in which the detective met Salome on the mean streets of the vice district, adapted to film noir and neo-noir, with the "object" or "thing" brought into focus, retaining in the forensic aesthetic the remnants of literate metaphysics, primed for the transition into electracy and the emerging categorial image.

The immediate contribution of this consultation is to modify, revise, the Frontier myth, to release it from its forensic state of mind, by calling attention

PHOTOGRAPH: BARBARA JO REVELLE

 DOI:10.1163/9789004534643_014

to another possibility already present in the scenario. The invention is jazz. The point of transition might be a figure such as Clint Eastwood, who played both gunfighter and hard-boiled detective, and who also is a jazz fan, using jazz soundtracks in his "Dirty Harry" films, as well as directing the biopic *Bird*, dramatizing the life and death of Charley Parker. Jazz in any case is generally recognized to be the "sound of the city" as such, heard in its accelerated rhythms and dissonant unresolved chord progressions. Because of this emblematic status jazz is the soundtrack of choice for the noir mode. One of the more famous episodes of *Miami Vice* featured a bluesy jazz soundtrack.

> Perhaps the greatest seeming difference between *Miami Vice* and film noir lies not in any particular element in the imagery but instead in the soundtrack. *Miami Vice* audio style and its relationship with image are much closer to music videos than they are to film noir. In "Calderon's Demise," the two flashbacks are done without any dialogue whatsoever. In each, a rock song fills the soundtrack and suggests an interpretation of the images. As Crockett and Tubbs travel to the Bahamas, Russ Ballard sings "Voices." This song, with lyrics about looking to the future, accompanies images from past episodes. In "Smuggler's Blues," the entire narrative was suggested by a song written by Glenn Frey. In that instance and in others less literally, the song precedes the image. Images are constructed to "fill in," as it were, the soundtrack. (Butler, 302)

This foregrounding of an MTV style in *Miami Vice* marks a point of transition from literacy into electracy, in which the expository and narrative modes of literacy become subordinated to the musical imaging of electracy. Our proposal for updating the Frontier myth, in other words, as a means to transvalue the values guiding foreign and domestic policy formation, is to promote a jazz worldview from supporting to leading role in the imagined community, especially since the ambition is to construct a global wisdom. The argument is that in fact jazz already is playing in a global state of mind, but, as is also the case with quantum physics, American common sense has not yet internalized and made native these versions of reality. Jazz (unlike "Acephale") fulfills Slotkin's primary criterion as the basis for a new myth, which is that the source of the myth be authentically a part of the community history (to be persuasive a myth must be grounded in the actual experience of a culture and society). Jazz and the cowboy are often mentioned together, as it happens, as the two most original inventions of American culture. In his attempt to isolate what is "American" about America, John Kouwenhoven composed the following list of items manifesting this feeling or quality. "1. The Manhattan skyline; 2. The gridiron town plan; 3. The skyscraper; 4. The Model-T Ford; 5. Jazz; 6. The Constitution; 7. Mark Twain's writing; 8. Whitman's *Leaves of Grass*; 9. Comic strips; 10. Soap operas; 11. Assembly-line production; 12. Chewing gum" (Kouwenhoven, 124).

This list as a kind of "heap" of items already gestures towards an electrate "definition" of a jazz worldview. The important point for Miautre, specifically, however, is the genealogy of jazz as invention, explicitly associated not only with the Black Atlantic and African-Americans in general, but with the history of Haiti in particular. In terms of chora as a category, gathering into a network the default topics and their moods encountered in a zone, the Haitian traders in our allegory unfold into the history of jazz.

Learning from New Orleans

The scene of impoundment or impasse does not call for answers, but for testimonials at this preliminary stage. Our policy issue concerned Haitian traders. In the frame of the EmerAgency, what do we make of this "answer" to Revelle's burning question (how do things stand with us)? At the limits or borders of the solutions supplied "up until now" by the Western tradition with respect to democracy and every other question, our consultation with the River revealed an unappreciated resource. What is this haitian-in-me? We are in any case no longer "merely" jewgreek, but jewgreekcreole. What does this creole dimension give us to work with? The continuing syncretism of civilizations is well underway and has already accomplished much. The Miami River is haunted by a New Orleans precedent. The Haitian guests arriving in Florida, dead or alive, are uncanny harbingers of what we already will have become.

The crisis of the city that motivated the formation of the Frontier myth included an identity dimension, which became condensed and displaced in the scenario into the problem of virility, and the traditional hierarchy valuing mind over body. A jazz remake of the Frontier inverts that value while continuing to address the problematic of virility. Haiti plays a prominent role in the story. "The questions of how Haitian Voodoo came to the continental United States, and the question of why jazz originated in New Orleans, are in fact parts of the same question. Jazz and rock 'n roll would evolve from Voodoo, carrying within them the metaphysical antidote that would aid many a twentieth-century Westerner from both the ravages of the mind-body split codified by Christianism, and the onslaught of technology. The twentieth century would dance as no other had, and, through that dance, secrets would be passed" (Ventura, 1985: 120). What happened in music may continue in consulting and in every other discourse and institution, as "we" become jewgreekcreole. Syllogism may become syllojazz.

The image tuning the zone is that of an impounded Haitian boat, piled high with its cargo of used mattresses. The image works categorially by means of emotion (felt) in the profound sense of atmosphere or mood. A function of this image as part of the deconstruction of literate method is to interrupt and supplement the instrumental mood of "anomaly" and "solution" with an "instrumental" jazz solo (a puncept). Conventional consulting works in a kind of booster mood of progress. Inventional consultants, rather, frame problems in the mood of "the blues" in a jazz idiom. The individual tuning of a problem zone specifies the image within this frame. To open "problem" with its empirical state of mind and forensic atmosphere to the realm of Arts and Letters in this syncretic frame is to integrate the practical politics of public policy formation with the emotional wisdom of the Afro-Caribbean musical and dance traditions (related with the syncretic religious traditions of the Black Atlantic) whose different names in different regions and nations reflect the different nuances of creolization that have occurred historically. Because Haiti freed itself so early from colonial rule, African religion there – Vodou (Voodoo) – survived more completely than in any other New World setting. In every setting, however, the conditions of colonialism set in motion a syncretism among European, African, and Indigenous cultures that has produced a similar conversion of the master to the slave culture that occurred in the previous great syncretic moment in the Western tradition, when the Roman Empire converted to Christianity.

In the United States, the creolization of the Judeo-Christian-Greco-Roman West with the Black Atlantic has developed in the mood of blues. The related mood in Brazil, for example, is the Samba feeling of "saudade"; in Argentina the tango feeling is "mufarse." Each of the moods has its own nuance specific to its national and cultural deployment, and each has been appropriated within the myth of origins of its respective society. An important step in mythography, to move from "America" to a global culture, is to make explicit this shift in our default mood from "forensic" to "blues," in a way that calls attention to the international character of the syncretism. The image apparatus renegotiates all borders. If it is true that Brazil plays soccer in a Samba style, then it should equally be recognized that other nations may play in a jazz style. The EmerAgency goal is to fashion a consulting practice from the lessons of "blutopia," combining the two major impulses of African-American music: "a utopian impulse, evident in the creation of imagined places (Promised Lands), and the impulse to remember, to bear witness, which James Baldwin relates to the particular history of slavery" (Lock, 1999: 3). "Blues" is not feeling "bad," but what you do about bad feelings.

> I have built a cultural narrative about the transformation of Western consciousness through African-American music by focusing on color and

alchemy. In *Shadow Dancing in the USA* (1985), Michael Ventura arrived at many of the same conclusions by constructing a historical narrative about jazz and rock 'n 'roll that focused on a return to the body through the 'loa' or spirit possession of the musician and listener/dancer. The bluesmen are the secular carriers of the Hoodoo or Voodoo religion which is a hybrid of traditional West African religions and Christianity. As such, bluesmen and women tend to spirit and psyche in a manner handed down from Africa to Hoodoo; spirit possession. Where the Northern soul, from shaman to Christian priest, operates dissociatively, leaving the body to travel the spirit world, the African priest, the Hoodoo conjurer, and the bluesmen ask the 'loa' to enter bodies and possess them. It is through this possession that the 'loa' is known and expressed.Blues lyrics have tremendous breadth, within which are two core streams, depression and libido. The depressive quality of the blues is the most recognizable to the majority of people, an example of which is the lament for being "done wrong" by a lover. But the blues is also highly sexual and at times exuberant. There are many "happy" blues about good time fun, including the joys of food and dance. The blues is about the passions of the flesh. Where the pathological Western mind tended to create dirty movies out of this blue, blacks turned these libidinous hurricanes into the huge body of art called the blues.... It is fascinating to note that in addition to the lyric content of

PHOTOGRAPH: BARBARA JO REVELLE

> the blues, its music and performance practices also worked to transform the dissociative consciousness. Harmonically, the blues is based on the tritone interval which is considered the most dissonant of all intervals and thus full of tension. In Renaissance times this interval was known as the "Diabolus en musica" and for a time its use in performance was actually illegal. In the blues almost every chord contains the tritone, even the tonic. The problem of union of the opposites is managed in the blues by permeating the music with the tension of the tritone. The Christian Devil is everywhere, the dream of salvation nowhere. (Diggs)

Mythography

The invention we are documenting within Miautre chora, arising in the same historical circumstances that produced the cowboy and Salome, is the jazz performance. One purpose of the consultation is to test the poetics of myth production abstracted from Slotkin's history of the Frontier myth, not to "complete" this particular myth, but to set the process going again and to democratize it. The experiment is to define the way in which jazz functions as a state of mind guiding not just music but any performance of practical reason (politics and ethics).

Historical Experience

"Myth" in Slotkin's definition is an historical experience promoted to the status of precedent and guide for further action in the future. The foreign policy of President George W. Bush, for example, in the second war with Iraq, was a literal application of Teddy Roosevelt's Western scenario to politics. The argument of American "exceptionalism" is that the unique experience of life in the frontier conditions of the American continent created a special destiny for the United States. The "strenuous life" was identified as that key quality productive of the American character or "virtue" in the Frontier myth. The Jazz myth, or jazz revision of the Frontier myth, represents the voice of the people identified in the original version as the crowd, the dangerous classes. The exceptionalism of the African experience in America is rooted in the emancipation from slavery – another story of survival in the extreme conditions of something more than a merely strenuous life. The clarinetist Sidney Bechet mythologized this event in his autobiography. "'All that waiting, all that time when the song was far-off music, waiting music, suffering music; and all at once it was there, it had arrived. It was joymusic. It was Free Day... Emancipation... That music, it wasn't spirituals or blues or ragtime, but everything all at once, each one.' The

PHOTOGRAPH: BARBARA JO REVELLE

significance of the jazz musical act is here defined by its relation to emancipation" (Gebhardt, 72).

> Bechet thus identifies a "mood about music, a kind of need to be moving.... You just can't keep the music unless you move with it." In this sense, the "birth of jazz" was constituted through the development of musical practices that expressed the ex-slaves' attempts to give meaning to their passage from the slave system to industrial capitalism. That is to say, it was the jazz musician who, through his or her attention to the tempo and the timing of the act, embodied the trauma of the slave event, acting as a cipher for the unconscious realization of black Americans' collective proximity to freedom and, likewise, a critical measure of the distance traveled by the whole of American society from slavery. Bechet links the birth of jazz to a recurrent memorialization of the slave experience that, in turn, reconstitutes the movement from Africa to the New World as the mark of jazz's fundamental originality. The making and remaking of jazz is thus the symbolic means of denying the motions of death and violence, of combating the destructive power of the state and its capitalist mode of production. In other words, under new conditions of industrial production and mass migration, the black musician is increasingly endowed with the cosmic powers of a historical reversal. As a consequence of this reversal, Bechet's

> jazz myth identifies, in black musical practices, a new feeling for life from which freedom would emerge as a principle of expression. (58–59)

Dissemination through Popular Culture

The frontier experience was put into narrative and circulated through the society in literary forms, including journalism, dime novels, literary novels, and ultimately cinema. The institutionalization of jazz began in the dimension of entertainment associated with vice. As commentators such as Ralph Ellison insisted, while it was true that jazz was played in the sporting houses of Storyville, New Orleans' red light district, that was not where the music was learned. Whether it was the more sophisticated "houses" of the city vice district, or the country whiskey dances (site of blues performances), jazz was good time party music meant for dancing. The music addressed the body directly, dance being, as the commentators note, the most democratic, least mediated form of entertainment. The clearest example of the transformation of society in the twentieth century, they note, is in dance (Crease, 70).

> There had always been popular music in America since the inception of some form of theater in the eighteenth century, but jazz was particularly unnerving because it threatened completely to engulf the business, the aesthetic, the purpose the very consciousness of music-making and music listening. With records, the beginnings of radio, and the urbanization of Americans, particularly blacks, jazz may have appeared more threatening and more pervasive than it really was. It was surely indicative of a frightening change, for if the rise of the nineteenth-century sentimental woman's novel could be called the feminization of American culture, the rise of jazz was in a real sense the primitivization of American culture.... Once again, as had become the common cultural dialectic in American life since the middle of the nineteenth century, it was city versus small town, agrarian values versus urban trends. It was just these forces of opposition that clashed during the struggle over Prohibition in the late nineteenth and early twentieth centuries. In the instance of Prohibition, small town agrarianism, and the conservative Christian reformism instinct won. In the case of jazz, it was the urban, secular, more liberal element of the culture that won. It is no accident that these two gigantic cultural movements should meet head-to-head in the 1920s, when jazz entered popular culture in a big way, largely through recordings and the burgeoning nightclub and speakeasy business (the latter made possible, in large part, by Prohibition), when the grandest social reform movement since abolitionism and Reconstruction was in effect. (Early, 406)

Jazz and cinema are invented at the same moment, and as part of the same institutionalization of electracy in entertainment, as the practices implementing and exploiting the features of the new recording technologies. The pattern by now is clear: jazz rehabilitates the body from the era of sin; the creole instructs the jewgreek on how to live through pleasure-pain.

Theorists

The theorists of the Frontier myth were Theodore Roosevelt, representing the Progressives, and Frederick Jackson Turner, representing the Populists. A similar division or conceptual polarity between elitism or separatism and universalism or assimilation is found also in the Jazz myth, in the figures of Leroi Jones (Amiri Baraka) and Ralph Ellison (respectively). "Jones's *Blues People* marks the junction: it was the first history of jazz written by an African-American, but it was also the last book Jones published before he became Amiri Baraka. Novelist Ralph Ellison objected to the militant spirit of Blues People. Ellison had been one of the most insightful and devoted commentators on jazz since the publication of Invisible Man in 1952, but his review of Jones's book turned out to be a farewell gesture to the musical (and social) world of an older generation" (Rasula, 67).

> For Jones, the sense of the relation between the virtuosity of the act and ideology of the jazz life is one of an irreducible social antagonism, and the terms in which we must grasp the dynamics of this antagonism are those of the blues impulse. Jones states the problem in these terms: "The blues impulse was a psychological correlative that obscured the most extreme ideas of assimilation for most Negroes, and made any notion of the complete abandonment of traditional black culture an unrealizable possibility. In a sense, the middle class spirit could not take root amongst most Negroes because they sensed the final fantasy involved." In one's analysis, the blues impulse was the barrier to the jazz act's assimilation to the American "progressive" ideology, even though the pressure to assimilate was already a possibility inherent in the act. We can reformulate Jones's proposition in this way: through the blues impulse, the real antagonism of racial and class struggle in the United States was manifested symbolically in the effort to prevent assimilation, but in the sense that assimilation had already taken place. The external opposition of whites to blacks, the very form of the antagonism between black and white Americans, was already internal to the jazz act. (Gebhardt, 14)

It is this problematic of the relationship between the dominant order and the subordinated "other" that is undergoing transvaluation in the imagined community.

The Iconic Gesture of Jazz

Jones/Baraka's argument, making a point that is central to choragraphy and electracy, is that the blues is not so much a music as a state of mind, an attitude, a mood, and that this is the dimension that white critics ignore. "The catalysts and necessity of Coltrane's music must be understood as they exist even before they are expressed as music. The music is the result of the attitude, the stance. Just as Negroes made blues and other people did not because of the Negro's peculiar way of looking at the world. Once this attitude is delineated as a continuous though constantly evolving social philosophy directly attributable to the way the Negro responds to the psychological landscape that is his Western environment, criticism of Negro music will move closer to developing as consistent and valid an aesthetic as criticism in other fields of Western art" (Baraka, 142). Jazz and blues are not synonyms, of course. At the level of the music itself, jazz is a hybrid, a composite of ragtime from Vaudeville, spirituals from the Black Christian church, and blues from folk life of the Delta. At the level of the blues impulse, however, "blues" names the fundamental mood of a worldview, the equivalent of "saudade" for Brazilian samba, or "duende" for Spanish deep song. As one commentator put it, blues is the treasure and jazz is the diver. In the PBS series, "Martin Scorsese Presents the Blues," another commentator observed, "blues is the roots, and the rest is the fruits." A related difference between jazz and blues proper is that the former does not involve the problem of "authenticity." Once invented within a particular community in singular conditions, jazz is available for everyone, as its spread around the world has demonstrated, and this spread is the key to its promise for the EmerAgency.

PHOTOGRAPH: BARBARA JO REVELLE

To formulate the Jazz myth we need to articulate the scenario into which the history is condensed and made mobile within the common sense of a cultural value system. What is the equivalent for jazz of the showdown for the cowboy, or Salome with the head of John the Baptist for the femme fatale? As is the case with the Frontier scenario, the Jazz scenario evolved from an original invention of an iconic type to a contemporary version. We may trace this type through several of its defining gestures or poses, in order to establish first the concrete vehicle of the scenario as an allegorical emblem. The first iconic scene is that of the bluesman, whose prototype is Robert Johnson, whose legendary meeting with the Devil at a crossroads is an expression, Julio Finn explains, of Johnson's initiation in the religion of Vodou (Hoodoo). The legend is important to the allegory of "Crossroads," for the Haitian traders sitting on the mattresses piled high on the deck of the impounded vessel link to Legba, Master of the Crossroads.

> In Voodoo the god of the crossroads is Legba, one of the most powerful in the whole of the pantheon. "No consideration of the roles played by African deities in new World religions should omit mention of Legba." Legba is a capricious and often malicious character, who delights in double-crossing those who put their faith in him. Courlander describes him as: "One of the most important of Haitian loa, frequently the first invoked in any service. He is the guardian of the gateway, the crossroads and the highway. He is a divine trickster who is both feared and loved, a force that is capriciously malevolent or benevolent. He is the intervener between mankind and the deities. It is in this sense that he "opens the gate," providing a channel of communication between man and loa. In Dahomean lore, Legba's mischief is often in the realm of sex.
>
> In Haiti, certain charms, called *garde carrefours*, are made with earth taken from the crossroads. Metraux explains how these pacts are made: "The Voodooist seeking co-operation of a god in order to achieve some ambition or simply wishing to put himself under the god's special protection, can make the god a formal proposal of marriage. Henceforth, they will have a common destiny." In black religions, to invoke the gods is to offer oneself for possession by them. If Johnson made a pact at the crossroads with Legba, in essence, this is what he was doing – a claim which he was by no means alone in making. Ledell Johnson, brother of guitarist Tommy Johnson, explained in *Living Blues* how his brother had struck his deal at the crossroads: "Now, if Tom was living, he'd tell you. He said the reason he knowed so much, said he sold hisself to the Devil. I asked him how. He said, 'If you want to learn how to play anything you want to play

PHOTOGRAPH: BARBARA JO REVELLE

> and learn how to make songs yourself, you take your guitar and you go to where the road crosses that way, where a crossroad is. Get there, be sure to get there just a little 'fore 12:00 that night so you know you'll be there. You have your guitar and be playing a piece there by yourself. A big black man will walk up there and take your guitar, and he'll tune it. And then he'll play a piece and hand it back to you. That's the way I learned to play anything I want.'" The evidence here presented proves that, first, the lore of making a pact with the Devil at the crossroads had come to the Delta area as part of the Hoodoo religion. (Finn, 217)

Houston Baker proposes the image of the black blues singer at a railway junction as "a trope for the blues as a forceful matrix in cultural understanding." That the crossroads is best emblematized as a railroad crossing, he says, is because "the dominant blues syntagm in America is an instrumental imitation of train-wheels-over-track-junctures. This sound is the 'sign,' as it were, of the blues" (Baker, 8). The crossing as matrix emblematizes the mediating function of chora. "The 'X' of crossing roadbeds signals the multidirectionality of the juncture and is simply a single instance in a boundless network that redoubles and circles, makes sidings and ladders, forms Y's and branches over the vastness of hundreds of thousands of American miles.... The singer and his production are always at this intersection, this crossing, codifying forces, providing resonance for experience's multiplicities. Singer and song never arrest transience – fix it in 'transcendent form.' Instead they provide expressive equivalence for the juncture's ceaseless flux. Hence, they may be conceived as translators" (7).

George Lipsitz uses the image of the crossroads as a metaphor for the contemporary condition of the public sphere in a post-industrial information economy. "For many musicians around the world, the 'popular' has become a dangerous crossroads, an intersection between the undeniable saturation of commercial culture in every area of human endeavor and the emergence of a new public sphere that uses the circuits of commodity production and circulation to envision and activate new social relations" (Lipsitz, 12). His first example is the Haitian group Boukman Eksperyans and their effort to recover Vodou practices for the work of emancipation against their cynical exploitation by politicians such as the dictator "Papa Doc" Duvalier (8). Especially relevant to Miautre as chora is Lipsitz's interest in the role popular music is playing in creating a global public sphere.

> A century ago, the combined effects of state building, urbanization, and industrialization transformed popular perceptions about change over time, making history the constitutive problem of the age of industrialization. Today the ever expanding reach and scope of electronic, computer chip, fiber optic, and satellite communication imposes a rationalized uniformity on production and consumption all over the world, making *place* the constitutive problem of the post-industrial era....
>
> Today, shared cultural space no longer depends upon shared geographic place. What Henri Lefebvre called "theatrical or dramatized space" becomes increasingly important as a substitute for the lost public sphere of the industrial city. New discursive spaces allow for recognition of new networks and affiliations; they become crucibles for complex identities in formation that respond to the imperatives of place at the same time that they transcend them. (6)

Robert Johnson died young, poisoned during an evening at a jook joint by a jealous husband of a woman with whom Johnson was having an affair. As the music evolved from blues into jazz, and from the setting of the country whiskey dance to the New Orleans saloon and sporting house, the figure of the masculine hero remained constant, as in the character of Buddy Bolden, considered to be the first jazz horn player. The iconic gesture of jazz, and the jazz hero in general, remain within the liberal ideology of the progressive era, ironically perhaps, portraying the meritocracy of the individual (Romance) hero.

The theme of virility is spelled out in descriptions of Bolden's career, including explicit links between musical and sexual power. "In classic mythmaking mode the chapter dwells on the prowess of the King, on some of the oldest meanings of virtue. 'Soon Kid Bolden became King Bolden,' the chapter spelling out an implicit coronational narrative. But the question of 'virtue' gets ironic very quickly. The implicit pun turns sardonic and open as the text explores Bolden's habits, which were as immoderate 'for women as for hard liquor and hot music.' It appears that Bolden, like many a Romance hero, was equally good with all these outlets for prowess. If he could draw a crowd by sticking 'his horn through a hole in the fence and the people came rushing,' he could take three women along on a parade, 'all three satisfied.' Jazzmen completed a process that had begun in Bolden's time, turning the barber into an icon that centered potency" (Garber, 83).

The Ellison/Baraka tension between universalist assimilation and black nationalism is evident in the contrast between the image of two African-American trumpet players, two stances or personas. Arthur Knight drew the contrast in a discussion of the documentary film *Jammin' the Blues*, referring to the repeated image of a grinning black man. "This last image – especially as embodied in Louis Armstrong's performance persona – was probably the most widely disseminated image of black musicianship in the world, and it had become a key image in the reformulation of the look of mid-1940s jazz, an image associated with the 'tomming' that bebop formed itself against. Writing in the journal Conjunctions about Miles Davis, who codified on trumpet the 'cool' playing style that Lester Young originated on tenor saxophone, Quincy Troupe recalled the meaning of not smiling for an audience. 'Besides the magisterial, deep-cool hipness of his musical language, the aspects of Miles that affected me most were his urbane veneer and his detached sure sense of himself as royalty, as untouchable in a touch everything world. Miles's refusal to grin in front of white audiences like so many other black entertainers made a statement to me'" (Knight, 39). The jazz life became codified in popular culture in a narrative referred to simply as "The Story." "The Story goes like this: a musician of genius, frustrated by the discrepancy between what he can achieve and the crummy life musicians lead because of racial discrimination or the demand that the music be made commercial, or because he has a potential he can't reach), goes mad, or destroys himself with alcohol and drugs. The Story might be a romance, but it is a valid one." (Vance Bourjaily, in Gabbard, 2002: 344)

The Jam Session

Emerging from this entire history, one particular scene condenses the import of the Jazz myth for an imagined community: the jam session – historically the sessions at Minton's Playhouse that produced the invention of bebop. What the Cabaret Voltaire was to the European avant-garde, Ralph Ellison declared, Minton's was to the American vanguard (Ellison, in O'Meally, 452).

> Henry Minton also provided, as did the Clef and Rhythm clubs, a necessity more important to jazz musicians than food: a place in which to hold their interminable jam sessions. And it is here that Minton's becomes most important to the development of modern jazz. It is here, too, that it joins up with all the countless rooms, private and public, in which jazzmen have worked out the secrets of their craft. Today jam sessions are offered as entertainment by night clubs and on radio and television, and some are quite exciting, but what is seen and heard is only one aspect of the true jam session: the 'cutting session,' or contest of improvisational skill and physical endurance between two or more musicians. But the jam session is far more than this, and when carried out by musicians in the privacy of small rooms (as in the mural at Minton's), or in such places as Halley Richardson's shoeshine parlor in Oklahoma City – where I first heard Lester Young jamming in a shine chair, his head thrown back, his horn even then outthrust, his feet working on the footrests, as he played with and against Lem Johnson, Ben Webster (this was 1929) and other members of the old Blue Devils Orchestra – or during the after hours in Piney Brown's old Sunset Club in Kansas City; in such places as these, with only musicians and jazzmen present, then the jam session is revealed as the jazzman's true academy. (453)

The bebop style of jazz developed in the jam sessions at Minton's, like the behavioral style and practices of the detective before it, created a means of perceiving and domesticating the shock of the industrialized city.

> Out of encounters in the streets of New York came local versions of the wisdom and agility required of all fleet-footed inventors. "Coolness helped to keep our values warm," says Ellison, "and racial hostility stoked our fires of inspiration." Not even the North could abide the kinds of interracial freedom the musicians cultivated; a palpable dissidence kicks the best of the music. The shock of relocation was 'handled' but the common language that musicians developed –styles of dress, music, drugs, and speech homologous with the structures of their experience. The requisite cool of the northern city dweller was unattainable unless negotiated through style....

PHOTOGRAPH: BARBARA JO REVELLE

> The various elements of bebop style were thus part of a new generational responsiveness to the northern city, particularly 1940s New York, a place distinguished less by its capacity to shock than by its ability to make little seem shocking. What evolved in turn was an aesthetic of speed and displacement – ostentatious virtuosity dedicated to reorienting perception even as it rocked the house. Every instrument became immediately more mobile, everything moved. (Eric Lott, in O'Meally, 460–461)

The point to make for the Miautre chora is that jazz is a potential voice within the Frontier myth, speaking from the side of the "other," of the crowd, but in a way that shared many of the features of the dominant version. The famous "double consciousness" of the Negro hypothesized by Du Bois, for example, in the black version of the "man who knows Indians" (it is the black who knows whites). Both instances of this ability to cross over a bar or frontier opposition is moreover a capacity precisely associated with the trickster figure. The gunfighter and jazzman both are tricksters. James Naremore, commenting on some designs by the director Vincente Minnelli, indicates how the primitivism attributed to the jazz life became syncretized with mainstream popular culture. "Like the folkloric artists, Minnelli relied on a kind of primitivism, explicitly associating blackness with sexuality, instinctiveness, and the

PHOTOGRAPH: BARBARA JO REVELLE

Freudian unconscious. At the same time, however, he promoted an uptown face of jazz, tied to contemporary fashion and big-time entertainments. In this context blackness began to signify both wildness and sophistication. The African imagery was as 'stereotypical' as any other cultural code, but it seemed attractive and denatured by parody or playful quotation; moreover, because it was regarded by audiences as vanguard, it tended to problematize the distinction between the savage and the cultivated Naremore, 1995: 178).

Similarly, the opposition of values dramatized in the Western, such as Wister's *Virginian*, was repeated within black sub-culture in the same terms of gender, class, religion, and morality. "Powdermaker's analysis supports the conventional view of a Black Belt society divided rather rigidly into a 'church culture' and a 'blues culture.' The all-too-easy conclusion that there were two cultures – one predominantly female and middle class which adhered to white middle-class proprieties, the other predominantly male and lower class which lived to make the most of the moment, valuing whiskey, women, and easy money – supposes that movement from one culture to the other bespeaks a divergence in desires, values, and behavior leading to severe feelings of guilt. Such a portrait may well describe the anxiety of black Americans becoming assimilated into white society, but it does not accord with life down home. If there were 'respectables' and 'no-'counts,' they were forced together by the same outside pressures of white discrimination" (Titon, 17).

World-View

The Jazz myth extracts jazz from its history and condenses it into a scenario that functions as an allegorical emblem. The jam session is the vehicle or image of this emblem. What is the tenor, the meaning or signified, of the image? We need to specify for jazz a metaphysics (a view of how the world is, the nature of reality); a morality (how one should act in such a reality); a state of mind or mood (the intuitive feeling that communicates immediately the lessons of the worldview). As Baraka insisted, jazz draws upon the blues impulse, which is a mood, a state of mind. This blues view is that the world is a hard hostile place. As Julio Finn states, "the bluesman sees his people caught up in a terrifying vortex, a nightmare reality the mechanism of which is arranged so that nothing in it can ever come out right for them – hell man, he don't want to hear about the ten black pop stars who've 'made it' – the bluesman is concerned with twenty-five million crushed, exploited, troubled people! People whom history has branded, people whom those in power have condemned" (Finn, 201). Finn rejects "The Story," pointing to the life of Robert Johnson to refute the romanticism of white critics who compare Johnson's blueslife to the stories of Keats, and James Dean – heroes dying young. "There was nothing whatever 'romantic' about being black in Mississippi in 1911, the year in which Johnson was born. Nor was there any particular luck in being born into a family of cotton-pickers. And it was decidedly a misfortune to have had the previous father of that family run out of town by a white lynch mob, on account of having dared to talk back to some rednecks, who followed up this good deed by getting the wife and children evicted from their land" (211).

Ralph Ellison takes up this point in his review of Richard Wright's autobiography, *Black Boy*, credited as the best account of the blues impulse, which of course is not confined to music. Indeed, commentators are in agreement that jazz (the blues feeling) may be found anywhere, in anything. "What after all is this taking new thing, that, condemned in certain quarters, enthusiastically welcomed in others, has nonchalantly gone on until it ranks with the movie and the dollar as the foremost exponent of modern Americanism? Jazz isn't music merely, it is a spirit that can express itself in almost anything. The true spirit of jazz is joyous revolt from convention, custom, authority, boredom, even sorrow – from everything that would confine the soul of man and hinder its riding free in the air" (J. A. Rogers, in Meltzer, 8). The important point is not to confuse the joy of the jazz moment with the blueslife that produced it. "We have here a problem in the sociology of sensibility that is obscured by certain psychological attitudes brought to Negro life by whites. The first is the attitude which compels whites to impute to Negroes sentiments, attitudes and insights which, as a group living under certain definite social conditions, Negroes could not humanly possess" (Ellison, in O'Meally, 557).

If the reality of the situation in which African-Americans found themselves was this nightmare of racism, what is the morality that follows from such a realization, as represented in the jazz mood? The moral imperative of the blues impulse in jazz is the opposite of ataraxy or withdrawal recommended by the ancient sages. "The blues is an impulse to keep the painful details and episodes of a brutal experience alive in one's aching consciousness, to finger its jagged grain, and to transcend it, not by the consolation of philosophy but by squeezing from it a near-tragic, near-comic lyricism. As a form, the blues is an autobiographical chronicle of personal catastrophe expressed lyrically. And certainly Wright's early childhood was crammed with catastrophic incidents" (553). Ellison's version of the morality of the blues is "that they at once express both the agony of life and the possibility of conquering it through sheer toughness of spirit. They fall short of tragedy only in that they provide no solution, offer no scapegoat but the self" (562).

Albert Murray makes a similar point about the blues impulse: "so it's a matter of affirming life in the face of adversity. Because my central image of the blues is that there are invisible blues devils that beset you on all sides, and they are an embodiment of entropy. You can't destroy them but you can push them back, you can hold them at bay" (Murray, in O'Meally, 575). The blues morality is that even in the depths of a lowdown life never to give up, to refuse victimage, but to embrace "marronage," that is, "fugitivity." The term is a neologism invented by Aime Cesaire, to associate a literary style of openness, experimentation, and formal innovation with a radical politics. Ironically, perhaps, the blues has its own version of "westering." "The poem ["The verb 'Marroner'/ for Rene Depestre, Haitian Poet"] announces and enacts its poetics under the sign of a neologistic verb. Cesaire invokes the history of fugitive slaves in the Caribbean, the runaway Africans known as maroons who escaped the plantations and set up societies of their own. The French noun for this phenomenon, marronage, is the basis for the word, the verb marroner, Cesaire invents, an act of invention exemplifying the independence for which the poem calls" (Nathaniel Mackey, in O'Meally, 518). The term has been translated loosely as "escape like slaves." Deleuze and Guattari's "line of flight" theorizes marronage. Marronage, often romanticized by European intellectuals, opens a passage between the Frontier and Jazz.

The Jazz Scenario

Albert Murray clarified that his proposal to take the blues as a model of conduct for the modern hero is based not simply on the blueslife or jazzlife, but on the music itself. The best model for survival techniques, for resilience in conducting a successful life, is the practice of improvisation, exemplified in blues and jazz performance. "Blues musicianship requires this, this, this, and this. Not just the ability to read the notes, but the ability to create in the presence of peers and betters, under the pressure of all tempos, with inverted chords and whatever. That's a good metaphor for me. That's a perfect metaphor for the kind of person you would like to be in life. If you could just handle everything like that, you'd be ready, because you'd be ready for the changes that are bound to come, the unforeseen changes" (Murray, in O'Meally, 577). This shift away from the romance hero, The Story with its hangover of progressive era ideology, to the musical performance, is the move we are following to arrive at the image itself as a category organized through mood. Nicholas Gebhardt discusses the jazz performance itself as a struggle between the institutional ideologies of the historical situation that produced African-American musicians on the one hand, and the virtuosity of playing in the moment in an act which is its own fulfillment, confirming the value of non-productive labor. In reading Gebhardt's review of the act of improvisation in a jazz ensemble, we should keep in mind that this scenario is to be extended in the Jazz myth to a model of conduct for policy formation as well as for personal conduct as modeled in entertainment forms.

> Paul Berliner's analysis of spontaneity adds an additional qualification to how the making of the improvised act is constituted through its seductive motions. For him, the effort that goes into making and remaking the act is determined by a musician's or a group's "habits of musical thought." These habits constitute the outer limit of what is musically possible during any one performance, and circumscribe the kinds of musical phrases that are practiced and learned prior to the performance. It is the structuring of those habits – how and in what order notes are played – that form the pathway from merely "playing the changes" to acts of spontaneous' improvisation. The act's spontaneity results from an amorous or impassioned moment of breakthrough, that point when a player "can be affected by his or her own notes." The improvised act's character is thus defined and valued by the unexpected potency (and pleasure) experienced by the musician on hearing what he or she in fact played. In this way, the spontaneity of the act confronts the musician with his or her capacity for musical activity and, in the process, opens up new regions of creative action, new powers of repetition. (Gebhardt, 149)

Several important points are suggested here. The first clarification involves giving a name to the mood of jazz proper, even with its blues impulse: *joy*.

PHOTOGRAPH: BARBARA JO REVELLE

There is considerable agreement on this term to name the feeling that is achieved through the jazz act (in whatever medium or practice). It is momentary, and may not be sustained, but motivates a repeated return to this activity. The narrative trajectory of *Bird*, the life of Charley Parker, traces the path of the hero's blueslife, in a version of the four locations or positions of a semiotic square: from rural farm setting to its contrary, represented by the white suburbs where he attempted to live for a time with his White wife and child; to the contradictory, the mean streets of the urban inner city; to its complement, the low stage of the jazz club (Birdland itself), where the music was made. In Eastwood's film, the central image, the symbolic core of the story, is that incident in the biography when Charley was just learning his craft. He took the stage on a kind of "open mike" night to play with a group. When he went into his solo, he stopped listening to the ensemble and destroyed the jazz act. To get his attention, the disgusted drummer threw one of his cymbals at Charley, and its crashing skid across the floor of the club brought an end to the session.

Parker did not achieve the epiphany of joy that year, but, like Robert Johnson, he disappeared for a number of years while he learned the craft of jazz. There is no story about the Devil tuning his saxophone at a crossroads. Instead, it was while improvising on a standard show tune, "Cherokee" (significantly for us, a pop rendition of a stereotyped "Indian" sound), that Parker discovered his peculiar style of playing that contributed fundamentally to the

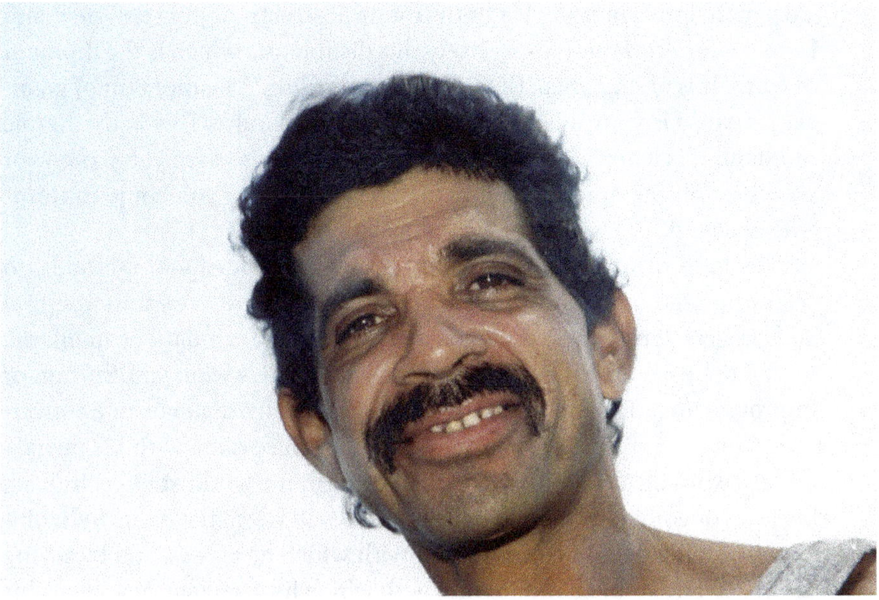

PHOTOGRAPH: BARBARA JO REVELLE

invention of bebop. Here is the crucial innovation that jazz contributes to the Frontier myth, as it mutates into the Jazz myth. The difference may be seen in the contrasting scenarios of the two versions: the confrontational showdown of the Western gunfighter; the collaborative jam session of the jazz musician. The proposal of the consultation, then, is to shift the priority of the common sense scenario guiding public policy (and entertainment) from the gunfighter showdown heroics to the jazz ensemble playing in the background, or on the soundtrack; and to shift the form of the scenario from narrative teleology to jazz collaborative improvisation.

> The convention of playful option-taking is what I call "the blues idiom" and jazz music. I define the jazz musician as one who approaches or creates or plays all music as if improvising the "break" on the traditional twelve-bar blues tune. Which brings me not only to improvisation and the creative process but also to blues idiom procedure and the 'jam session' as the representative anecdote for life in the United States. By improvisation, of course, I most definitely do not mean "winging it" or making things up out of thin air. The jazz musician improvises within a very specific context and in terms of very specific idiomatic devices of composition....
>
> The break is a device which is used quite often and always has to do with the framework in which improvisation takes place... The break is an

> extremely important device both from the structural point of view and from its implications. It is precisely this disjuncture which is the moment of truth. It is on the break that you "do your thing." The moment of greatest jeopardy is your moment of greatest opportunity. This is the heroic moment. It is when you establish your identity; it is when you write your signature on the epidermis of actuality. That is how you come to terms with the void.
>
> This kind of improvisation is applicable to educational methods, to scientific method, to inventions. Your knowledge of chordal structural and progression is your knowledge of the experience data of mankind; it is all the wisdom and mistakes of the ages. That is what you "riff" on or improvise on.... The musician is always engaged in a dialogue or a conversation or even argument – not only, as in a jam session, with his peers – but also with all other music and musicians in the world at large. Indeed, his is an ongoing dialogue with the form itself. He achieves his individuality by saying "yes and also" to that with which he agrees, and by saying "no," or "on the other hand" to that with which he disagrees. Yes, all of this is as applicable to science and engineering and political procedure as it is to music and literature and all of the other arts....
>
> Improvisation, as it functions in the blues idiom, is something that not only conditions people to cope with disjuncture and change but also provides them with a basic survival technique that is commensurate with and suitable to the rootlessness and the discontinuity so characteristic of human existence in the contemporary world. However such improvisation proceeds, when it succeeds it puts you on good terms with life. It generates an atmosphere of well-being and celebration. It is music to have a good time with. (Murray, in O'Meally, 112–113)

The purpose of Miautre is to provide a practice based on this blues idiom as a support for a virtual civic sphere, by means of which the public improvises on the "standards" of the imagined community, to reeducate collective common sense, and bring the Frontier myth up to date. To participate in the EmerAgency consultation on Miami, through its online chora, Miautre, is to perform civic life in the mood of jazz.

Real Time requires Be-Bop reason, ad hoc improv decision. Another title for "Crossroads": Skeeversville Blues. What is Jazz as myth? An attitude, an orientation, a direction, bringing into policy formation a feeling universally appreciated: Joy. It is a promise of happiness.

CHAPTER 14

Vortex

A purpose of the consultation on the Miami River is to design and test the logic of a categorcal image, "Crossroads." The process involves extracting the "thing" from the literate apparatus to refunction it within electracy. We located a threshold of transition from one apparatus to the other within the scenario of the Frontier myth in its contemporary form (the forensic aesthetic), in a cinematic close-up on a detail evoking a mood. The mattress-thing in "Crossroads" is the detail that evokes a categorial mood (Stimmung). The relay for the new practice of the virtual civic sphere (Miautre) is jazz improvisation, as a figure for flash reason, deliberation in Real Time in an Internet public sphere. For now the point is to note the connection of jazz with the figure of the ragpicker in the default topics collected within the Miautre chora. As the emblem to show the relation of jazz to the larger context of modernism, Alfred Appel used the figure of the ragpicker, explicitly associated with the poetics of collage-assemblage.

> As singers, Armstrong, Waller, Teagarden, and Holiday typically had to modify or tear apart and rebuild poor or mediocre Tin Pan Alley material in a procreative manner analogous to the ways in which modernists such as Picasso begot paper collage, wood assemblage, and metal sculpture. "I'm king of the ragpickers!" Picasso proclaimed gleefully around 1930, after ha had created *Woman in a Garden,* his first welded tin-and-iron sculpture, proof that machines do not rule. Picasso's proposed memorial to the open-minded poet Guillaume Apollinaire (1928) visualizes a royal ragpicker's best weapon, its two concave frontal wire "feelers" representing a powerful antenna/radar/satellite dish/magnet combination that won't miss a thing, especially since it's grounded by six legs. With or without such divining rods and antennae, a King or Queen of the Ragpickers is a representative figure, confronting and ordering chaos – our goal, simply said, in life if not art. (Appel, 8)

The camera puts the ragpicker's art into the hands of ordinary citizens, who internalize into their worldview a photographic sensibility. Choragraphy is a practice for this orientation, which is synesthetic, syncretic, including text as well as picture and sound as supports for reasoning in this electrate way. What remains to be clarified is exactly how this practice forms categories – its

 | DOI:10.1163/9789004534643_015

epistemology, ontology, cosmology. We are working now in the mystorical register of Career (Theory), putting the FRE consultancy in its default historical context. We are not outside the zone but within its field in the same way as the other figures whose default moods we have traced. The proper context for our interest in the categorial image is the tradition of the search for a perfect or universal language.

> It is at this point that we can bring together the two paths we have been tracing – those of writing and light – which is in fact the historic junction at which the camera stands in the nineteenth century. It is as a species of natural writing, writing by God's own hand with the very fingers of his light, that photography was able to claim its unique proximity to truth in representation.... In this manner the camera entered nineteenth-century consciousness, not simply as a new mode of representation, but as a new language of truth, at a time when the natural claims of language as a vehicle for truth were themselves being tested. This situates a vital historical cleft. Belief in the transparency of language had been disturbed by the invention of the printing press, and the spread of written "vernacular" languages, including the first translations of sacred texts such as the Bible and the Koran. By the nineteenth century, the potential gap between sign and meaning had become an object of theoretical concern, culminating in the work of Saussure, Nietzsche and Freud. But even as the concept of the sign within which logocentrism had maintained its coherence became increasingly problematic, photography (and later cinematography) became new repositories for the old dream of a universal language. (McQuire, 31)

The dream of a universal language began as a practical pedagogy in manuscript culture and its reliance on mnemonic techniques to support its oratorical performances. In the Renaissance scientific method and numeracy coexisted for a time with the memory theater tradition, but eventually the two modes went separate ways, with Descartes, Bacon, and Ramus rejecting the memory palaces which were appropriated by the hermetic movement and became associated with the Cabala, magic and mysticism. The epistemology and cosmology of this mnemonic hermeticism was the prescientific one that most civilizations embraced, the system of correspondences between the macrocosm and microcosm, "as above, so below." Michel Foucault in *The Order of Things* traced the history of the separation of Western knowledge from this correlational or mimetic metaphysics, with the first break in the seventeenth century with the establishment of the Classic episteme, and the second in the nineteenth century with the Modern episteme. He itemized the logical features of correspondence – the prose of the world – that disappeared from official or mainstream learning in these periods. "Up to the end of sixteenth century, resemblance played a constructive role in the knowledge of Western culture. It was resemblance that largely guided exegesis and the interpretation of texts; it was resemblance that organized the play of symbols, made possible knowledge of things visible and invisible, and controlled the art of representing them. The universe was folded in upon itself: the earth echoing the sky, faces seeing themselves reflected in the stars, and plants holding within their stems the secrets that were of use to man" (Foucault, 17).

PHOTOGRAPH: BARBARA JO REVELLE

Foucault isolates four figures that determine the knowledge of resemblance. *Convenientia:* "This word denotes the adjacency of places more strongly than it does similitude. Those things are 'convenient' which come sufficiently close to one another to be in juxtaposition" (18). *Aemulatio:* the relation of emulation enables things to imitate one another from one end of the universe to the other without connection or proximity: by duplicating itself in a mirror the world abolishes the distance proper to it." *Analogy:* In this analogy, convenientia and aemulatio are superimposed. Like the latter, it makes possible the marvelous confrontation of resemblances across space; but it also speaks, like the former, of adjacencies, of bonds and joins. Its power is immense, for the similitudes of which it treats are not the visible, substantial ones between things themselves; they need only be the more subtle resemblances of relations." *Sympathies:* Sympathy is an instance of the Same so strong and so insistent that it will not rest content to be merely one of the forms of likeness; it has the dangerous power of assimilating, of rendering things identical to one another, of mingling them, of causing their individuality to disappear – and thus of rendering them foreign to what they were before" (19).

PHOTOGRAPH: BARBARA JO REVELLE

In electracy, dissociated from its hermetic claims, resemblance reenters metaphysics as what has been called "liquid architecture," to provide the operating features of image categories. These features were maintained during the epistemological exile of the signature tradition by the arts, and especially by two inventions related to universal language, based on misunderstandings of Egyptian hieroglyphics and Chinese ideograms. "Many of the most important intellectual figures of the sixteenth and seventeenth centuries devoted themselves to the problem of hieroglyphic writing and, later, to that of Chinese ideograms. The contemporary trend towards 'Egyptomania' and the fashion for emblems and imprese in European culture is symptomatic.... The fact that non-European civilizations had been able to achieve the systematic representation and communication of concepts by means of hieroglyphs or images instead of the letters of the alphabet not only seemed to confirm the feasibility of the ars memoriae and Lullism, but also suggested the possibility of devising a universal language which could be 'read' and 'understood' independently of the differences between contemporary languages, nationalities or historical situations" (Rossi, 78).

Michel Beaujour traced the survival of this tradition in the mode of the "self-portrait," from Montaigne to Michel Leiris. "By presenting psychoanalysis as a practice that solves, dissolves, and so forth, all the signs thanks to which Renaissance esoteric thought – with a Pico della Mirandola or a John Dee, for example – undertook to reconquer the magic powers supposedly held by antiquity, by becoming the Horapollo and Poliphilo of neuroses and hysteria, Lacan promises more than Leiris himself thinks he can deliver. But the hieroglyphs, blazons, labyrinths, charms, enigmas, and the rest (one would willingly add emblems to this list) are in fact part of the gear deployed by Michel Leiris in his self-portrait, and his work, moreover, concerns hermeticism and equivocations, about which one is hard put to decide whether he 'solves' them or whether he maintains them with all of rhetoric's resources to produce an adorned *copia*, of which the Asianism of his double, Dr. Lacan, would be, as it were, the playful hyperbole" (Beaujour, 200).

Mystory is a "self-portrait" in this sense. "Unlike humanism, the self-portrait seeks to be an individualized patchwork using for its own purposes bits and pieces borrowed from the vast humanist system of which the subject partakes as a 'cultured' man. It is no longer a question of standardizing the individual's memory to fit a cultural model but, on the contrary, of working with fragments that do not conform to the stereotype and out of which the subject can fashion an idiosyncratic ensemble of metaphors where he will find himself (again) or get lost" (197). Chorography is related to this tradition as an updating of the art of memory.

> Although today many are still of the opinion that images provide a means of communication that can overcome language barriers, the explanation of the way in which images can accomplish this by now takes one of two forms: the Egyptian and the Chinese way. The Egyptian way today belongs only to art history.... Any language of images is based on the alleged fact that images exhibit some properties of the represented things. Yet in any representable thing there will always be a multitude of properties, and there are infinite points of view under which an image can be judged similar to something else. We can see this by looking at the various versions of a semiotic apparatus (if not a true language) which remained alive

> for centuries and which flowered in the same period when the western culture was looking for perfect visual languages: the arts of memory. An art of memory establishes at its expression-plane a system of *loci* (that is, of places in the literal sense of the word) which may be imagined as the rooms of a building or palace, or as an urban street or square. This system of *loci* is destined to house a set of images, drawn from the same iconographical field, which will play the role of lexical units. The content-plane is given by a system of *res memoranda*, in other words, of things to be remembered, usually belonging to the same conceptual framework. In this way, an art of memory is a semiotic system. For instance, the system of grammatical cases is expressed (and thus recalled) by the different parts of the human body. Not only is this a case of one system expressing another system; it is also a case where the two planes are (in Hjelmslev's sense) conformal. It is not arbitrary that the head stands for nominative, the chest, which can receive blows, stands for accusative, and the hands, which possess and offer, stand for genitive and dative. (Eco, 1995: 170)

The figure who combines all these historical trends – scientific, hermetic, mnemonic, emblematic and ideographic – is Leibniz. "Leibniz's tendency to transform his *characteristica* into a truly blind calculus, anticipating the logic of Boole is no less shown by his reaction to the discovery of the Chinese book of changes – the *I Ching*" (284). Both the scientific and hermetic movements featured classifications of knowledge, with each claiming a correspondence between its mode of representation and reality itself. The design of Miautre has something to learn from both the poetics produced as spin-offs of the search for a universal language – emblems and ideograms. Let us consider first the Chinese relay, in order to place image technology in the context of the desire that is also part of an apparatus. From the *I Ching* we may learn how an image metaphysics works, how an epiphany becomes predictive.

The Web of Changes

A self-consciousness about the relativity of the categories organizing our experience is one symptom of the transformation of the apparatus of language from literacy to electracy. Michel Foucault is just one of the more prominent theorists to refer, for example, to the "Chinese Encyclopedia" anecdote found in the work of Jorge Luis Borges. To clarify the shift in categorization that we are exploring it is worth recalling Foucault's discussion, which is used to introduce his own history of categorization in the era of print. The limit of our own "order of things" is more easily apprehended, Foucault said, when juxtaposed with another system of thought that seems "impossible," unthinkable, from our point of view: "This passage quotes a 'certain Chinese encyclopaedia' in which it is written that 'animals are divided into: (a) belonging to the Emperor, (b) embalmed, (c) tame, (d) sucking pigs, (e) sirens, (f) fabulous, (g) stray dogs, (h) included in the present classification, (i) frenzied, (j) innumerable, (k) drawn with a very fine camelhair brush, (l) et cetera, (m) having just broken the water pitcher, (n) that from a long way off look like flies.'"

> The monstrous quality that runs through Borges's enumeration consists in the fact that the common ground on which such meetings are possible has itself been destroyed. What is impossible is not the propinquity of the things listed, but the very site on which their propinquity would be possible. The animals "(i) frenzied, (j) innumerable, (k) drawn with a very fine camelhair brush" – where could they ever meet, except in the immaterial sound of the voice pronouncing their enumeration, or on the page transcribing it? Where else could they be juxtaposed except in the non-place of language? Yet, though language can spread them before us, it can do so only in an unthinkable space.... Absurdity destroys the *and* of the enumeration by making impossible the *in* where the things enumerated would be divided up. Borges adds no figure to the atlas of the impossible; nowhere does he strike the spark of poetic confrontation; he simply dispenses with the least obvious, but most compelling, of necessities; he does away with the site, the mute ground upon which it is possible for entities to be juxtaposed. (Foucault, 1973: XVI–XVII)

Foucault explained that Borges filled the missing space or site with the Western myth of China: "In our dreamworld, is not China precisely this privileged "site" of "space"? In our traditional imagery, the Chinese culture is the most meticulous, the most rigidly ordered, the one most deaf to temporal events.... There would appear to be, then, at the other extremity of the earth we inhabit, a culture entirely devoted to the ordering of space, but one that does not distribute the multiplicity of existing things into any of the categories that make it possible for us to name, speak, and think" (xix). Deleuze proposes this Borgesian/Chinese "and" as a constitutive feature of the time-image (and hence of the image category).

> Film ceases to be "images in a chain... an uninterrupted chain of images each one the slave of the next," and whose slave we are (*Ici et ailleurs*). It is the method of BETWEEN, "between two images," which does away with all cinema of the One. It is the method of AND, "this and then that,"

> which does away with all the cinema of Being = is. Between two actions, between two affections, between two perceptions, between two visual images, between two sound images, between the sound and the visual: make the indiscernible, that is the frontier, visible (*Six fois deus*). The whole undergoes a mutation, because it has ceased to be the One-Being, in order to become the constitutive "and" of things, the constitutive between-two of images. The whole thus merges with that Blanchot calls the force of "dispersal of the Outside," or "the vertigo of spacing": that void which is no longer a motor-part of the image, and which the image would cross in order to continue, but is the radical calling into question of the image. (Deleuze, 1989: 180)

The Asiatic Mode of Production

This weakness in the category system (metaphysics) of a civilization motivated the work of the "suture," providing a closure or completeness effect. Zizek's Lacanian account of this quilting point that closes a metaphysics is one of the theoretical supports for the EmerAgency project to replace the ideological upholstery buttons (Nation, God, Money) with the sinthome (the singular vinculum produced mystorically). The object @ (Thing), he says, "is not the subjective supplement which sustains the objective order, but the objective supplement which sustains subjectivity in its contrast to the subjectless objective order" (Zizek, 2001: 65).

> In order to produce the effect of self-enclosure, one must add to the series an excessive element which "sutures" it precisely insofar as it does not belong to the series, but sticks out as an exception, like the proverbial "filler" in classificatory systems, a category which poses as one among the species of a genus, although it is effectively just a negative container, a catch-all for everything that does not fit the species articulated from the inherent principle of the genus. In traditional Marxism the unfortunate concept of the "Asiatic mode of production" played a somewhat similar role: this concept functioned as a kind of container for the inassimilable rest, i.e. it was not effectively defined by a set of positive determinations, but Marxists put into it that which did not fit any of the other modes of production, so that we find in it ancient China, Egypt, etc. (65)

This category of the "Asiatic mode," turns out to be a pivot, the point of this reversal into a different category system altogether, in which, with the support of the digital prosthesis, a new kind of object ("the object on account of which

PHOTOGRAPH: BARBARA JO REVELLE

'objective reality' is forever inaccessible to the subject") achieves classificatory power (for example – a mattress). The image category is precisely not conceptual and therefore does not rely on the methods of definition (distinguishing essences from accidents in a hierarchy of species and genus). The photographic holistic nature of chora makes the infinity of properties and points of view potentially available in a scene a positive condition. The task of this consultation is to review precisely the categorial operations of the Asiatic mode. As Francois Jullien explains, for example, the "monstrous" list – the Borges list being a typical example – is in fact ordered by an attunement (a state of mind called *shi*) that is self-evident to a native of the civilization.

> It is as if the Chinese using the lists would have no need to derive a more abstract concept from the material instances, as if they had no need for any theory over and above what they could themselves instinctively and actively feel to be the pertinence of *shi* through the cases listed. For them, *shi* was a "practical" – indeed the most practical – term and one to be considered as such. *Shi* is self-evident, permeating whatever field one considers, and as soon as one is exerting oneself effectively and educating oneself in an apprenticeship, the very idea of explaining *shi* becomes pointless, even harmful to anyone using it. To consider doing so could only occur to someone uncommitted, unconcerned (from the

perspective of his own particular logic) someone merely reading the text. (Jullien, 112)

"Shi" means disposition, circumstance, power, potential, referring to the tendency or propensity of the energy of change operating in any situation. It entered Western culture in the twentieth century through such practices as *I Ching* divination, Feng Shui, and Tai Chi. It turns out the Borges' list resembles those produced to render the fingering positions for playing the lute, or the moves for a martial art. David Sudnow's description of playing jazz piano stops just short of this Chinese rhetoric of positions. "An index and fourth finger touch two closely spaced black keys, in a chromatic way of being. The keyboard is a setting of places, with measurable dimensions. The hand is an 'organ' with measurable dimensions. The knowing relationships obtaining between them, the way the hand finds itself correspondingly configured to fit dimensions of this keyboard, involves a mobile hand engaged in a course of action" (Sudnow, 58). The relevant point for choragraphy is that we may learn how to tap into the categorial moods operating in our own culture by noticing the postures and gestures of the body (as in the time-image).

PHOTOGRAPH: BARBARA JO REVELLE

In his book on the Chinese mood (state of mind) of *shi*, Francois Jullien placed Foucault's reading of Borges' "Chinese Encyclopedia" in the context of the general difference between Western and Eastern metaphysics (category formation).

> [Western reasoning] first seeks to adopt a commanding position that provides a theoretical perspective ordering all the material to be organized. This makes abstract thought about it possible, resulting in a vantage point from which one can usually derive some classificatory principle of homogeneity. Chinese reasoning, in contrast, seems to weave along horizontally, from one case to the next, via bridges and bifurcations, each case eventually leading to the next and merging into it. In contrast to Western logic, which is panoramic, Chinese logic is like that of a possible journey in stages that are linked together. The field of thought is not defined and contained a priori; it just unfolds progressively, from one stage to the next, becoming more fertile along the way. Furthermore, the path along which it unfolds does not exclude other possibilities–which may run alongside temporarily or intersect with it. By the end of the journey, an experience has been lived through, a landscape has been sketched in. Not everything is visible and unequivocal, as in a Western picture; rather, the view unfolds like a Chinese scroll in which a path running up a mountainside (and thereby giving it consistency) appears at one point, then disappears around the hill, to reappear even further on. (Jullien, 1995: 124)

This context gives added meaning to Raymond Williams' assessment of the effect of James Joyce's *Ulysses*, due to the internalization of the action: "In a way there is no longer a city, there is only a man walking through it" (cited in Donald, 1999: 5). Chorography attempts a hybrid or syncretic method combining Western and Eastern (non-Western) reasoning (or should we say: alphabetic and ideogrammic logic). The important thing is the realization that chora appropriates, incorporates and revises any and all metaphors of "place" as logic, rhetoric, poetics.

The value of Foucault's archeology as a relay is its study of the experience of order and change at a collective historical level – the same experience that chorgraphy approaches from the side of the individual. The discussion of Borges also suggests what to do with "Haiti," which serves as our "China." In our case, however, a chora is formed out of a specific geography – the Miami River zone – and whatever is found there, certainly as heterogeneous as the items on Borges's list, is gathered into a belonging and held together by the attunement of an atmosphere, a mood, recorded in a photograph with a mattress punctum (the Miami *shi*).

Wisdom

The Chinese *Book of Changes* is organized by image clusters rather than by concepts. The clustered "heaps" (the six lines) collected within each hexagram cohere into a system through the shared civilizational mood of *shi*. One way to discern the shape or borders of our Western habitus is by means of a cross-cultural comparison of our conceptual logic with pre-scientific modes of wisdom such as divining. The value of augury (which treats reality as a text) for the virtual city is that online sites, unlike the actual urban environment, are designed as "forests of signs." Divining was as central to Western civilization as to any other, as Foucault reminds us, as a basic practice associated with the correlational epistemology replaced finally by Classic representation. It is important to study the principles of divining that provide the "vehicle" of or relay for our interface metaphor for a choral approach to information design. Placing divination in a grammatological frame means to consider it as a mnemonic practice, central not only to the oral apparatus, but also to written cultures, especially in the chirographic or manuscript phase of the technology. The practical function of divination is to relate individual experience to collective cultural memory. Oracles work by providing a cognitive map that enables individuals to bring to bear on personal difficulties the stored wisdom of the civilization. The heuretic strategy is not to reproduce a particular divination system but to extrapolate from the general operational logic of divination as well as from a particular example, to design from a divining relay a practice informed by our modern values and beliefs.

The basic elements of oracular knowing include a symbol, chance and a critical situation, whose combination produces an image of fate (Karcher, 107). "Anything that intersects our thoughts and feelings in an unusual way at an important moment, activating that part of our imagination that sees behind the apparent, is an omen" (23). Perhaps the best known example of divination practiced by a canonical figure in the Western tradition is that of St. Augustine, who made use of two devices during the spiritual crisis that led to his conversion to Christianity. In the emotional state of struggling with his crisis he overheard a child repeating a refrain, "Take it and read! Take it and read!" To receive overheard phrases as an oracle was known as "consulting Hermes." To open an important book (Vergil's *Aeneid*, the Bible) at random and take what is read as an oracle is known as "sortes." Augustine opened his copy of the Testament at random and encountered the following passage: "Not in rioting and drunkenness, not in fornication and wantonness, not in strife and envying; but put on the Lord Jesus Christ and think not of fulfilling the lusts of the flesh." "Instantly," he wrote in his *Confessions*, "a light of confidence darted into my heart and all the darkness of doubt vanished" (51). The confidence was due to the fact that the passage addressed exactly the issue motivating Augustine's hesitation, which was his attachment to the "flesh."

The system used as the relay for designing our interface metaphor is the *I Ching*, the Chinese Book of Changes. Although the Miami theoria suggests a hybrid between poststructuralism and Afro-Caribbean practices, the context frames this encounter in the modernist history of syncretism in which vanguard artists sought alternatives to Western hyperrationalism in all manner of non-Western forms and modes. In any case, the Yoruba culture's *Ifa* divination system, a major source for Santeria and Vodou, formally resembles the *I Ching*. In our grammatological frame the problematic "primitivism" is replaced by syncretism. There is no pretense that the interface metaphor is "authentic," pure, or correct. The aim is not access to traditional Chinese or African wisdom, but to contemporary Miami wisdom. The EmerAgency hybrid is a construction extrapolated from certain essential features of both epistemologies and practices (divination and poststructuralism). At a formal level, divination fulfills the function of cognitive mapping, or of mystory,

> In its practice, the Yoruba system of divination that is the source of much of the Afro-Caribbean practices is quite a bit like the Chinese system contained in the *I Ching*. With the *I Ching*, to discover a response to your question, you throw yarrow sticks or flip coins to indicate one of 64 hexagrams. A coin flipped six times has 64 possible results of this chance operation. If for example, you ask, "Should I leave this love affair?" and your coin comes up heads-heads-tails-heads-tails-tails, you have the hexagram Chien or "Development (Gradual Progress)," and the response to your question begins "A tree on a mountain develops slowly according to the law of its being and consequently stands firmly rooted." They say that divination works best if you come with a "burning question," for then the oracle's enigmatic responses will work the way so-called projective therapies work. (Hyde, 110)

The *I Ching* is a relay that shows us a possible poetics for making our own divination system. Espen Aarseth expresses the fit between the *I Ching* and the Internet in strictly formal terms.

> Possibly the best-known example of cybertext in antiquity is the Chinese text of oracular wisdom, the *I Ching*. Also known as the Book of Changes, the existing text is from around the time of the Western Chou dynasty (1122–770 B.C.) and was written by several authors. The *I Ching* system also inspired G. W. von Leibniz, who developed the binary mathematics used by today's digital computers. The *I Ching* is made up of sixty-four symbols,

> or hexagrams, which are the binary combination of six whole or broken ("changing") lines. A hexagram contains a main text and six small ones, one for each line. By manipulating three coins or forty-nine yarrow stalks according to a randomizing principle, the text of two hexagrams are combined, producing one out of 4,096 possible texts. This contains the answer to a question the user has written down in advance. (Aarseth, 9–10)

Of special interest for Miautre is that the *I Ching* demonstrates an alternative solution to the metaphysical and methodological problem that Plato solved by inventing chora in *Timaeus*. As Jullien explained, when Plato introduced a third term (the receptacle) to resolve the question of how being and becoming interacted, this step in turn entailed a fourth nature, "agent," from which followed the whole cause-effect approach to reality that distinguishes the Greek from the Chinese metaphysics (Jullien, 252).

PHOTOGRAPH: BARBARA JO REVELLE

> Why this logical necessity for a third principle conceptualized as a "substratum or subject"? As Aristotle declared, contraries "cannot act upon each other," "do not change into each other," and "are mutually destructive," In logical terms, they are mutually exclusive. In contrast, the entire Chinese tradition insists that contraries both oppose each other and "contain each other mutually: within yin there is yang, just as within

> yang there is yin".... They together form all that exists. There is thus no need to posit a "third term" to support their relation. Even the regulating principle does not exist over and above the two contraries; it simply expresses their harmonious relation. The two contraries form on their own a self-sufficient configuration; the propensity that stems on its own from their interdependence orients the process of reality (251).

Another account of Chinese cosmogony, however, blurs the clear distinction posed by Jullien by describing "emptiness" as functioning in a choral manner. "This median emptiness, itself a breath, issues from the original emptiness, from which it draws its power. Median emptiness is necessary for the harmonious functioning of the yin-yang pair; it attracts the two vital breaths and draws them into the process of reciprocal becoming. Without it, yin and yang would be in a relationship of frozen opposition. They would remain static substances and be formless. It is clearly this ternary relationship between yin and yang and median emptiness that gives birth to, and also serves as a model for, the ten thousand existents" (Cheng, 50).

The important point is that the Chinese demonstrate a categorial practice different from the conceptual and topical ontology invented by the literate Greeks. "The matter of interest seems to be the configuration formed by chance events in the moment of observation, and not at all the hypothetical reasons

PHOTOGRAPH: BARBARA JO REVELLE

that seemingly account for the coincidence. While the Western mind carefully sifts, weighs, selects, classifies, isolates, the Chinese picture of the moment encompasses everything down to the minutest nonsensical detail, because all of the ingredients make up the observed moment" (Wilhelm, XXIII). Jullien adds that "the essential difference is that Greek thought introduced order into becoming from the outside (on the basis of numbers, ideas, forms), whereas in Chinese thought order is conceived as lying within becoming; it is what makes becoming a process.... Chinese thought became sensitive early on to the regular, spontaneous fecundity stemming simply from the alternation of the seasons" (218).

Descriptions of the relationship between *yang* and *yin* evoke the winnowing action of the Eleusinian choral metaphor: "This is what the *I Ching* refers to as the real secret of the change, or what is called by Chuang-tze the dust flying like a wild horse. *Ch'i* [shi] which sometimes goes up and at other times comes down, is the beginning of motion or rest. What goes up is the light, *yang* part, what comes down is the heavy, *yin* part. It can consolidate or dissolve in the form of wind and rain, snow and frost, mountains and rivers, and myriad other things" (Carson Chang in Lee, 56). In our context, the cross-cultural comparison shows that the Chinese version prioritized the relational action of space over the definition of things, establishing the correspondence between outer and inner "emptiness." In any case, chora as electrate category may learn from the oracle a rhetorical skillset for thinking in the Moment of Now.

That which resisted measure and hence control the Greeks called "necessity" – "residue that belongs to the irrational – accident, fortune, monstrosity, or any other unintelligible manifestation of necessity. Becoming is once and for all identified with 'matter' and there is no escape from this fixing of essences" (217). The value of the accidental detail is a major point of difference between Greek and Chinese cosmology. Jullien extends this comparison into a series of contrasts to distinguish Chinese reason from its Greek counterpart: not a deductive sequence, but a global insight; not problem, but process; specifically not the problem of being, but the capacity to function; not confrontation, but correlation; not events, but change; not a heroic journey, but a field unfolding; not metaphysics, but divination.

Hexagrammatology

The chief insight to emerge from Jullien is that the fundamental organizational unit of Chinese thought expressed in the *I Ching* is the *situation*, with a certain tendency or propensity. The world is approached as a continuously changing or flowing series of circumstances manifesting *shi* – a configuration of energy, such that Chinese metaphysics is frequently used as an analogy to explain Quantum physics, that is, a world in which "things" are not essences or substances but secondary, emergent features of a holistic background field. The Classical Chinese constructed an order for this world, whose prototype is the *I Ching*. The legendary origin of *The Book of Changes* provides further resonance for the Miami River zone, in that it began as a vision, a revelation of the relations between Heaven and Earth given to several sages at the Yellow River *in the form of a mysterious map*. The map was derived from a pattern of spots on the back of the visionary dragon-horse, that supplied a blueprint for the *I Ching* (Lee, 33).

The map was converted into groups or "piles" of lines – trigrams – that, in the manner of archetypal ideograms, figured eight "germinal situations" that contained virtually every arrangement of circumstances in experience. The emphasis here was not on predicates but predicaments. The eight trigrams extended by a combinatorial pairing produced sixty-four archetypal situations. Of interest in our context is a certain resemblance to Greek category formation, in that the germinal situations are identified by a set of attributes and properties. Most handbooks explaining the system include a chart providing the following information for each of the eight germinal trigrams (with just one example tracked here to illustrate): "NAME: Ch'ien; TRIGRAM [three solid or yang lines]; IMAGE, Heaven; QUALITIES: creative, strong, light of day, firmness; DIRECTION: Northwest; RELATIONS: Father; PARTS OF BODY: Head; TIME OF YEAR; Early Winter; TIME OF DAY: Daytime" (Wing, 15). The trigrams were arranged into two different sequences historically, one moving through the sixty-four combinations giving a configuration and its opposite (Heaven is followed by Earth), and the other following the cycle of the seasons. The importance of the sequences is that a situation has a direction, a "grain," a tendency. In general, once one knows which situation one is in, the *I Ching* sequence constitutes a kind of "entailment" (hence the effect of foresight or prediction).

Miautre focuses on several aspects of the system, in order to extract a poetics that may be used to generate equivalents (a wisdom oracle for a global public sphere). First, the parts of the system are itemized, condensing into a "genre" description of features that evolved over many generations historically. Formally a hexagram consists of: a name; the hexagram itself; the primary judgment (the text defining the overall situation and interpreting what action is appropriate); the subordinate judgments for each of the six lines. In some editions there is also a picture selected from Chinese art to illustrate the situation. In addition to the immediate features of the hexagrams, there are several commentaries, adding layers of philosophical interpretation, and other documentation. There is also the claim that the yin-yang exchange references an historical conflict between two dynasties of ancient China. A key feature is that the underlying proportional analogy organizing the system is that of a landscape passing through the seasons (grounded in an agricultural civilization).

The advice derived from the *I Ching* expresses the world view of Classical Chinese civilization. The wisdom of the book, as Jullien explains, reflects the long debate in Chinese philosophy between the naturalist (realists) and the moralists over whether events unfold automatically, spontaneously, or whether individual virtue may play a significant part. The Taoist-Confucian debate correlates in certain respects with the Protestant-Catholic debate in Christianity over whether salvation is predetermined or may be influenced by good works. The basic theme of the advice reflects what Jullien calls "efficacy" – the enlightened person learns the tendency at work in a given situation, waits for the right moment (similar to Greek *kairos*), and acts in a way that exploits the propensity to his or her advantage. The consistent message in any case is patience, since the metaphysics of yin-yang dictate that any tendency eventually must reverse itself. The recommended strategy usually is to avoid confrontation or rhetoric, and achieve one's aims indirectly through manipulation of circumstances (live as a "courtier").

Ideogram

The insight derived from studying the *I Ching* as a relay for a Miautre interface motivated the design. Visitors to the site do not simply use an existing and completed divination system, but would participate in the creation of a contemporary system, within the framework of choragraphy. An immediate function for Miautre is to formulate and disseminate a practice for conducting this experiment, based on an electrate category formation. There is a direct connection between the writing systems of the Greeks and the Chinese and the respective category formations (metaphysics) developed in each civilization (alphabet-concept, ideogram-*shi*). The heuretic premise is that the principles of image reason have already been invented in "pure" form within the experimental and vanguard arts, participating in the "paideuma" of the emergent apparatus that began forming early in the nineteenth century. In this grammatological frame modernism looks somewhat different from the story that it tells about itself. What remains to be accomplished (and this may be one of the pedagogical services of Miautre as an educational enterprise) is the conversion of *collage/montage* into a logic to be taught, learned, and applied alongside the dialectics that have spread to every dimension of modern life from their origins in Classical Greece.

The exemplary figure for us among the many possible candidates is Ezra Pound, not only because of his poetics of the ideogram, relevant to the *I Ching* relay, but more fundamentally because of his choice of "logo" to represent his poetics: the Vortex. He got the idea for this image (which lent its name to the Vorticist movement in England) from reading scholarly accounts of Classical Greek cosmology, such as John Burnet's *Early Greek Philosophy*. The passage that was the source of Pound's inspiration was by Leukippos, but we recognize in it the allusion to *Timaeus* and Plato's most important metaphor for chora.

> The worlds came into being thus. There were borne along by "abscission from the infinite" many bodies of all sorts of figures "into a mighty void," and they being gathered together produced a single vortex. In it, as they came into collision with one another and were whirled round in all manner of ways, those which were alike were separated apart and came to their likes. But as they were no longer able to revolve in equilibrium owing to their multitude, those of them that were fine went out to the external void as if passed through a sieve; the rest stayed together, and becoming entangled with one another, ran down together, and made a first spherical structure. This was in substance like a membrane or skin containing in itself all kinds of bodies. And, as these bodies were borne round in a vortex, in virtue of the resistance of the middle, the surrounding membrane became thin, as the contiguous bodies kept flowing together from contact with the vortex. And in this way the earth came into being. (Bell, 152)

PHOTOGRAPH: BARBARA JO REVELLE

The original context of the image of the sieve or winnowing basket –as part of a divination ritual in the Eleusinian Mysteries – is also specifically relevant to Pound, who used the scene of the *nekuia* or descent into Hades from Homer's *Odyssey* as the framing structure for the *Cantos*. The ritual of descent was the mystery celebrated at Eleusis. "The conclusion one can draw from these Eleusinian references is that Pound's decision to begin the *Cantos* with a translation of Homer's *nekuia* involves the adoption of the Eleusinian ritual of spiritual death and rebirth in the form of a journey to the Underworld. In other words, the *Cantos* require of the reader a reinterpretation of the *Odyssey* if the parallel between the two works is to be properly understood. For in the *Cantos* Odysseus is a spiritual voyager whose principal acts are the conquest of Circe and the visit to the Underworld. These acts parallel the central ritual acts of the hierophant at Eleusis, but are by no means Odysseus' most important actions in Homer's *Odyssey*" (Surette, 55).

With this context Pound evokes perhaps the key turning point in the origins of the literate apparatus – the moment when Plato borrowed the Eleusinian scene to represent chora within an ontological (literate) mode of categorization. Pound "makes it new" by returning to this moment in order to take chora (sacred space) in a different direction. In doing so he is participating in the anti-Platonism of his own milieu. At the same time, he exploits the context to indicate his intentions, relevant also to chora. Odysseus descends into Hades to consult with Tiresias, a consultation that Pound extends to his entire epic, which is a search through history to identify the qualities of greatness that appear from time to time in certain places, through the efforts of certain individuals, to produce strange attractors in civilization. He is just one among the many philosophers and artists attempting to escape from the limits of Western metaphysics in the early twentieth century – specifically from the confines of abstract conceptual and topical categories. "The opposition between abstraction and sensation is a prominent feature of Modernist poetics. Pound and Eliot, like the philosophers [Bergson, James, Nietzsche], assume that instrumental conventions displace us from immediate experience. Pound incessantly attacks our propensity to substitute conceptual abstractions for concrete sensations, and makes the precise rendering of immediate experience a cornerstone of his poetic program" (Schwartz, 6). A contrast between Vortex and Topos supports the formulation of chora as the "the places of invention" for digital memory.

Reflecting the spirit of his age, Pound turned to non-Western sources for help with inventing his alternative practice. His reading of Ernst Fenollosa's studies of the Chinese Ideogram (undertaken during the Vorticist years) provided the

PHOTOGRAPH: BARBARA JO REVELLE

relay for a non-conceptual, non-topical, image-based way of gathering heterogeneous materials into sets.

> In tables showing primitive Chinese characters in one column and the present "conventionalized" signs in another, anyone can see how the ideogram for man or tree or sunrise developed, or "was simplified from," or was reduced to the essentials of the first picture of man, tree, or sunrise. [Thus, man + tree + sun = sun tangled in the tree's branches, as at sunrise, meaning now the East]. But when the chinaman wanted to make a picture of something more complicated, or of a general idea, how did he go about it? He is to define red. How can he do it in a picture that isn't painted in red paint? He puts (or his ancestor put) together the abbreviated picture of
>
> ROSE CHERRY
>
> IRON RUST FLAMINGO
>
> That, you see, is very much the kind of thing a biologist does (in a more complicated way) when he gets together a few hundred or thousand slides, and picks out what is necessary for his general statement. Something that fits the case, that applies in all of the cases. The Chinese "word" or ideogram for red is based on something everyone KNOWS. (Pound, 1960: 21–22)

The analogy with biology, alluding to Louis Agassiz (who revolutionized the pedagogy of the life sciences by having students dissect actual specimens), is misleading to the extent that it suggests that Pound's ideograms are inductive. Their mode of inference rather is "conductive," an inference based on aesthetic coherence operating with the holistic gest of epiphany. The more apt analogy is with cinema, especially the montage methods of Sergei Eisenstein, which were invented out of a similar exposure in his case to Japanese arts and writing. Fenollosa's description of the Chinese ideogram as a "verb," an action, rather than as a thing, included reference to photography that clarifies the connection of Vorticism with the emerging image apparatus. "'Things,' wrote Fenollosa about 1904, are 'cross-sections cut through actions, snapshots'" (Kenner, 146). Kenner also described Joyce's epiphany as a "snapshot," related to the "aesthetic of glimpses" at work in Pound (35).

The point in our context is that the Vorticist engages in the same task that Aristotle undertook when he invented the "thing" as a certain kind of entity, structured by essences and properties. This discussion of reality in terms of action snapshots is a symptom of the emergent "social machine" of electracy. The poet now is looking for something else in experience other than essence or substance, and produces a different means of coherence, another kind of entity – the Vortex. "By tying the concept to its constituent particulars, the ideogram keeps us from detaching the abstract form from concrete sensation – the 'leaf' from individual leaves, or 'redness' from particular red objects. And by displaying its own etymology, the ideogram prevents us from assuming an absolute distinction between concept and figure, metaphysical truth and metaphorical imagination: we are less prone to forget that our abstractions arise from the swift perception of relations" (Schwartz, 89). Photography makes ideogrammic reason practical; choragraphy makes it available as a rhetoric, the basic device for image metaphysics (category construction).

"Vortext"

Apparatus theory is not a technological determinism (there is nothing "new" about "images"). Rather, it is a matter of efficacy (timing). Pound's logic of the "Luminous Detail," which he said already existed among the best scholars and historians, supplies a logic to go with the equipment – the imaging equivalent of "definition." "Luminous Details are the transcendentals in an array of facts: not merely 'significant' nor 'symptomatic' in the manner of most facts, but capable of giving one 'a sudden insight into circumjacent conditions, into their causes, their effects, into sequence, and law'" (Kenner, 152). Among the multitude of facts available just a "few dozen" of these "interpretive details" give the intelligence of a period: "'They govern knowledge as the switchboard governs an electrical circuit'" (Kenner, citing Pound, 153). Pound learned from Fenollosa's how to arrange the details in a way that made them luminous. "For the ideogram sketches a process, seizes some continuous happening – the movement of attention through an eye, the flow from roots to branches within a tree – and fixes it, like Buckminster Fuller's knot, with three or four minimal vigorous spatial gestures. Thus the word, freed from evanescent sound, transcends the moment of utterance and reutters itself in a vibrating field of force where, augmented by neighboring words, particulars rush *from* and *through* and *into*" (160).

Vortex dynamics offer a principle for configuring the dataflow of a "mirror world," where information rushes *from, through, and into*. Pound returns to the moment of chora before Aristotle constructed "things" out of "mere heaps" of properties. "According to Aristotle, Plato has no good way of distinguishing between a participant [in the realm of becoming] changing and a participant being replaced because he ultimately has no clear way of regarding the participant as anything other than a 'mixture,' that is, as anything other than a mere heap, rather than a thing" (Mann, 34). Pound described the *Cantos* as an epic, but in fact he jettisoned for all practical purposes every conventional organizing schema except one – the heap. The commentators note that the great difficulty of the *Cantos* is due to their lack of the conventional sources of coherence – exposition, plot, character, even musical form. Instead, it is an epic structured by figure alone – by a haiku-esque juxtaposition of disparate fragments. It is an enormous heap of details gleaned from culture and nature.

> Pound reverses the process [of essentializing] by juxtaposing phenomena in order to reveal through their relations the invisible idea. From the relations of the seen arises the unseen form, "the *forma*, the immortal *concetto*, the concept, the dynamic form which is like the rose pattern driven into the dead iron-filings by the magnet itself, but separate from the magnet. Cut off by the layer of glass, the dust and filings rise and spring into order." The force of the magnet is akin to that force which orders sphere, great bass, monad, and right reason.… The important point is that the accumulated data will at one point cease to be just a 'heap' of detail. As in a flash, the whole subject or phenomenon will appear, in the form of an image, in the reader's consciousness. What Pound avers here is that revelation is a process, the final state of a cumulative agglomeration of detail. (Gefin, 38)

This reduction to a minimum level of literate order is at the same time a turn in a new direction. The Vortex image is a step in the direction of an electrate categorization (chora). "By the 'image' I mean such an equation [one that "causes form to come into being"]; not an equation of mathematics, not something about *a, b,* and *c,* having something to do with form, but about *sea, cliffs, night,* having something to do with mood. The image is not an idea. It is a radiant node or cluster; it is what I can and must perforce call a VORTEX, from which, and through which, and into which, ideas are constantly rushing. In decency one can only call it a VORTEX. And from this necessity came the name 'vorticism'" (Pound, 1965: 152). The updating of vortex within grammatology might be expressed in the term "vortext." The psychogeographic map of the zone is a *vortext.* "Testimonial" attains categorical or abstract force through this ideogrammic effect (the pattern appears in a flash through the accumulative repetitions of mattresses).

The synchronic quality of time in Pound's vision addresses the Real Time crisis posed by Virilio. Pound wrote the *Cantos* as a witness to history, "live" so to speak, identifying a pattern of events that recur and that give shape to time. The metaphors of turbulence used to characterize Vortex suggest its affinity with "strange attractors," that come and go within time that flows like a river, with its chaotic eddies (one of Pound's images for VORTEX). A center of energy emerges, develops, reverses, declines. "Pound's heuristic device is always the subject-rhyme. To elucidate the Italian New Birth of circa 1500, he compares it with the America of circa 1770. Specifically, Jefferson and his successors building a nation are rhymed with Malatesta building the Tempio" (Kenner, 423). A Vortex has to be "somewhere on the map" (238). Malatesta's Rimini was a Vortex. Pound thought that the London of his own moment could be a Vortex. "A Vortex is a circulation with a still center: a system of energies drawing in whatever comes near ('Energy creates pattern'). So you could have a National Vortex, likewise when it came to *virtu,* a personal Vortex ('Vortex Lewis'; Vortex Pound'; Vortex Gaudier-Brzeska'). And the personal Vortex since 'we do not desire to cut ourselves off from the past,' will draw into the artist's personal dynamism his usable arraying of past Vortices" (239).

The personal witnessing is an essential part of the image, in that the poet understands the present moment through identification with historical peers. Pound understands London and its situation in the history of the British Empire through a juxtaposition (resemblance) with Sextus Propertius in the situation of the Roman Empire. Similarly the possibility of the future is divined by this intuition of pattern (it is a time-image). Kenner stated the risk of the method: "a poem about history, taking note of recurrences, could go on endlessly, like wallpaper" (Kenner, 376). Reed Dasenbrock noted the disaster that befell the *Cantos,* in which Pound was using his new category to demonstrate that Mussolini's Italy was the next embodiment of the ideal city. The

PHOTOGRAPH: BARBARA JO REVELLE

generalizability of the Vortex as a category, rather than as a feature of Pound's style, is due precisely to the fact that the vision Pound created by its means was falsifiable. "Pound cannot impose a form on his poem; it must find its form in the form of history. In short, Mussolini must bring order out of chaos for Pound to bring order out of his chaos. Pound has linked the destiny of his epic to the destiny of his implicit epic hero: his epic poem is literally being written *in media res*" (Dasenbrock, 213). The Vortex is a logic, not a style, or it is a style of logic. Gefin selects Canto IV as a good example of the type of juxtaposition he distinguishes as the "cumulative ideogram,"

> in which Pound juxtaposes certain mythological elements (such as the cannibalism in the story of Tereus and Philomela and the metamorphosis of Actaeon into a stag) with stories from Provence (the eating of Cabestan's heart and the 'metamorphosis' of Peire Vidal into a wolf running through the forest). Again, in Pound's hands the individual values and characteristics of the "factual atoms" (ideogrammic components, pattern units, sub-wholes) are not blurred or distorted in order to make them fit his scheme. Pound does not intervene subjectively to say that "this *is* that" or one thing *is like* another thing. Pound establishes not analogies or similarities but relations, or relations-in-process, in a language "beyond metaphor." The "new angle" Pound aims at in his juxtapositions

> inheres in the changes and the metapmorphoses mutually effected by each particular unit, and these changes bring about the creation of a new whole which is simultaneously perceptual and conceptual. (Gefin, 39)

A zone becomes choral when the field of heterogeneous items suddenly flash into a constellation, revealing the circulation centered around the attractor (jouissance) like a hurricane over Miami.

Miautre online, the virtual city, invites netizens to participate in a Vortex-Miami serving as a category of an electrate metaphysics. In electracy metaphysics is live (classification on the fly).

PHOTOGRAPH: BARBARA JO REVELLE

CHAPTER 15

Choramancy

Interface Metaphor

How to put Miautre online? The usefulness of hybrid divination in the original consultation suggested that divination might provide a good interface metaphor for chora consulting. Since research showed the similarity of the Chinese and Yoruban oracles, and since we were not adopting pure divination but syncretizing divination epistemology with poststructural poetics, we adopted the Chinese system as the relay. To call attention to our appropriation of *The Book of Changes* structure of a core of eight hexagram/archetypes, generating a total of 64 situations, but that supplements the Taoist and Confucian philosophies informing Chinese wisdom with the values of the industrial West, we named the oracle the Ka-Ching (the onomatopoetic word evoking the sound of a cash register, often heard punctuating the banter of sports casters). The hexagrams of the *I Ching* provide a relay for learning how an image becomes categorial, since each hexagram represents one situation that functions as an archetype. The hexagram is legible and applicable intuitively to an infinite number of specific circumstances through the mood or atmosphere that it conveys.

The I Ching is to category formation what acupuncture is to medicine: nobody knows exactly why it works, but its utility is obvious to those who try it.

The syncretic status of "consulting" the river means that the querent not only understands his or her situation through the policy allegory, but also that the values directing the policy are open to renewed attention. While an advantage of divination as an interface metaphor is its familiarity due to the New Age movement in contemporary culture, making its operation more or less intuitive, the purpose of creating a contemporary divination system is not to reproduce, but to make explicit, the dominant values and hence open them to examination. "The ultimate postmodern irony," Zizek observes, "is the very moment when, at the level of the 'economic infrastructure,' 'European' technology and capitalism are triumphing world-wide, at the level of 'ideological superstructure,' the Judeo-Christian legacy is threatened in the European space itself by the onslaught of the New Age 'Asiatic' thought, which, in its different guises… is establishing itself as the hegemonic ideology of global capitalism" (Zizek, 2001a: 12). In the context of Zizek's warning that in its popular form this

 | DOI:10.1163/9789004534643_016

"Asiatic" trend amounts to a "what me worry?" abuse of the wisdom of "letting go," it is important to emphasize the syncretic nature of the EmerAgency hybrid of the two kinds of "consulting" (scientific and magical). The wisdom of the traditional divination systems, including the Chinese/Yoruba model, is based on the philosophies of premodern, agrarian societies. To be adequate to the "wisdom" of a post-industrial post-colonial global society, choramancy necessarily is updated to represent processes that did not exist in the era in which divination evolved. In the same way that the *I Ching* includes Taoist and Confucian commentaries on the hexagrams, the Ka-Ching must provide commentaries based on contemporary Western theory and experience.

William Gibson's use of Voodoo possession as a metaphor for cyberspace provides a precedent and context for the psycho-divination hybrid of the consultation. To perform the Copernican revolution in consulting in which we do not explain the River's problems, but they explain us, the testimonial includes the meaning of "consultation" as it is used in divination. The consultation of the Miami River – the poetic encounter with the haitians in Miami – may be understood as a continuation in digital media of the syncretism that began with the coming of haitians to New Orleans that led eventually to the invention of Jazz. To say that Miami will be to hypermedia what New Orleans is to music gives an idea of the ambition, the larger scope of the experiment with choramancy. The cultures that relied on divination as a decision procedure considered as "external" many things that literate culture assumes to be "internal" to the self. In short, to consult the Miami River is to write the (political) unconscious.

The relevance of Vodou in particular to electracy (and to the Internet as a practical institution) is suggested by the novels of William Gibson, who invented the term "cyberspace" in the prototype of cyberpunk fiction, *Neuromancer*. Gibson read in his doctor's waiting room a *National Geographic* article about Haitian religion. He recognized at once the value of spirit possession as a metaphor for the experience of "jacking-in," a metaphor that he introduced in the sequels to *Neuromancer* – *Count Zero* and *Mona Lisa Overdrive*. What is it like to think with the total knowledge of "the matrix," the Unconscious, the Symbolic order, which is extimate for me? *It is like spirit possession.* Michael Benedikt edited the collection entitled *Cyberspace*, crediting Gibson's vision in *Neuromancer*, however much it might be tainted by its origins as an adolescent male fantasy, as a motivation for serious research aimed at inventing the institutional practices for the new technologies. This passage from Gibson's COUNT ZERO explains the metaphor.

> *"Bobby, do you know what a metaphor is?"*
> *"A component? Like a capacitor?"*

"No. Never mind metaphor, then. When Beauvoir or I talk to you about the loa and their horses, as we call those few the loa choose to ride, you should pretend that we are talking two languages at once. One of them, you already understand. That's the language of street tech, as you call it. We may be using different words, but we're talking tech. Maybe we call something Ougou Feray that you might call an icebreaker, you understand? But at the same time, with the same words, we are talking about other things, and that you don't understand. You don't need to." He put his toothpick away.

Bobby took a deep breath. "Beauvoir said that Jackie's a horse for a snake, a snake called Danbala. You run that by me in street tech?"

"Certainly. Think of Jackie as a deck, Bobby, a cyberspace deck, a very pretty one with nice ankles." Lucas grinned and Bobby blushed. "Think of Danbala, who some people call the snake, as a program. Say as an icebreaker. Danbala slots into the Jackie deck, Jackie cuts ice. That's all."

"Okay," Bobby said, getting the hang of it, "then what's the matrix? If she's a deck, and Danbala's a program, what's cyberspace?"

"The world," Lucas said. (Gibson, 1987)

An irony: blocked at the port, Haiti comes in through the (cyber) screen.

One difficulty of the interface design is how to extrapolate from the relay (the traditions of the emblem and ideogram) to our own system – a project for designing a global wisdom. One option is to organize Miautre as a complete divination system, treating each of the denizens (the people Revelle interviewed, such as Simon Lubin) as embodying archetypes, perhaps more in the manner of the major arcana of the tarot, with the different properties of the scenes (the mattresses for example) constituting the minor arcana. A second (preferred) option is to consider Revelle's Testimonial as having established *one situation*, and that the purpose of Miautre is as a relay guiding the construction of more situations (the project taken up by John Craig Freeman in his subsequent work). The first step in this latter approach is to revise the prototype Testimonial into a germinal situation. The germinal situations codified in the hexagrams achieve their coherence through a unifying mood (Wing, 14). The hexagrams gather materials into sets not by means of logic and definition but by means of art or style. "Through the dynamism produced by this configuration, a wider meaning emerges from the theme and spreads like an aura, pervasive yet elusive. What both the Chinese and Westerners call the 'poetic atmosphere' is for them created by *shi*" (Jullien, 130).

The design task involves understanding Miautre as both *one situation* and a complete system in itself. The preliminary step with respect to the former consideration was to translate the FRE Testimonial into a kind of hexagram, with a

PHOTOGRAPH: BARBARA JO REVELLE

Name (giving the situation conceptual status); Logo (reflecting a plan for placing a set of situations into a system); Judgment (articulating the circumstances of double binds in the scene of impoundment); Commentary (unpacking the ethical subtleties of poststructural theory and the effects of syncretizing with Vodou); Picture (the photograph "Crossroads" and its contextual scenario). The possible format and style were suggested by two hexagrams: 52 and 12. Hexagram 52 is "Meditation" ("Stillness") chosen because the painting illustrating it showed a boat moored on a river. "Meditation is to turn one's back so there is no consciousness of the Self. Should he go into the courtyard without noticing anyone, it is not a mistake. Stillness (mountain) upon stillness (mountain) forms the condition for Meditation. An enlightened person, therefore, does not allow his thoughts to go beyond his situation" (Wing, 125). The editor comments on the painting used to illustrate the mood of the situation, a depiction of a boat on a river: "The occupant of this 'Boat on a River' seems heedless of his surroundings in his relaxation and meditation. Yet we see that the boat is not floating aimlessly on the water but tied firmly to the shore. Art historians date this album leaf to the Ming Dynasty of the fifteenth century, a time when China was withdrawing from external contact. The Imperial Court delighted itself with a life full of splendors, wherein closeness to nature, detachment and contentment were the rule of the day" (124).

The mood of the FRE Testimonial, however, corresponds more closely with hexagram 12: "Stagnation." "Stagnation is brought about by inferior persons. There exists no advantage – even to persons of exceptional character and virtue. The superior is going; the inferior is coming.... The forces in nature are in a state of perfect and undiscriminating impasse" (45). The mood of "Stagnation" is illustrated by a depiction of a legendary incident that involves a border: "Lao-tzu, the famed Taoist philosopher, has decided to leave the stagnation and corruption of the empire and become a recluse. As he is passing through the last barrier, he is stopped by a guard and the demand is made that he leave the text of his *Tao-Te-Ching*.... Although Shang Hsi depicted here an event in antiquity, the cart, clothing, and headgear are all of the fifteenth-century Ming era, a time when internal rot and rebellion were beginning to weaken the empire" (44).

The point is not to replicate the Chinese situations, but to look for the generic situations in our own historical conditions. It is deceptively easy to translate Testimonial into the format of a hexagram, in order to get a sense of Miautre as a categorial situation. The translation, however, demands that one sort out the details of the scene to select which attributes are "typical" (if not "essential"). "It would seem impossible to deal with today's life situations through the use of only sixty-four hexagrams. However it is a mistake to identify the sixty-four hexagrams with sixty-four actual situations. The hexagrams do not represent actual or phenomenal situations (in Kantian terms), but germinal situations, each of which might be actualized in many different ways" (Lee, 98). There are many ways in which one may be "impounded," "detained." It is more difficult to write the primary and secondary judgments in a way that expresses the historical conditions of contemporary America.

Transvaluation of Values

The crisis of the virtual (post-industrial) city announced by Paul Virilio requires a rehabilitated mythography. Egents become distributed sages, responsible for identifying and constructing a contemporary wisdom consultancy. Miami "Virtue" revisits and updates the scenario formulated in response to the first crisis of the city, during the rise of industrialization. Shelley's "Triumph of Life" (the poem used as the exemplary text in a Yale School introduction to deconstruction), expresses what was happening to virtue, to the status of the hero, and to the place of the individual in relation to world historical forces during that epoch. "Rousseau's self-immolating attempt to gain protection from the contradictions of temporal life makes him a proper choice – one who 'with the weight/

Of his own words is staggered' – to mediate Shelley's vision of the grim 'Triumph of Life' over all happiness and all heroism. The specter of Life in Shelley's poem asserts with fresh force the myth of some vast power irreversibly at odds with human self-projection. This power runs riot through nineteenth-century mythmaking. It is variously conceptualized – nature, passion, society, race, the nation, the World-Spirit, the world market, the Immanent Will, evolution, necessity, history. Whatever its form, it is, compared with older mythologies of Fortune or Providence, as much more unremitting and humanly unanswering as the coercions of a world-spanning industrial-commercial society were proving to be in comparison to earlier civil systems" (Berthoff, 64–65).

Prior to the turn into industrialization there was a certain human scale to the transvaluations of value, however profound and sweeping the changes might be for society as a whole. These changes include adding the theological virtues to the cardinal moral virtues; courtly love as an adaptation of the military virtues to Christianity; Machiavelli's desublimation of *virtu* in *The Prince* (described as doing for practical reason – prudence – what Descartes did for pure reason); the inversion in the status under capitalism of "acquisitiveness" (*pleonexis*) from vice to virtue, and a similar inversion within science in the status of the vice of *acedia* into the virtue of curiosity. The very categories of virtue and vice, good and evil, were relativized by the time of Nietzsche and Freud into expressions of the will to power of the hegemonic order of a given society. This insight reveals the limitations of all previous moralists, as Alasdair MacIntyre noted.

> The appeal to a universal verdict by mankind turns out to be the mask worn by an appeal to those who physiologically and socially share Hume's attitudes and *Weltanschauung*. The passions of some are to be preferred to the passions of others. Whose preferences reign? The preferences of those who accept the stability of property, of those who understand chastity in women as a virtue only because it is a useful device to secure that property is passed only to legitimate heirs, of those who believe that the passage of time confers legitimacy upon what was originally acquired by violence and aggression. What Hume identifies as the standpoint of universal human nature turns out in fact to be that of the prejudices of the Hanoverian ruling elite. Hume's moral philosophy presupposes allegiance to a particular kind of social structure as much as Aristotle's does, but allegiance of a highly ideological kind. (MacIntyre, 215)

In European cinema, following World War II, according to Gilles Deleuze, an image practice emerged capable of reasoning within the scale of "the triumph of Life over all happiness," in the conditions of the industrial city in which individuals found it impossible to transform circumstances into situations through the power of intention. In such circumstances, Deleuze said, makers must become seers, voyants, a mode of understanding supported by the cinematic time-image. This technique, invented within such film movements as Italian Neo-Realism, including Pasolini's poetics of writing with the objects of reality, suggests how choral images function as theoria. "The character has become a kind of viewer. He shifts, runs and becomes animated in vain, the situation he is in outstrips his motor capacities on all sides, and makes him see and hear what is no longer subject to the rules of a response or an action. He records rather than reacts. He is prey to a vision, pursued by it or pursuing it, rather than engaged in an action" (Deleuze, 1989: 3). One of his examples is the film *The Lonely Woman*, which "follows a female tourist struck to the core by the simple unfolding of images or visual cliches in which she discovers something unbearable, beyond the limit of what she can personally bear. This is a cinema of the seer and no longer of the agent" (2).

> A purely optical and sound situation does not extend into action, any more than it is induced by an action. It makes us grasp, it is supposed to make us grasp, something intolerable and unbearable. Not a brutality as nervous aggression, an exaggerated violence that can always be extracted from the sensory-motor relations in the action-image. Nor is it a matter of scenes of terror, although there are sometimes corpses and blood. It is a matter of something too powerful, or too unjust, but sometimes also too beautiful, and which henceforth outstrips our sensory-motor capacities. *Stromboli:* a beauty which is too great for us, like too strong a pain. It can be a limit-situation, the eruption of the volcano, but also the most banal, a plain factory, a wasteland. In Godard's *Les carabiniers* the girl militant recites a few revolutionary slogans, so many cliches; but she is so beautiful, of a beauty which is unbearable for her torturers who have to cover up her face with a handkerchief. And this handkerchief, lifted again by breath and whisper, itself becomes unbearable for us the viewers. In any event something has become too strong in the image. Romanticism had already set out this aim for itself: grasping the intolerable or the unbearable, the empire of poverty, and thereby becoming visionary, to produce a means of knowledge and action out of pure vision. (18)

Lacan observed a prototype of this beauty in Antigone, as the mark of *Até*, the calamitous blindness of conduct explored in tragedy. It exposes the condition of waste and loss inherent in experience. The banality of the mattress bears this mark in Miautre, witness to an excess of desire beyond the reach not only of the action-image but of action as such in the conventional terms of instrumental reason. The task of the Ka-Ching is to identify the qualities of character relevant to a heroism of postindustrial life in which situations exceed the resources of embodied experience.

Happiness Blues

The turning point in modern virtue is registered in philosophy in Diderot's *Rameau's Nephew*_dramatizing a dialogue between representatives of the old and the new values – the Philosopher speaking for the tradition of the "honest soul," and the Nephew speaking for the modern alienated man of *ressentiment*. "It is not Diderot-*Moi*, not the *philosophe* with his archaic love of simple truth and morality, with his clearly defined self and his commitment to sincerity, who, for Hegel, commands esteem. Rather, it is Rameau, the buffoon, the flattering parasite, the compulsive mimic, without a self to be true to; it is he who represent Spirit moving to its next stage of development" (Trilling, 44).

> Shakespeare unabashedly uses material and social establishment and what it is presumed to assure in the way of order, peace, honor, and beauty as emblems of the spiritual life, as criteria by which the sufficiency of the inner condition may be assessed. He conceives the self in terms of states and activities which imply achievement and reward, such states as innocence, such activities as repentance and atonement, such achievements and reward as redemption, "a clear life ensuing," and even – how astonishing it is! – happiness.
>
> It is this vision of life that Hegel means to discredit when he speaks with condescension and even contempt of Diderot-*Moi*. The "honest soul" is rejected by Hegel because it is defined and limited by its "noble" relation to the external power of society, to the ethos which that power implies. Nobility has been bourgeoisified in Diderot-*Moi* but not essentially transmuted. And we in our time, at least as readers, are in essential accord with Hegel's judgment. We reject the archaic noble vision of life because we desire to escape the limiting conditions which it imposes. Our commitment is to such freedom as is to be found in the exigent spiritual enterprise which, in the English translation of the *Phenomenology*, goes under the name of "culture."
>
> Culture, as Hegel idiosyncratically defines it, is the characteristic field of experience of the base self; it proposes the activity by which the disintegrated, alienated, and distraught consciousness expresses its negative relation with the external power of society and thereby becomes "Spirit

> truly objective," that is, self-determining… This breaking up of everything real, although of definitive importance in the career of Spirit, is anything but a happy activity. The experience of the self in culture is fraught with pain; it entails "renunciation and sacrifice." (42–43)

Here is a point of entry for the EmerAgency, for a distributed adjustment to common sense, a realignment of the correlations across the registers of culture. For, as Trilling observed, "it would be absurd to say that the lives we actually live are controlled by the present-day repudiation of the old visionary norm. As householders, housekeepers, and parents we maintain allegiance to it in practice, possibly even in diffident principle. But as *readers*, as participants in the conscious, formulating part of our life in society, we incline to the antagonistic position" (41). The contradiction or disjunction may be seen in the difference between the nostalgia of William J. Bennett's campaign to restore the values of the "honest soul," in such collections as *The Book of Virtues* and *The Moral Compass*, in contrast with Slavoj Zizek's declaration, reflecting on 9/11 (the most contemporary example yet of Virilio's general accident), that "the betrayal of desire has a precise name: happiness" (Zizek, 2002: 58). "'Happiness' belongs to the pleasure principle, and what undermines it is the insistence of a Beyond of the pleasure principle. In a strict Lacanian sense of the term, we should thus posit that 'happiness' relies on the subject's inability or unreadiness fully to confront the consequences of its desire: the price of happiness is that the subject remains stuck in the inconsistency of its desire. In our daily lives, we (pretend to) desire things which we do not really desire, so that, ultimately, the worst thing that can happen is for us to get what we 'officially' desire. Happiness is thus inherently hypocritical" (59).

A counter-intuitive possibility? The name of the impasse is "happiness." Image categories make states-of-mind directly accessible to collective experience.

Our failure to understand collectively the disconnect between desire and happiness is a source of the ruin in reality that stumps conventional consulting. Here is a function for the Jazz myth – to modulate the global state of mind from "happiness" to "blues." The proposal is that choramancy and the categorial image make it possible to include desire and the sublime object in electrate wisdom, as the ultimate way in which common sense is augmented in Miautre. The aporia of the Frontier myth revealed in the consultation may be unstuck by blocking (in multiple senses) "happiness" with "blues," understood as a state of mind in which it is impossible to ignore the base life theorized in Europe as "formless" (Bataille) and lived in modern cities. What some see as the vulgarization of Western mores may be understood more affirmatively in terms of what Ralph Ellison described as a jazz approach to life. "I believe that Hemingway, in depicting the attitudes of athletes, expatriates, bullfighters, traumatized soldiers, and impotent idealists, told us quite a lot about what was happening to that most representative group of Negro Americans, the jazz musicians – who also lived by an extreme code of withdrawal, technical and artistic excellence, rejection of the values of respectable society. They replaced the abstract and much-betrayed ideals of that society with the more physical values of eating, drinking, copulating, loyalty to friends, and dedication to the discipline and values of their art" (Ellison, in O'Meally, 120).

Host, Guest, Parasite

Rameau's Nephew embodies that position in the genealogy of virtue associated with the "parasite" as "noise" in the network of "hospitality" (host-guest-parasite). This role/function (actant in narrative theory) replaces "nurse" in the contemporary metaphor of "chora," as J. Hillis Miller suggested in his contribution to the collection on "The Triumph of Life." Miller provides a further theoretical significance for the notion of "frontier" that is deconstructed in Miautre.

> Deconstruction does not provide an escape from nihilism, nor from metaphysics, nor from their uncanny inherence in one another. There is no escape. It does, however, move back and forth within this inherence. It makes the inherence oscillate in such a way that one enters a strange borderland, a frontier region which seems to give the widest glimpse into the other land ('beyond metaphysics'), though this land may not by any means be entered and does not in fact exist for Western man. By this form of interpretation, however, the border zone itself may be made sensible, as Quattrocento painting makes the Tuscan air visible in its invisibility. The zone may be appropriated in the torsion of the mind's expropriation, its experience of an inability to comprehend logically. This procedure is an attempt to reach clarity in a region where clarity is not possible. In the failure of that attempt, however, something moves, a limit is encountered. This encounter may be compared to the uncanny experience of reaching a frontier where there is no visible barrier, as when Wordsworth found he had crossed the Alps without knowing he was doing so. It is as if the "prisonhouse of language" were like that universe finite but unbounded which some modern cosmologists posit. One may move everywhere freely within this enclosure without ever encountering a wall, and yet it

> is limited. It is a prison, a milieu without origin or edge. Such a place is therefore all frontier zone without either peaceful homeland, in one direction, land of hosts and domesticity, nor, in the other direction, any alien land of hostile strangers, "beyond the line."
>
> "The place we inhabit, wherever we are, is always this inbetween zone, place of host and parasite, neither inside nor outside. It is a region of the *Unheimlich*, beyond any formalism, which reforms itself wherever we are, if we know where we are. This "place" is where we are, in whatever text, in the most inclusive sense of that word, we happen to be living. (Miller, 231)

A symptom of alienation from our own agency in a commodified reality is that we fail to recognize this experience of constraint as a "measure" (a ratio governing movement through information). The Mattress vinculum locates just this frontier virtual measure or limit, a border at once inside and outside, serving as the groove through which the chora coheres.

The hospitality dilemma reminds us that there is always a third party mediating any pairing, such as that between host and guest. In a chapter entitled "Rat's Dinner" Michel Serres began with a section called "Diode, Triode," to explain his point that a dialogue has three positions, three functions. He is working a (macaronic) puncept on "parasite," meaning noise, guest, and leech, in an exploration of the mise en scene of the concepts of information. Here is the link by which chora gathers topoi into a holistic category – a conductive repetition of signifiers across the cultural encyclopedia. Serres figures the requirements for invention, creativity, in the vocabulary of tourism and hospitality.

> Given, two interlocutors and the channel that attaches them to one another. The parasite, nesting on the flow of the relation, is in third position. Up to now, this model was adequate; it was the elementary link of the system. But now, the positions change. The guest becomes the interrupter; the noise becomes interlocutor; part of the channel becomes obstacle, and vice versa. Questions: who and where is the third man? These questions have fluctuating answers, functions of noise, time, and of the new relations of equality or similarity between the terms. The same and the other change places with the third.... A diagram is needed where the branches are not determined and where the cuts are not specified. (Serres, 1982: 53–54)

Deleuze finds this same functionality of the third mediating position in Foucault. "Even Foucault needs a third agency to coadapt the determinable and determination, the visible and the articulable, the receptivity of light and the spontaneity of language, operating either beyond or this side of the two forms. It is for this reason that Foucault said that the grappling implies a distance across which the adversaries 'exchange their threats and words,' and that the place of confrontation implies a 'non-place' which bears witness to the fact that the opponents do not belong to the same space or rely on the same form" (Deleuze, 1988: 68). This third dimension supports "the thought of the Outside." This mediating agency is in the family of chora, to be applied to an Internet public sphere and deliberative politics in Real Time.

Serres' information model is applicable to our electrate revision of consulting, addressing the crisis of the Time-accident. The triode diagrams as a "Y," the branch in the channel, that Serres calls the "joker," is a logical object. The joker is wild; its value is "Whatever." Wherever it comes in a series, "it permits it to bifurcate, to take another appearance, another direction, a new order. The only describable difference between a method and 'bricolage' is the joker. The principle of bricolage is to make something by means of something else, a mast with a matchstick" (Serres, 160). In dream logic the series is all jokers (162). In practice, the entity that may be counted on to play the joker is my personal experience, on condition that it be put into a series, made into a sinthome for a mystory. "His novelty is having injected his life into the produced object, instead of drawing his life from the chosen object. The only novelty is my improbable life" (170). The joker operates as a poetic figure. "This is not this; this is something else. This chain breaks away from redundance, identity or repetition. An object has to be found that can be spoken of in this way. Or a subject, it matters little. It is thus a vital experience that the rejected child never be himself. It is also a cultural constraint that a woman must metamorphose. It is a social experience that the one who is sacrificed is anyone. But it is especially a Judaic invention, an explosive novelty in the Fertile Crescent, that the one who is sacrificed is substituted, that suddenly, the victim is something else: a goat, a kid, but also the beginning of a completely other series. I shall call this object the joker" (160).

Here is what happens to the wilderness in the Jazz myth. The haitian is wild (a trump). One does not ask "Why haitians?" but looks for the Y circuit that they create.

Aporia

Over its long history the *I Ching* attracted many commentaries, some of which are considered to be an integral part of the book ("The Ten Wings"). We imitated this layering by looking for a match between the Miami mood and our disciplinary discourse. The experience of recognition passed from the uncanny to eureka when we "dialectized" the impounded haitians and the impasse of Barbara and Ron with Jacques Derrida's account of aporia (gathering the popcycle discourses of Family, History, and Career). In this way the zone not only answered Revelle's burning question, and revealed the attunement of the city, but gave guidance at a theoretical level regarding policy formation. The haitians impounded on the Miami River constitute a mise en scene of a theory. This scene shows both a specific answer to the official question of the inquiry (what is to be done about the image crisis?), and a theoretical answer to the question of method as such in electracy.

Exceeding Movement

The tour of a problem engages the basic metaphor of "method," referring to a way or path. The figurative passage through the steps of method and the following of hyperlinks online are mediated by Revelle's literal drifting through the zone. In the Greek tradition Philosophy began with a sense of wonder growing from an initial difficulty (aporia) resulting from conflicting arguments. The initial state of ignorance with respect to this difficulty was compared by Aristotle to a man in chains (translations of aporia include "difficulty, question, problem, with no way out") (Peters, 22). The response to the aporia was the application of "method" – an exploration of various "routes," "diaporia," a canvassing of previous opinions, leading to a proof, or solution, "euporia, lysis, a loosening of the chains," which is the heart of the philosophical procedure (22). Behind "Method" as a concept and practice (invented by Plato in *Phaedrus*), abstracted from a mise en scene of following a path, is a story of emancipation (the jazz story).

The River is a "mirror" in which the consultant's experience of identity is mediated by the image of the other. This encounter with the other is an act of identification, an experience of "recognition" whose original event is emblematized in the jubilation of an infant seeing its own reflection in a mirror (the original "eureka"). Choragraphy appropriates this experience of (mis)recognition as the locus of the choral tension between being and appearing, knowledge and meaning. The method folds together narrative with vision, text with image, to take advantage of the semiotic expansion of Aristotle's original moment of *anagnorisis* or revelation in tragedy into the semiotic "square of veridiction." The relevance of the semiotic square to choragraphy is the premise that chora is comprehension operating not by concept nor percept but act. Chora "does something" in the thymic or passional dimension of intelligence.

> The modern story requires considering the processes that unfold on the cognitive level, whether it be a question of observation, information,

> persuasion or interpretation, trickery, illusion, lies, or secrets. Greimas takes up this requirement (which has its origin in the dramatic function of "recognition" in Aristotle and also in the well-known analysis of the trickster figure in anthropology) through a series of audacious methodological decisions. In the first place, he quite openly splits "doing something" into "doing something pragmatically" and "doing something cognitively," where the latter branch sets up the acting subject as a noological subject distinct from the bodily subject. Next he apportions this cognitive doing something between two poles: persuasive doing something and interpretive doing something. The essential advantage of so treating the cognitive dimension in terms of doing something is that it allows him to submit the operations of knowing anything at all to the same transformation rules that govern action properly speaking. In this way, the inferences from appearance to reality, which interpretation consists of, are forms of doing something capable of being inscribed on a narrative traversal just as the other forms of doing something are. (Ricoeur, 1985: 54)

This conflation of action, knowledge, and narration suggests the possibility of a mode of thought experienced as a narrative of a place (such as the Miami River). Choragraphy finally is not a narrative, but a region in which a multitude of narratives intersect. This function requires the addition of another dimension to Greimas's semiotics. An important difference between instrumental convention and EmerAgency invention is this opening of the discourse of consulting to the "square of veridiction," hence adding input from the trickster to that of the scientist (not opposition but supplement). Everything depends upon the possibility of the testimonial writing a *recognition scene*. A consultation shows and unfolds this experience as doing something not only cognitively, but affectively: doing something well, or in a certain manner. The psychological dimension of this narrativized moment of recognition includes topologically the collective outside position from which "I" (the consultant) is "recognized" with the Other. In the event of the consultation, the position of the subject within the societal whole is registered, mapped, along with the positions possible in the system of values organizing that scene (Miami). The challenge of Real Time to choragraphy is that the egent must access the narrative development instantly from within the moment of decision. The augmented narrative in the digital prosthesis supplies the egent with this default wisdom.

In deconstructive consulting the problem-solution approach to method is not replaced but is put in tension with its unthought alternative – the pair "aporia-impossibility." The theoria tours the zone with this sense of both "way" and "no-way" (opposition not as a static dualism but as a dynamizing generator).

The narrative of a testimonial is organized not by conventional conflict, but by impasse. It is the impasse between belief and knowledge, that needs a chora.

> What was going to be at stake in this word [aporia] was the "not knowing where to go." It had to be a matter of the nonpassage, or rather from the experience of the nonpassage, the experience of what happens and is fascinating in this nonpassage, paralyzing us in this separation in a way that is not necessarily negative: before a door, a threshold, a border, a line, or simply the edge or the approach of the other as such. It should be a matter of what, in sum, appears to block our way or to separate us in the very place where it would no longer be possible to constitute a problem, a project, or a projection, that is, at the point where the very project or the problematic task becomes impossible and where we are exposed, absolutely without protection, without problem, and without prosthesis, without possible substitution, singularly exposed in our absolute and absolutely naked uniqueness, that is to say, disarmed, delivered to the other, incapable even of sheltering ourselves behind what could still protect the interiority of a secret. (Derrida, 1993: 12)

The mapping between "Crossroads" (the photograph and its title) and aporia constituted a new understanding of how to go on, in that it revealed the impasse as threshold (Derrida playing Confucius in the Ka-Ching). The explanatory effect of this figure moved first from the image to the theory: the deconstructive aporia became intelligible in the context of a social, lived embodiment of an aporetic condition of the river. The crossroads X or chiasm served Derrida from the beginning as a figure of a doubling and crossing at work in language and history. The X does not symbolize the unknown, he said, but is "a kind of fork (the series 'crossroads, quadrifurcum, grid, trellis, key, etc.') that is moreover, unequal, one of the points extending its range further than the other: this is the figure of the double gesture, the intersection" (Derrida, in Gasche, 1986: 172).

Aporia is for the inventional consultant what problem is for the conventional one. These two "ways" are not oppositional but complementary (literacy + electracy). A tension between problem and aporia is produced by refusing the calculative, analytical loosening of the bonds of contradiction, and instead intensifying and multiplying them. "Method" as metaphor is the passage, way, or path (odos, methodos, Weg) from problem to solution. The deconstruction of consulting works precisely on this metaphorical dimension. The point of departure of articulating this difference with method is to experience the difficulty itself as such, to witness it and meditate with it. The function of the choral image ("Crossroads" in this case) is to evoke this feeling of identification (the haitian in me). "The question of knowing what it means 'to experience the aporia,' indeed to put into operation the aporia, remains. It is not necessarily a failure or a simple paralysis, but sterile negativity of the impasse. It is neither stopping at it nor overcoming it. (When someone suggests to you a solution for escaping an impasse, you can be almost sure that he is ceasing to understand, assuming that he had understood anything up to that point.) Does one then pass through this aporia? Or is one immobilized before the threshold, to the point of having to turn around and seek out another way, the way without method or outlet of a Holzweg or a turning that could turn the aporia – all such possibilities of wandering?" (Derrida, 1993: 33).

This theoretical context shows that the relationship between the macroosm and microcosm in choragraphy is dialectical, with explanatory effects moving in both directions. Limit as such may not be understood, but must be undergone. "Crossroads" gives not a solution to but an image of aporia, and in so doing "figures" the experience, thus composing a map. It is just such a map that is needed in order to tour a problem, in order to rehabilitate global common sense. The title for Miautre as vortext is "Aporia." Within the larger setting of an online divination system, "Aporia" is but one situation, whose functionality depends ultimately on its placement within a system of other choral situations. If the Aporia vortext were the answer to a burning question, what would it mean as a parable? The juxtaposition with Chinese epistemologies and cosmology reveals finally that what we found in Miami was prepared in advance as the default condition of Western reason. In contrast with a preference for detour and obliqueness characteristic of Chinese civilization at all levels, Jullien observed in Classical Greece, and hence in the West in general, a commitment to confrontation – an "antilogy" – found in every sphere of life from logic to military strategy. "In this face-to-face antagonism, can we not detect not simply a point of departure but a chosen position continuously at work in Western logic?" (Jullien, 2000: 48).

Here is the map. Do you find yourself at an impasse? Welcome to the Western imagination. An invitation: who will compose another chora?

Open Virtue

What about the heroes if not of postmodernity then of electracy, keeping in mind the change of identity formation that is part of a shift in the apparatus? Electracy, that is, emerges within the industrial city, at the same time that the Frontier myth is being formulated. Berthoff points out where to look for the change underway in our own moment when he locates the core of "virtue" that he takes to be (with Trilling) the true topic of literature. From the entire history of virtue in the Western tradition he extracts the quality of human vitality: "The great subject of imaginative literature is, in a word, virtue, broadly understood as the vital presence and force within creaturely being wherever met and consequently, too, within historical experience. Virtue and (to complete the formula) the odds against it, including those inwardly generated, as it acts to maintain itself as a living force" (Berthoff, 15). This sense of virtue is not concerned with moral virtue and vice nor an inventory of behavioral excellence. It resonates with Pound's "vortex" and the notion of mood or Stimmung, in a way that is most useful for moving on in an epoch reduced to the passions (emotivism).

> *Arete*, the Greek equivalent, through identifying first of all the character of an arms-bearing male leadership, was also regularly used for the nonhuman and inorganic –in Herodotus it is applied to animals, objects, and the land itself – besides serving Homer to name the special excellence of the home-keeping wife Penelope… In German and Slavic speech, where classical derivations are more obviously hybridized and imported, the same range of meanings is distributed over a diversity of terms that appear to have undergone comparable transformations. In German it may be *Mut*, in its older and folklorish connotations of *mettle* and mood-determining *feeling* as well as – more restrictively – *courage*, that comes nearest matching the attributive range of *virtus* and *arete*.
>
> In such ways a physical or physiological component has remained active within the broad concept of virtue; a component which overrides the domain of what is conventionally teachable and cannot be separated from the primal mysteries of birth, blood, sexuality, genius, or creaturely bias. The Classical and Renaissance doctrine of the humors directly responds to this understanding, as do later pseudo-doctrines like phrenology and physiognomy. It is this irreducible physical presence and consequence, this element of the constitutionally given, that distinguishes virtue from conceptions of moral goodness or from such behavior as properly results from correct training. It appears that every systematic moral calculus must eventually give way to this element or learn how to accommodate its consummating and transforming power. (18)

I noted previously a word available to name this life force or physical vitality that Berthoff might have used to avoid some of the confusion resulting from his efforts to abstract a core signification of "virtue" from its encyclopedia of historical uses. Promoted by Marcos Novak as the quality of liquid (digital) architecture, the word is "duende," as elaborated in the writings (essays and poetry) of Garcia Lorca, and which has as synonyms all the mood terms associated with the syncretic musics of the Black Atlantic (Samba's *saudade*, Tango's *mufarse*, Jazz's *blues*). Here is the "virility" at the heart of the modern crisis. Here is the name for what was missing, what had been drained from the Western tradition by its denial and abjection of the body. Jazz and the blues have been described as a gift given to civilization during the potlatch of colonialism, making it possible to think through the body just at the moment in our history when we need it most. Berthoff calls attention to this event in its lexical manifestation. "One of the livelier lexical events of recent times was the double reactivation of 'soul,' among American blacks, and 'head,' among heads, as unitary terms for the whole self, fully alive and reattached to an ultimate power source (as in Allen Ginsberg's 'ancient heavenly connection to the starry dynamo in the machinery of night')" (Berthoff, 58).

Hubert Dreyfus is just one of those who has worried about the consequences for society and for individuals if the body is missing from cyberspace, if the apparatus is appropriated by the old Neoplatonic dream of transcendence. The key question for educators in electracy is whether the bodily presence required for skills and culture may be supported by the Internet (Dreyfus, 49). The answer to this question from our default position within the paradigm of our discipline – the poet as modern hero – is that the practices of choragraphy and choramancy are designed to bring the body (in its force as duende) into a virtual public sphere initially through an aesthetic engagement with the world.

PHOTOGRAPH: BARBARA JO REVELLE

The artists at the very least invented the logic of imaging guiding EmerAgency consulting. The avant-garde arts tested over the last century and a half the formal devices of collage-montage-bricolage that are native to electracy, providing the new ratio by which to construct meaning.

In the larger context of the crisis of the information economy, what is it possible to say within the episteme of the FRE career discourse, with respect to the project of establishing the virtue of electrate identity?

> And through the nineteenth century art has as one of its chief intentions to induce in the audience the sentiment of being, to recruit the primitive strength that a highly developed culture has diminished. To this end it proposes a variety of spiritual exercises, among which are suffering and despair and cosmic defiance; conscious sympathy with the being of others; comprehension of the processes of society; social alienation. As the century advances the sentiment of being, of being strong, is increasingly subsumed under the conception of personal authenticity. The work of art is itself authentic by reason of its entire self-definition: It is understood to exist wholly by the laws of its own being, which include the right to embody painful, ignoble, or socially unacceptable subject-matter. Similarly the artist seeks his personal authenticity in his entire autonomousness – his goal is to be as self-defining as the art-object he creates. (Trilling, 99)

The FRE proposal for extending democracy to the virtual city is to connect netizens with an experience of the Open, in order to activate this sentiment of being crucial to the judgment of well-being within the civic sphere.

If the core attitude (orientation) of literacy is problem solving (codified in cognitive science schema theory), what is the equivalent for electracy? This core experience has been described under a variety of terms. Rilke called it the Open, and Heidegger took his cue from Rilke for his account of the Clearing. Bergson called it Time and Foucault called it the Outside. These accounts emphasize that the Open is not thinkable or sayable within the conventions of instrumental calculation. It does not lend itself to analysis but operates in a quantum manner, as mood or Stimmung. Art has always drawn upon it, but it is also relevant to practical reason as well as scientific and political problem solving. Proust wrote one of the best accounts of the Open as aesthetic event. It is the moment when Swann first hears Vinteuil's Sonata, that one phrase or motif "that had just been played, and had opened and expanded his soul, just as the fragrance of certain roses, wafted upon the moist air of evening, has the power of dilating our nostrils."

PHOTOGRAPH: BARBARA JO REVELLE

> And so, hardly had the delicious sensation, which Swann had experienced, died away before his memory had furnished him with an immediate transcript, summary, it is true, and provisional, but one on which he had kept his eyes fixed while the playing continued, so effectively that, when the same impression suddenly returned, it was no longer uncapturable. He was able to picture to himself its extent, its symmetrical arrangement, its notation, the strength of its expression, he had before him that definite object which was no longer pure music, but rather design, architecture, thought, and which allowed the actual music to be recalled. This time he had distinguished, quite clearly, a phrase which emerged for a few moments from the waves of sound. It had at once held out to him an invitation to partake of intimate pleasures, of whose existence, before hearing it, he had never dreamed, into which he felt that nothing but this phrase could initiate him; and he had been filled with love for it, as with a new and strange desire. (Proust, 160)

As the semiotician Algirdas Greimas explained in his study of the passions, "before hearing Vinteuil's phrase, Swann is an ordinary individual, without ideals, without a project, subsisting intellectually and emotionally on little things, void in an insignificant world. Vinteuil's phrase is the figure of this initial operative subject, for it will indeed summon a place, designate in his mind the zone where, as

Proust writes, Odette's name will be inscribed" (Greimas, 14). If the place opened by the aesthetic experience makes it possible for Swann to fall in love, this "love" may be generalized (as it has been from Plato to Heidegger) as "philo-Odette" (in Swann's case) as well as "philo-Sophia": it is the opening to the world that allows things to matter for the subject. Greimas makes explicit the interdependence of the two orders of "value" in experience: the structural and the axiological (formal and moral). "In other words, for signification to occur and for tensivity to be stabilized, the only solution for the operative subject is to categorize the loss of the object, and that is why negation is the first discrete operation. It is only on this condition that, by the introduction of discontinuity into continuity, behind the shadows of value, the subject will be able to have knowledge of the object. Without contradiction, summons simply determines a pure singularity in the tensive continuum and does not succeed in making signification occur. This is why, after having seemed 'singular' and irreducibly individual Vinteuil's phrase appears as a network of contrasts, internal negations, to be discovered and uncovered, and ends up being the sign of an absence, an absence whose existence Swann never suspected, and which will give meaning to his life (15).

The egents' testimonials hold in external storage a rendering of this system. It takes choragraphy to map such a place for a collective memory. Choral consulting produces in the egent the equivalent of Swann's experience of Vinteuil's Sonata. Indeed, the goal is to approach the Open not as appreciation (only) but as making (as making from the point of reception). The egent experiences the Open by composing a time image.

Swann's adoption of the musical phrase as a personal measure making experience possible is the touchstone scene for almost every theorist of the past century, from Benjamin to Deleuze. The instruction is elementary, already commonsensical, but until now not actionable as practical reason. Hannah Arendt proposed that Kant's aesthetic judgment provided the ratio extendable to ethics and politics. Here is an agenda for the EmerAgency.

I'm Not Stuck; I'm Dancing

The no-way of aporia is not an end or stopping point of thought but the opening of another way that may be for electracy what method was for literacy. The tension between problem and aporia (a productive relationship of conceptual and image categories) suggests the need to identify a "movement" that is for event what method (the road or path, whether clear or blocked) is for concept. The metaphor for aporia as "way" comes from the Afro-Caribbean scene for which the Haitian harbingers are a metonym. Referring to the "places of aporia in which I have found myself regularly tied up, indeed, paralyzed," Derrida explained that he "was then trying to move not against or out of the impasse but, in another way, according to another thinking of the aporia, one perhaps more enduring" (Derrida, 1993: 13). The idiom of this alternative way is the phrase "il y va d'un certain pas [it involves a certain step/not; he goes along at a certain pace]" (6). This deconstructive "step" remains open to creolization. Our consultation returns to the metaphorical root of "method" to imagine this jewgreekcreole state of mind as a "certain way of walking." "For me the aporia is not simply paralysis but the aporia or the non-way is the condition of walking: if there was no aporia we wouldn't walk, we wouldn't find our way; path-breaking implies aporia. This impossibility to find one's way is the condition of ethics" (Derrida, 1999: 73). Our practices emerges from the blocking of these two steps (French theory and Creole practice). Choragraphy and choreography do converge, after all.

Make an impossible recommendation.

To shift the Miami River from the category of problem to that of aporia, to look not for its solution but for that in it which is *impossible* for us, is the advice, relevant to sinthome as choral category. The impossible reveals the metaphysical border, the negative dimension of our identity. The zombie is the suture, the upholstery tack in the mattress of the procrustean bed that ideology made for us, and in which we lie. What does this fact about my situation make impossible? One consequence of staying on the side of aporia is to realize that the haitians represented not a problem that could be solved, but *a state of mind* for rethinking all problems, and "problem" as such. The generalizability of this discovery remains to be tested: The problem is integral, not anomalous (the outside is inside). The block in pathic knowledge represented in the aporia evoked in "Crossroads" is grasped as an ontological "freeze frame." "It is like a freeze-frame which both indicates its basic (or bass) position in the polyphony of chaosmic components, and intensifies its power relative to them. Thus it does not constitute a degree zero of subjectivation, a neutral, passive deficient, negative point, but an extreme degree of intensification. It is in passing through this chaotic 'earthing,' this perilous oscillation, that something else becomes possible, that ontological bifurcations and the emergence of coefficients of processual creativity can occur" (Guattari, 1995: 81–82).

Ports of Opportunity

The Miami River is a port, a crossroads, an interface joining North and South America, as much a part of the Caribbean basin as it is of the United States (a lounge in the Orlando airport is named "the Floribbean"). The haitians waiting on their impounded, burned, substandard, sinking, mortgaged, repossessed boats, with their used cargoes collected in warehouses that have been looted, vandalized, rain-damaged, scattered, auctioned off, given away, these temporary denizens of the river, these guests, show and reveal, give the location of the port as door, passageway, and opportunity. The goal of the theoria is to find the door (threshold) in a zone. It will always be blocked, and that is how one finds it (but this claim is what remains to be replicated or falsified by other theorias). "'Opportunity' comes from the Latin 'porta,' which is an 'entrance' or 'passage through.' The word is associated with doors and entranceways (portal, porch, portico), and an 'opportunus,' then, is what offers an opening, or what stands before an opening, ready to go through. For the Romans, a 'porta fenestella' was a special opening that allowed Fortune to enter. The Greek root is 'poros,' which is a passageway for ships but also any passageway, including one through the skin, that is, a pore. 'Poroi' are all the passages that allow fluids to flow in and out of the body. A pore, a portal, a doorway, a nick in time, a gap in the screen, a looseness in the weave – these are all opportunities in the ancient sense. Each being in the world must find the set of opportunities fitted to its nature" (Hyde, 1998: 46).

As Lewis Hyde elaborates, tricksters hang around 'poroi' and are involved in both closing and opening the way. "Kairos" is the name for that hypericon of opportunity, the timing of shooting the shuttle through the rising and falling warp threads. Hyde links 'poros' and 'kairos' to the "lucky find." "The fabric of social and spiritual life contains, shapes, and binds our lives; a lucky find tears a hole in the cloth so that we can slip free.... Against all this – against the nets of reciprocity, the grid of ethics, the weave of fate – we have the find which is a lucky break, a hole in the surrounding cloth. The lucky find is an opportunity, a pore or penetrable opening in an otherwise closed design" (133). Hyde's interest in the trickster stories found throughout the world is the common theme they express regarding the importance of maintaining a balance between the essential and the accidental. The man who knows Indians, and the double consciousness of the African in America, are examples of a trickster capacity for occupying both sides of a frontier.

Hyde calls attention to the inadequacy of Aristotle's codification of categories (a major step in the invention of the "concept," and of metaphysics itself). "In the *Categories* Aristotle argued that we ought to be able to look at any group of things, say 'apples' or 'human beings' or 'jazz recordings,' and determine what is essential to membership in the category and what isn't.... Accidentals are present by chance, essentials by design. Accidentals are changeable and shifting; essentials are stable. The real significance of a thing lies with its essences, not with its accidents" (97). Baudelaire famously defined "modernism" as shifting attention to the opposite side of this pair, signaled by attending not to the eternal dimension but to the circumstantial, ephemeral fashion of the day. The trickster tradition also counters this preference given to essences. The ability to work with contingency constitutes the special craft of the trickster and points to a paradox of image categories.

Sublime Hospitality

That tourism is so intimately involved with the hospitality industry makes it a good site for testing the capacity of Entertainment as the institution supporting the emergence of electracy. Derrida's introduction of the language of hospitality into the history of democracy demonstrates the practicality of thinking tourism and theory together (theoria). The history of Western thought from Aristotle to the present theorized political gathering into communities by means of the logic of exemplarity, and the exemplum for political relationship is friendship. The model includes an oppositional interdependency between the concepts of friend and enemy (hence politics has been condemned to melodrama). The friends have always been conceived of as men, specifically as brothers, with the relationship generalized and made universal historically in the "fraternity" of the French Revolution (the Rights of Man).

Alasdaire MacIntyre showed that "friend" as a bond of community has disappeared from modern experience. The revolutionary Jacobin virtues of liberty, equality, fraternity were the final attempt to save this sense of a common responsibility, their obsolescence proven by the resort to terror to impose friendship as a public collective practice in a social order in which it survives only in the private realm. "The true lesson of the Jacobin Clubs and their downfall is that you cannot hope to re-invent morality on the scale of a whole nation when the very idiom of the morality which you seek to re-invent is alien in one way to the vast mass of ordinary people and in another to the intellectual elite. The attempt to impose morality by terror – the solution of St. Just – is the desperate expedient of those who already glimpse this fact but will not admit it" (MacIntyre, 221).

The Haitian is neither enemy nor friend. A part of the effect of "trouvaille" in bringing Derrida into the vortext was his opening up certain pivot words (choral words), showing an affiliation between "river" and "arrivant" – a term that could be used to describe the Haitian boat people.

> What takes place with the aporia? What we are apprehending here concerning what takes place also touches upon the event as that which arrives

> at the river's shore ["arrive à la rive"], approaches the shore, or passes the edge – another way of happening and coming to pass by surpassing. All of these are possibilities of the "coming to pass" when it meets a limit.... What is the event that most arrives? What is the arrivant that makes the event arrive? The new arrivant, this word can, indeed, mean the neutrality of that which arrives, but also the singularity of who arrives, he or she who comes, coming to be where s/he was not expected... and such is hospitality itself, hospitality toward the event. One does not expect the event of whatever, of whoever comes, arrives, and crosses the threshold – the immigrant, the emigrant, the guest, or the stranger.... Such an arrivant affects the very experience of the threshold, whose possibility he thus brings to light before one even knows whether there has been an invitation, a call, a nomination, or a promise. (Derrida, 1993: 33)

At the historical level Derrida's image covers the Miami River itself as an embodiment of the problematic of borders, classification, categories: the question of what is inside and what is outside that is central to the experience of epiphany, attunement, extimacy, goes to the heart of Western metaphysics. The denizens of the river are figures of the other, are the arrivants (those who come), caught within this problematic space or place of a frontier, limit, edge. Our own destiny may be read (is revealed) in their "situation," and epiphany is precisely the art of articulating situations, which suggests that the anagogical dimension may be rendered popularly through the allegory of the river. A tour of the problem/aporia writes the hole of our situation (how things are with us). It is not that the fate of God-Is-Able replaces Derrida, but that the juxtaposition (the combination of tourism and theory) is mutually illuminating.

The juxtaposition of the river image and aporia constitutes a picture theory, an emblematics. What is revealed is that "division," the way things are sorted into kinds at every level and scale, is mutating now, so that our Frontier scenario must similarly be revised. In the politics and ethics of hospitality the "enemy" is replaced by the stranger, foreigner, immigrant, as figures of the other (but for a generalized other who need not be literally foreign). The exemplary relationship for thinking this post-conventional democracy is the host/guest relationship, relevant to a politics of well-being. As one of the commentators stated in the Ken Burns documentary on Jazz, civilization may be summed up in one word: "welcome." This word summarizes the difference between the Frontier and Jazz policies regarding the defense of civilization (First Strike, and Welcome, respectively).

To state his question Derrida formalized the aporetic sentence, which in turn draws upon the rhetoric of negative theology, to evoke a thought of that which is still to come, absolutely, and therefore lacks concept, shape, or eidos. The question is formulated using the structure "X without X." "The central question of this essay would be that of a friendship without hearth. Ultimately, a friendship without presence, without resemblance, without affinity, without analogy. Is an aneconomic friendship possible? Can there be any other friendship? Must there be another? Can one answer this question otherwise than with a 'perhaps' – that is, by suspending in advance the very form of a 'question,' and the alliance of the 'yes' – in order to think and to dream before them? And must not this reflection account for a certain end of Lysis, in its final leave-taking?" (Derrida, 1997: 155).

A quick adaptation of Derrida's X to our consultation (Florida tourism) might read: *a hospitality industry without commodification*. Applied to the policy question of the Caribbean Code, this formula proposes the suspension of the interdiction and boarding of Haitian vessels by the Coast Guard, to be replaced with the "boarding" of "haitians" at city expense: *learn about boarding from boarding*. Here we have one formulation of the impossible: a key to the city for everyone on the Rose-Marie Express, the God-Is-Able, the Deliverance Express. This sentence must be suspended, since it resonates with the one that Socrates made at his trial, when given an opportunity to name an alternative to the death penalty, after being found guilty of impiety: "free maintenance at the state's expense" was what he deserved, he said. Nothing radical here, an alien might reply who has been housed in one of the detention centers for those arriving in Florida without proper documentation.

Guest's Dilemma

The various conventional problems addressed by Miami River policy (such as immigration) cumulatively evoke the problematic of democracy and its future in the apparatus of electracy. Does virtual speed eliminate the urban condition? Derrida noted that the limits of the friend-enemy model of politics have been exposed by the decline of the Cold War. This context also answers those who doubt the potential of aporia as a guide for policy, since policy during the Cold War was guided by the paradox known as Prisoner's Dilemma. This dilemma, which originated as a parlor game, was at the heart of the invention of game theory by John von Neumann, and directed thinking about nuclear deterrence (Poundstone). Prisoner's Dilemma works best when the interests of the two parties are absolutely opposed, with each party pursuing its own interest at the expense of the other (exactly the premise of dialectical argument formulated in Aristotle's *Topics*). In fact, the policy conclusion of this dilemma was "impossible" and never enacted (the odds favored a nuclear first strike). In the post-Cold War moment of global terrorism, however, American foreign policy under George W. Bush officially embraced the statistical odds of first strike strategy.

The politics of the stranger, in contrast, suspends the problematic of interests, without expectation of reciprocity. It remains to be seen what sort of policy might be entailed by the aporia of hospitality, but its impossibility in principle is nearly the inverse of Prisoner's Dilemma. "For unconditional hospitality to take place you have to accept the risk of the other coming and destroying the place, initiating a revolution, stealing everything, or killing everyone. That is the risk of pure hospitality and pure gift" (Derrida, 1999: 71). Pure gift, in other words, constitutes the impossible necessity of ethics. What is at stake may be seen when reminded that during Taliban rule in Afghanistan Osama bin Laden was known as "the Guest."

Learn about the present from a present.

My Friend, My Other

The shift from a politics of friendship to one of hospitality motivates the name for the chora of the virtual city, a name that will distinguish it from Miami and Myami alike. At the level of the signifier, a series unfolds (creolizing French, Spanish, English) as follows: Miami, Mon Ami, Mi Amigo, My Friend, My Other, Mi Otro, Miautre. Against the rule of "belonging" that enforces language as a shibboleth, deconstruction opens language to the Creole remainder, to heterogeneity, dreamwork, the obliqueness of literature. The original shibboleth was the name of a river. Judging by the way Americans pronounce "Notre Dame" (noter daym), "Miautre" will be pronounced "my-odor." In other words, Miautre is "funky." Or as the poets say, the ideogram does not follow a path but creates one as it unfolds in language ("inveniere"). A macaronic pun motivates the link between Miami and this deconstruction of a politics of friendship. If we accept the luck of this series joining the words to the themes of friendship (Mi Amigo, Mon Ami) we produce a name for the Miami River as chora: MIAUTURE – a place to invent (virtually first of all) the politics of hospitality. In naming this virtual civic place we have already begun to avert the general accident of the Internet. The work of testimonial is not that of solutions but of witness. Miautre opens the virtual community.

Welcome.

AFTERWORD

Electrate Virtue

Gregory L. Ulmer and John Craig Freeman

Choragraphy Today

Why a new edition of Miami Virtue today? The FRE continued to develop choragraphy following the experiment at the Miami River in Florida, including the evolution of EmerAgency consulting into konsult as the genre of electrate education. The present condition of a General Accident (happening everywhere simultaneously, as Paul Virilio warned) calls for a review of resources and probes available to support well-being against disaster. Our present konsult on the U.S./Mexico border addressing Climate Change migration takes up choragraphy again, with Miami Virtue as a benchmark of learning and point of departure. We have resisted documenting a genealogy of choragraphy in Ulmer's theory over the past four decades, from Applied Grammatology (1985) to Konsult: Theopraxesis (2019), although it is worth noting the place of Miami Virtue in that history. The FRE began work on the river zone in the mid-1990s, with residency summer of 1998, with book and "Ulmer Tapes" completed 2001, but not published until 2012, by the online Small Cities Imprint, directed by Will Garrett-Petts. Choragraphy was introduced in Heuretics: The Logic of Invention (1994), retaining "under erasure" the original spelling of the ancient practice (chorography), the subsequent change of spelling indicating how different from the original practice was our version motivated by Derrida's grammatology. Miami Virtue documents the first field test of the practice, with further probes treated in Electronic Monuments ("Wishing-Y" konsult on 9/11, with Will Pappenheimer, 2005) and Konsult: Theopraxesis ("Murphy's Well-Being" with Jack Stenner and Barbara Jo Revelle, 2019). The genealogy would allow a review of the various devices of electrate pedagogy generated or appropriated along the way: mystory, the heuretic CATTt, scope of wide image, MEmorial, theopraxesis, konsult.

Any survey of the theory becomes more intelligible and useful correlated with a record of John Craig Freeman's investigations of choragraphy and electracy in digital arts.

 | DOI:10.1163/9789004534643_017

Imaging Place

In 1998, during the fieldwork at the Miami River for the first edition of *Miami Virtue*, John Craig Freeman joined Barbara Jo Revelle on location. Although Revelle's residency there was featured in the book, Freeman was experimenting with a new artistic practice which would be native to electracy and incorporate the theory that Gregory Ulmer was developing and the lessons learned during the project. This work became *Imaging Place*.

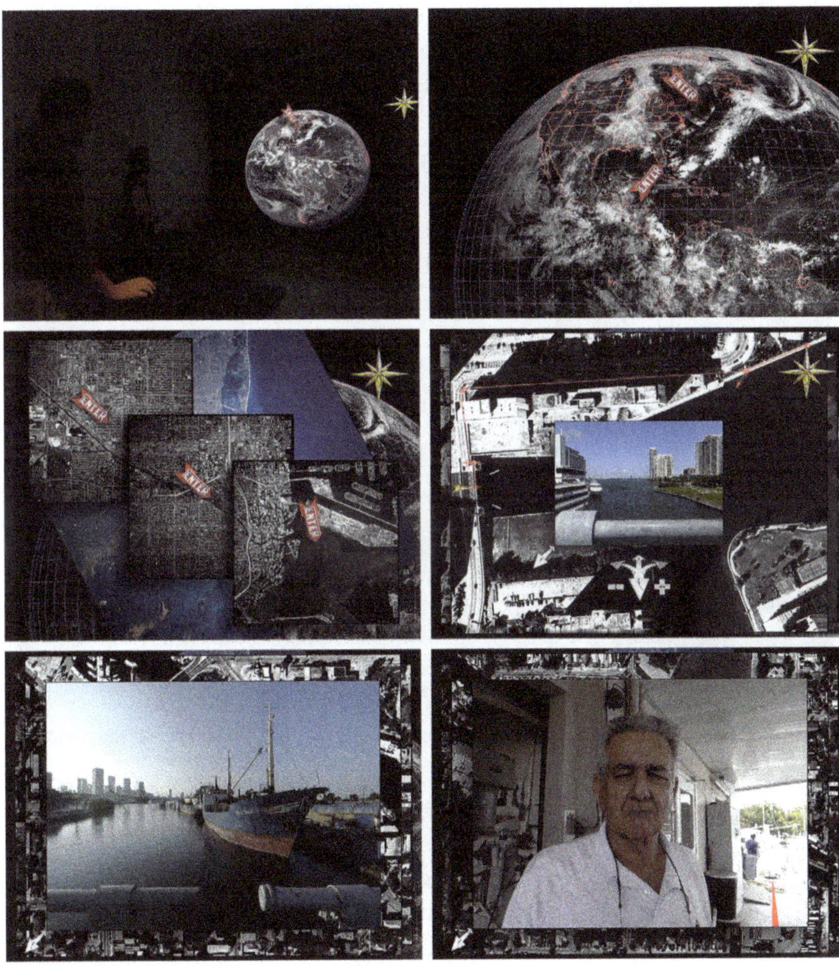

Imaging Place: Miami River, 1997, interactive place-based proto-virtual reality installation and screenshot sequence
PHOTOGRAPH: JOHN CRAIG FREEMAN

Imaging Place was a place-based proto-virtual reality artwork that combined panoramic still photography and video, with a game-like interface to document situations where the forces of globalization are impacting the lives of individuals in local communities. It was developed as a standalone application and exhibited extensively as interactive installation art between 1997 and 2007. In its final form, it included dozens of stories, with hundreds of individual locations from several countries across four continents. It still exists on a legacy computer running a legacy operating system.

Also, in 2001, wanting to collate the extensive theory Ulmer was developing in order to translate the ideas across media, Freeman sat down with Ulmer and recorded over twenty hours of video. This video archive titled The Ulmer Tapes was the reference point for the development of *Imaging Place* and all of Freeman's subsequent work (Freeman, 2001). Links to this video archive were included in the first edition of Miami Virtue, as it was published as an eBook (Ulmer et al., 2012).

Even the most recent work (in progress), titled *Climate Change Migration Stories*, makes use of the concept and the methodology emergent from the Miami River in 1997, as a strategy for encountering a subject location.

Climate Change Migration Stories: Border Crossing, Sasabe AZ, 2022, virtual scene for virtual and augmented reality
PHOTOGRAPH: JOHN CRAIG FREEMAN

Go to the place. Find out what is happening, what is true, what is concocted misinformation or fake news. Include the voices of the locals. Consult the oracles, and report back in the public square. In short, a contemporary version of the ancient Theoria, theory-tourism, which the FRE called Solonism.

Climate Change Migration Stories is an augmented reality public art project designed for exhibition in public squares and other publicly accessible spaces. Using new forms of virtual and augmented reality technology, this project documents stories of climate change migration, told by people who are directly impacted, based on their perspectives and world views.

The audience encounters the project through way-finding signage placed at the physical location, with information and instructions about how to experience the augmented reality content. Users guide themselves with mobile phones on a walking tour through a series of virtual scenes. The experiences consist of world scale, photorealistic scans created on location in and around the U.S./Mexico border.

Users walk through the physical site of the exhibition, as they are transported virtually to the border. When the user encounters a scan of a person, spatialized audio begins to play. The audio includes stories that were collected on location. These stories have been translated into both Spanish and English. The scenes also include ambient audio such as sounds of birds, traffic and music from street vendors' carts, making the experience more immersive.

Imaging Place made use of QuickTime VR, an early digital panoramic photography technology based on a largely command line software tool to seamlessly stitch individual photographs together to create an immersive image.

Imaging Place: Miami River, NW 17th Avenue Bridge from Sewell Park, 1997, panoramic photography for QuickTime VR
PHOTOGRAPH: JOHN CRAIG FREEMAN

In addition to specialized software, this technology required a custom tripod head that offset the nodal point of the camera lens to be directly over the pivot point of the tripod.

Although this type of photography seems strictly technical, Freeman applied concepts and techniques inspired by Guy Debord and the Situationist International outlined in Miami Virtue. Freeman drew a correlation of the nodal point of the panoramic image to the Situationists' notion of the plaque tournante. The panoramic simultaneously represented the centrifugal quality of the image radiating in all directions from the center of the camera lens nodal point, as well as a centripetal force acting as an attractor, drawing one to that specific location in the urban experience, and by extension, the user navigating *Imaging Place*. The act of planting the tripod acknowledged the psychogeographical vortex (node of ambient mood) present at that location.

Imaging Place: IMiami River, Swing Bridge, 1997, Barbara Jo Revell interviewing Captain Simon Lubin, interactive place-based proto-virtual reality screenshot sequence
PHOTOGRAPH: JOHN CRAIG FREEMAN

Freeman extended this technique by creating a multi-nodal network of panoramic images which he connected using dolly video clips from one node to the next, and then by developing a navigable digital interface using computer games as an interface metaphor. In an attempt to replicate the intention of the psychogeographical dérive or drift through the site and to capture that experience in a representational form, Freeman created an interactive artwork that allowed users to navigate the entire Miami River, to zoom into dozens of individual locations, from the multimillion-dollar condos along Brickell Avenue at the river's mouth, to the inland backwaters, which were once part of the Everglades.

In the work at the border, rather than panoramic photography, Freeman used a LiDAR (light detection and ranging) scanner. The device is capable of extracting a fully three-dimensional, world-scale model of a location. The scanner spins an infrared pulse light laser on its vertical axis while the entire device slowly rotates atop a tripod. Points in a digital XYZ coordinate space are created by measuring the distance of surfaces, based on the time it takes for the light to be reflected back to the device.

Climate Change Migration Stories: Nogales AZ, 2022, on location at the border wall with border patrol vehicle and LiDAR scanner
PHOTOGRAPH: JOHN CRAIG FREEMAN

Again, the mechanics of the scanner reflects the conceptual plaque tournante (hub). The resulting model can be used to create experiences for virtual and augmented reality. The scene loads with the user's feet located dead center of the vortex, represented by the occlusion shadow, the circle on the ground where the laser pulse is blocked by the base of the scanner itself.

Whereas *Imaging Place* applied a game-like interface that allowed the user to navigate complex multi-node scenes, the virtual and augmented reality equivalent allows the user to walk through the scene, making the experience inhabitable, an inhabitable monument. A series of placemarks, when tapped, advances the user from node to node. Moving through the space becomes an act of reasoning. With Ulmer's theory developed during the production of *Miami Virtue* still informing the recent work of Freeman and the FRE (Florida Research Ensemble), this afterword testifies to the continuity of two decades of development leading up to our current practice of konsult as the educational genre of the digital apparatus.

Climate Change Migration Stories: Sasabe Store, Sasabe AZ, 2022, virtual scene for virtual and augmented reality
PHOTOGRAPH: JOHN CRAIG FREEMAN

Aliveness

What is our present theoretical understanding of konsult as choragraphy? An inspiration for our experiment in digital education is Heidegger's invocation of the Presocratics and Anaximander in particular as part of his method of destruktion, returning to the beginnings of Philosophy in order to make a new start, the other beginning, to follow a path discovered by the Presocratics but abandoned or forgotten, in favor of the propositional method preferred in the Academy and thereafter in the Western tradition. The Academy proposed that Nature speaks Mathematics. The scientific logic initiated there found its end (telos) in the Gestell of technology, enframing the world as Standing Reserve for production (instrumental utility). Heidegger retrieved an alternative modality of truth, Aletheia (disclosive withdrawal), exemplified in the Ancient world explicitly in the Oracle at Delphi, which, as Heraclitus said, taught that when Nature "speaks," it neither reveals nor conceals but intimates. Nature speaks poetry as well as mathematics. Heidegger, concluding that the other beginning could be reopened by means of poetry and art, built his existential phenomenology relaying the works of modern poets, from Hölderlin to Rilke, and painters, Cézanne and Klee in particular. His essay, "The Origin of the Work of Art," is a resource for konsult, developing a theory of design by appropriating the word Riss, which means both rift and design, to name what he learned from the artist Albrecht Dürer: Art lies hidden in nature: he who can wrest it from her, has it. This "wresting" (rift) opens chora between Earth and World (nature and culture). Choragraphy maps the Open (gap, interval, hole).

The analogy guiding our experiment is between first and second Phusis, mathematics and aesthetics. In his OWA essay, Heidegger used the Greek temple as a primary example, treating it in aesthetic rather than mathematical terms, thus anchoring the analogy, given Anaximander's first beginning vision occurring on the site of the original temples in Ionia, the temple of Hera on Samos, for example. The method of destruktion operates by reoccupation (Blumenberg) or retrieval (McLuhan). We reoccupy the original questions of literacy, while seeking different answers. We need to encounter Being as it is given today.

The question concerns ontology. How does being (life) show itself in aesthetics? The significance of this framing should be noted: the assumption of electracy is that causality in the electrate apparatus is identified through the Anthropocene as humanity itself (human capabilities or virtues projected collectively in institutions). Embodied humanity is to electracy what God is to orality or Nature to literacy The Anthropocene is anthropomorphic, if not anthropocentric. The manifestation of ontology in art is the effect of aliveness (vitalism). The myth or legend of Pygmalion dramatizes this effect: The artist's desire projected into the created object brings it to life (Galatea), producing experience of presence. Virtual presence is the scene of ontology today. Mary Shelley's Frankenstein is in this tradition. This projection of desire as ontological force is manifested in social reality as fetish, appropriated from animistic primitivism by Marx and Freud, to theorize the effect of alienation in modernity. Through alienation, reification, objectification, humanity lost intuitive connection with its own agency, failing to recognize its own powers in the social and psychological conditions of the industrialized city. We know (scientifically) that human activity is responsible for Climate Change, but politically we don't believe it, and are unable to control the appetites in whose service the disaster unfolds. In fetishism we fail to recognize that the desired object (commodity) seemingly external to us, is a projection of our own spirit. Surplus value generated by commodities (Marx) created profit. Lacan modelled surplus enjoyment on Marx's insight to account for jouissance (attraction/repulsion), libidinal enjoyment as energy.

Study of aliveness in art has a long history, with Hogarth's S-curve line of beauty as an emblem. Lacan in his Sinthome seminar claimed the curves of his knot diagrams continued Hogarth's inquiry. The sexual connotations of the S curve resonate with the oscillating sine wave, the most fundamental movement of nature, whose paradigm is a wave breaking on the seashore (celebrated in the myth of Aphrodite born of sea foam). A basic exercise in the experience of aliveness in art is proposed by the comics artist Lynda Barry. In workshops with amateurs (people who had not drawn since childhood), Barry invited them to draw a car, and then to draw batman. The "primitive" (outsider) drawings were alive (the lines transformed formal properties into emotional expression). Drawings composed "correctly" (conventionally) Barry experienced as "dead."

A. J. Greimas in his semiotics proposed a diagram of the deep (actantial) structure of all narrative that maps how an object becomes animated by a subject's desire and passion. In our context Greimas's study of narrative is an ontology. Greimas found in narrative an articulation of the primary human faculties (which we designate with the portmanteau term "theopraxesis"). The dynamics of narrative develop in three axes the modalities of savoir (Theoria, knowledge), vouloir (Praxis, will) and pouvoir (Poiesis, imagination). In diagrams of the three axes, the upper level of discourse (communication), and lower level of story (performance), connect through the desire of the subject for an object of value. Events in story are appropriated in discourse to communicate a value (the object). The Subject's wanting the Object drives the narrative. The object may be anything (it is a "symbol" in any case). Greimas's example is that of a miser and his gold. The miser's passion projected into gold gives it a life of its own that controls the subject. This wanting expresses the faculties of Thymos in Plato's vocabulary (Passion, vouloir), and Feeling (Viscera, pouvoir). The drama works out whether or in what way the Subject does/not capture (con/dis/junction) the Object in conflict with an opponent. This dynamic of narrative Lacan identified as the Master/Slave relationship structuring the formation of the Subject relative to the Other of the Symbolic order. The ontological insight is that through passion of the subject an object becomes present, alive, subject in its own right, and like the Gingerbread Man tends to run off.

Peter Sloterdijk took up Francis Fukuyama's argument (The End of History and the Last Man), that the demand for recognition by people in history is thymotic, with the feelings of injustice and resentment as the source of rage accumulating in the contemporary global order. Greimas captured both forces and their interrelationship in his diagram: the relation of subject with object (desire) and opponent (passion). Michel Serres endorsed René Girard's theory of triangular desire as a synthesis of these two forces (want is socially mediated). In Lacan's terms, the subject is constituted by the desire of the Other. The analysis in terms of human faculties (capabilities) is that capitalism banked on savoir being able to buy off vouloir by satisfying the appetites of pouvoir. Or in Kantian terms, Knowledge and Desire allied against Will (moral freedom). This gambit has reached its limit and broken down, as Kant would have predicted, since that alliance works only for Beauty. In conditions of the Sublime, knowledge is overwhelmed and an alliance between Will and Imagination is necessary to restore human orientation without understanding. Our coming politics must renegotiate the attunement among the apparatus faculties/institutions. A shorthand reference for this argument would be that the rage of BLM protesters cannot be assuaged by gift certificates to Amazon, no matter how generous. Kant condensed into one word the demand of (thymotic) Practical Reason: Respect. The prototype is Jacques Derrida's respect for Nelson Mandela: the affect is compelled, not willed.

The metaphysics of attraction/repulsion, modelled in the tradition of the arts, must be taken into account as framing any social movement to reimagine our institutions, beyond recourse to rationalist social engineering. The modality of literacy (science) is neutral, confined to declarative propositions provable as true/false. Electracy engages with the other modalities conditioning experience: subjunctives, conditionals, counterfactuals in language, not to mention im/possibility. Phenomenology (in Husserl and Merleau-Ponty) replaced Cartesian savoir (I think, the Cogito), with existential "I can" (to be able, pouvoir). This "I can" (Ich kann) too optimistically assumes correlation among the three faculties, as if knowing, willing and making automatically aligned, a possibility refuted by history and imagined in literature. Narratives submit protagonists to three tests: competence, object value, measure of worth. We know what the problem is. We intend to fix it. We are (not) able. To be able is the modality augmented in electracy. The modality of to be able is dramatized in the axis of subject between donor (helper) and opponent. The donor actant embodies pouvoir, Classical Phronesis, personified in American mythology as "the right man with a gun." A prototype is the movie Shane. Shane is not the "hero" in actantial terms but the donor, helping the farmer hero overcome the evil rancher and his hired gunslinger. The fundamental object of value in American mythology, communicated in countless narratives, is the gun.

A question for electracy concerns this modality of being able, as foregrounded in Lacan's Seminar X, Anxiety. Anxiety/Fantasy relate dialectically (fantasy against anxiety), organized at the primal level by the Phallus (the resistance of im/potence, im/possibility). Aliveness in design sublimates the feeling of being alive. Conatus (conative map) is striving to persevere in one's own being. Ontology in electracy applies design aesthetics to theopraxesis of fantasy objects (phenomena of Second Phusis). Aletheia intimates what makes life worth living, guiding konsult as education. Laws organizing the choral interface of electracy are not those of God or Nature, but Fantasy, whose laboratory is art. Not the world as it should be (Philosophy) or is (History), but how we want it to be (Literature). Gerald Holton's thematic dimension of education, concerning how disciplines are created in the outsider imaginations of revolutionary science, rather than how they are verified in the disciplines of normal science, is a pedagogical version of this quarrel. This ancient quarrel among the faculties about which one was closer to reality (truth) must be mediated into collaboration in konsult. What remains uncertain is whether an apparatus capable of designing flesh augments happiness, as advertising claims. As psychoanalysis demonstrates, life/death metaphysics has a moebius (one-sided) structure. Such is the perversity of humanity as cause: death may appear as what makes life worth living (jouissance).

From Architecture through Information to Libido

Konsult seeks continuity relating microcosm and macrocosm, between local individual and global collectivity. Apparatus history indicates homology between human capabilities (body and three virtues or faculties) and institutions collectively realizing the capability (Home, Church, School, Corporation). The challenge is to design a conative (cognitive) map (choragraphy) of Existential Positioning to support and manage well-being relative to this homothetic system. The Other Beginning formulates our project as witness of Second Phusis by analogy with the invention of Philosophy observing First Phusis. Anaximander realized on the site of the new stone temples (6th century BCE) that Nature speaks mathematics. He devised the first gnomon, first "globe," first world map, first model of the cosmos, based on the cylindrical drums of marble crafted by the architects to erect the temple columns. Subsequently, in the Academy, architecture was adopted as the paradigm directing the new metaphysics of literacy, not only in the figure of the Demiurge as

Creator in Plato's Timaeus (source of the concept of Chora), but also in Aristotle's account of cause, of which there are four, on the model of the building of a temple: Material (stone); Formal (plan); Efficient (Architect); Final (purpose). In addition, Aristotle noted two auxiliary factors: Automaton (necessity); Tuché (fortune). A feature of rift design is to treat necessity as affordance or opportunity for invention (fortune). Vitruvius, author of the first written work on architecture (27 BCE), based his account on Aristotle's theopraxesis. The three principles of architecture are: Firmitas (the building proper); Utilitas (usefulness); Venustas (beauty) – the three capabilities, still observed today, as in Donald Norman's Experience Design: Useful, Usable, Desirable.

Reference to Vitruvius clarifies the macrocosm question, concerning how to understand the relationship between individual faculties and the institutions of apparatus. Konsult retrieves the tradition of Vitruvian Man, whose most famous rendering is by Leonardo da Vinci, inscribing the figure of a perfectly proportioned man within the shapes of Euclidean geometry. This icon expresses the principle of micro/macrocosm, the correlation and correspondence proportionally of humanity and nature. The assumption has been, from Vitruvius through Leonardo to Le Corbusier in modernism, that the proportions of the human body as organism, emblematized in Corbusier's Modulor device and expressed in the geometry of the Golden Section (Fibonaci sequence), is the proper measure relating culture and nature. To adopt the ratios of the human body (beauty) as proper for architecture (buildings and cities) is anthropocentric. The conditions of world created within electracy (the digital apparatus) challenges Modulor, seeking a new measure, a new anthropomorphic correspondence against sublime alienation disrupting post/industrial global civilization.

One response has been to reject "man" (the human) as measure, and to confound anthropomorphism and anthropocentrism (see Foucault's The Order of Things). Object Oriented Ontology (OOO), for example, gets it half right: nonhuman objects may become subjects, but their vitality comes from human energy. Or rather, OOO is an ontology of the other world, of Second Phusis, emergent virtually within media narratives, where everything and anything may be an actant (Benjamin Bratton adopts "actant" to characterize the operator position in the world as "stack"). The Bernoulli principle may offer a relevant analogy. The first attempts by men to fly fixed on an obvious but irrelevant feature of bird flight (the flapping of the wings = flapping of arms). This poetic conceit gave way to scientific analysis, the discovery of the physics of lift through airflow. The analogy suggests that Modulor should locate a different model of correspondence for EPS. A new paradigm for Second Phusis was proposed by Michel Serres, who noted the transformation of energy distinguishing electracy from literacy. Referencing the three states of matter (solid, liquid, gas), Serres reviewed the history of technology, from architecture in antiquity, through thermodynamics in early modernity, to information science today. The insight is that if Phusis was observed on the site of architecture, then Second Phusis must be oriented relative to our contemporary mode of energy – information – with the ambition of unfolding a metaphysics out of this paradigm, as Western Philosophy was unfolded out of architecture. The caveat is to avoid baby/bathwater errors; not to throw out anthropomorphism (Anthropos) with anthropocentrism. It is useful to recall our principle of reoccupation of old questions in search of new answers, as for example when we notice the usefulness of "information design" and "information architecture" as syncretic relays.

Serres outlined a way forward in noting "information" as shorthand for the fields of complexity, chaosmosis, autopoiesis, emergence and the like. Serres offers numerous emblems and analogies for understanding this ontology. The best model for the information paradigm, according to Serres, is weather. This figure translates electrate attraction/repulsion as high and low pressure systems that steer meteorological events tracked and reported daily in the news. Edward Lorenz, who formulated the Lorenz attractor to help define chaos, was a meteorologist. The butterfly effect is a useful way to understand self-organizing behaviors, the indeterminacy of emergent order, that calls attention to seemingly marginal details as potential forces of change. The insight directing our search for a focalizer is topological geometry, which precisely does not measure, is unconcerned with scale, and hence reimagines the Vitruvian ratio. The rubber sheet geometry of topology keeps track of invariant positions (originally dubbed by Leibniz as analysis situs).

Destruktion (alluded to in Derrida's deconstruction), Heidegger's method of tracking the descent of terminology, the passage of concepts through translation from one cultural history to the next, locates our vector. The relevant keyword is Edge. The history begins with Greek Daimon (daio related to division, split); Latin translation as Genius; Goethe translation into German as Grenze (Limit). This vector continues through Euler's discovery of edge as part of his invention of graph theory (network systems) through solving the puzzle of the seven bridges of Koenigsberg (Kant's home city) in the eighteenth century. An edge is constituted by the faces of forms meeting at vertices (tested on Plato's polyhedra). Lacan's adaptation of topological geometry and knot theory for studying and teaching psychoanalysis clarifies and confirms the continuity of this historical passage from Daimon to Edge (evoking threshold and border) for rethinking measure in every sense (including especially ethics and politics). Lacan explicitly characterized his account of the subject in terms of topological relations among three orders (Real, Symbolic, Imaginary) as an update of Aristotle's theopraxesis. Here is the key: the fundamental energy at issue is libido, for which "information" is a metaphor.

Lacan provides a map for our problem of electrate metaphysics, understanding how the human faculties and their apparatus-defining institutions may be coordinated for well-being. Lacan shows that the three orders (diagramed as a Borromean knot) are held together by a clasp constituted by the object @ (object cause of desire): the human body (Anthropos). How else might the field of the gaze be graphed (actor network)? Weather replaces the aesthetic body as electrate Modulor. Leonardo's Vitruvian Man may be replaced as emblem of measure by his studies of flow, sketches of turbulent water, the swirling eddies or vortices that fascinated him. Or is the Mona Lisa's smile the primary ontological curve, its subtlety winking at the dynamics of turbulence? Does Second Phusis (syncretized nature and culture) appear as turbulent design in disaster's scene of instruction? Our purpose in documenting disaster is to witness emergent ontology. Through disaster being is given as human (in)capability. The Other Beginning, the vector from Daimon to Edge, calls for a new axiology (theory of value) to direct collective theopraxesis.

Innervation (Facilitation)

Apparatus history evolves through a process of growth, destabilization, and adaptation, when institutions created to manage well-being collapse in the face of new conditions. Electracy emerges from the chaos created by the revolutions beginning in the late eighteenth century that overwhelmed the metaphysics of literacy. Among the challenges of this history is the realization that literacy includes individual selfhood and democracy as identity, the implication being that selfhood and democracy are unsustainable and are being displaced within electracy. Our program of coordinating faculties, however, shows not just that apparatus coexist in conflict through history, but that they may be attuned (Kant/Arendt hypothesis). Meanwhile, history repeats itself, with the corporation beginning to separate from control by the nation state, just as the state separated from the church in early modernity, with the implication that entertainment is becoming hegemonic in global civilization.

Walter Benjamin outlined the project of innervation central to Second Phusis (Bernard Stiegler and Mark Hansen continued innervation as a program in our own moment). Benjamin was hopeful that new media and popular culture, as he witnessed them in the first half of the twentieth century, were apprenticing the public for the transformation of the lifeworld created by industrialization.

In the same way that play "trains" children in preparation for adulthood, so does popular culture train citizens for thriving in a mutating lifeworld. "Innervation" refers to the human organism literally changing neurophysiologically to adapt human faculties to the conditions of existence. For innervation to be socially functional requires cultural inventions to support capability (to actualize potential: realization). Paul Virilio continues this argument, recalling Plato's warning that writing as an artificial memory would destroy living memory. What happened in practice, culminating in the age of print, was the adaptation of organic human memory away from direct memorization of information to the management of logic and rhetoric applied to texts in libraries. Virilio defined the challenge today relative to digitizing (digitalizing) of perception itself (ubiquitous imaging combined with real-time data transmission). Phenomenology (human experience of appearance) becomes writable ("deep fake"). Images transform from representations to simulacra (copies with no original). Here is the Automaton and Tuché of electracy. Electracy designs flesh (Merleau-Ponty's term for the digital category replacing Aristotle's substance), just as literacy writes speech. The existing capability, put into a machine, requires a relevant metaphysics (operating practice and institution).

It is important to recall that electracy opens a new chora with its own metaphysics, so that while there is remediation (reoccupation and retrieval) of parts of the established apparatus, the purposes of electracy are specific to itself. Science is and remains literate; religion is and remains oral; family is and remains paleo. Science, religion and family adopt computing but retain their original metaphysics (necessarily), now supplemented and displaced. The implication is that electrate innervation prepares for a new time-space-cause. It is not augmented literacy. The battles over screen-time within families is the site of emerging electracy, with consequences for human neurophysiology. As our theorists already documented, we don't have to wonder about the qualities of electrate categories of experience: they are modelled in popular media. Citizens receive apprenticeship in theopraxesis of electrate chora, the dimension opening in our apparatus, in enjoying the creation of animation arts from comics through cinema and television into digital media. At issue is the configuration of primary interface relating culture and nature, displacing writing and its institutions. The central role of digital animation and its powers of metamorphosis in exploring and mapping the topological rubber sheet and superflat conditions of simulacral flesh accounts for the importance of Walt Disney as the Plato of Electracy (Mickey Mouse our Socrates). Not that Disney invented animation or comics, but he is the paradigm codifying the possibilities of the inventions, especially in their extension through theming into the built (architectural) world projecting the powers of the corporation and entertainment into everyday life (disneyfication). The three circles of Lacan's Borromean knot diagram precisely the outline of Mickey Mouse.

Creating an electrate equivalent of the allegory of the cave in Plato's Republic is instructive. In Plato's education, students (prisoner) are released from their chains, turned (converted), led out of the cave to behold the true light, the sun (illumination). The allegory of electrate education is the cat & mouse chase cartoon, representative of all chase scenes so common in popular narratives. Ulmer proposed several other texts to fill this allegorical function (Joseph Brodsky's Watermark; the film version of The Wizard of Oz). While these texts are relevant, they don't capture the essence of the apparatus in the way that the animated chase scenario does. The ideal readymade allegory is not the adventures of Mickey Mouse, Tom & Jerry, Road Runner & Coyote and the like, but George Herriman's Krazy Kat, which in its obsessive scenario demonstrates the operations of a minimum system (universal motor), observable in everything from any game played with a ball through Lacan's drive diagram to the operations of transistors or René Thom's elemental catastrophes. The point for now is that such entertainment productions introduce netizens to the forces of the electrate dimension, the choral interface organizing the emerging lifeworld. If tragedy was the primal story of literacy, the joke (gag) is at the heart of electracy, evoking the fumisterie of Dadaism invented in the Cabaret academies of Montmartre in nineteenth-century Paris, culminating in the Cabaret Voltaire in Zurich during WWI, continuing in the Funk of hip-hop culture. Lenny Bruce is our Homer. What are the implications of this allegory?

Catastrophe

Lacan once reminded his seminar that humans are mammals. René Thom adds: we are predators, hence our fascination with the chase. Heidegger mentioned the predation cycle as informing Dasein (being in the world), and René Thom referenced it as paradigm in his semiophysics. Thom's catastrophe theory and theory of morphogenesis contribute to topological dynamics, relevant to choragraphy of electrate interface. What is it like to be immersed in electracy? How does one negotiate phase space structuring the zone between virtual/actual dimensions? The Cat & Mouse cartoon is an allegorical mise-en-scene of the interface. In the cartoon, the cycle always goes awry, at least from the functionalist perspective: the capture is always foiled. The ritual relating Krazy, Ignatz, Officer Pupp, and the brick, clarifies that the cycle as story is appropriated in discourse to communicate a value. Nature itself is instructive in this regard, manifesting its own version of game theory, the Master/Slave dance of desire. Roger Caillois in The Mask of the Medusa described the play in nature that Lacan called "gaze" and Deleuze and Guattari referenced as "rhizome": the relation of wasps and orchids was their paradigm, the networked relations organizing nature. Their paradigm is symbiotic (the orchid tricks the wasp into serving as its sex worker). Caillois's favorite example is the eyespots evolved by insect prey to deter predators. The play is between caricature (exaggeration) and camouflage (deception), with predator and prey applying both strategies.

Michel Serres invoked the same relationship in a less symbiotic mode, referencing the "parasite," exploiting the semantic field that in French includes "noise" as one of the meanings. His preferred parable is that of the Wolf, the Lamb, and the Shepherd. It is worth noting that the design of camouflage used by the military in World War I is credited to Cubism, which (inspired by Cézanne's passage) perfected the breakup of contour that make Gestalts recognizable. Plato introduced ideal Form by analogy with Eidos (shape) as Idea. Electracy subjects form to transformation, flow, metamorphosis. Predation is not a bug, as the saying goes, but a feature of existence, and its process is complexity itself. Konsult pedagogy apprentices egents for this mise-en-scene, with many relays available for study, from the duel between Sherlock Holmes and his Nemesis Moriarty, to Freud's Marriage Broker jokes. Our parables are lived daily at the U.S./Mexico border, the universal motor of migrant, coyote, ICE, and wall.

The parables allegorize the material, psychological, and social forces structuring electrate metaphysics, the axis attraction/repulsion of appetite and electricity. The predation cycle is driven by three forces (the same three powers documented in every area of culture), identified by Thom as hunger, fear, sex. Thom developed his elementary catastrophes articulating phase space in part to formalize the actantial diagram Greimas found structuring narrative, in order to chart patterns of change. In the anthropomorphic world of narrative the drives inform three orders of exchange whose disturbance triggers story: money (economy, hunger); language (communication, fear); kinship (family, sex). Roland Barthes's reading of the Balzac story, "Sarrasine" (in S/Z) shows how these exchanges constitute the Symbolic code, the dominant code among the five codes organizing narrative. We want to learn how life manifests itself in this interface (how flesh appears, in order that it may be rift designed).

According to Thom, a subject floods its territory with pregnance (Thom uses the Gestalt term with his own nuance). The field is enlivened with the subject's intentionality relative to its desire/need/demand (a cat's hunger). This field creates background noise (the parasite), which turns out to be necessary for perception. When a mouse-prey-object enters the zone of attention, it becomes salient, figure/ground shape. Its presence is sensed by an interference in the noise (noise amplifies the salience). The ideal event is triggered:

the cat chases, catches and eats the mouse, achieving satisfaction, allowing it to rest, and so the cycle turns. Thom diagramed this cycle as the Cusp Catastrophe. It is worth noting that the Anthropologist, Claude Lévi-Strauss, explained the formal operations of art in terms homologous with Thom's pregnance/salience. Estrangement in art design produces salience in noise (formal order out of chaos), just as the method of epoché suspension of the natural attitude in perception (habit) produce phenomena in the method of Husserl's Phenomenology. As McLuhan argued in his Laws of Media, all figure/ground relations are codified in the minimum system of tropology (metaphor, metonymy, synecdoche, irony). A challenge for konsult is to retrieve figure/ground discernment (daimon, edge) in formless conditions of inchoate ambiguity, metaphysical sprawl.

Thom argues that this event of "capture" structures morphogenesis as such, including quantum events of photon exchange at the subatomic level, or the transitive sentence (subject-verb-object) in language, not to mention any drama in which two subjects desire the same object (diagrammed as the Butterfly Catastrophe). Thom admits that in practice the event often misfires, as detailed through infinite variations in cartoon chases, as well as in the point-counterpoint of rhizomes and parasites in nature, or cops and robbers or courtship. Thom's semiophysics modelling how shapes emerge through growth and change, resonates with the universal motor or minimum system invoked by Buckminster Fuller and Gregory Bateson, among others. Lacan's diagram of unconscious drives explains why in the chase allegory the predation cycle must fail: the drive necessarily misses its aim, since its goal (the tension between aim and goal) is to keep the game going, that is, to live. Heidegger makes a similar point, regarding Dasein, being-towards-death as what is ownmost for humans. We fall out of the natural attitude and begin living (wake up) when opponents resist and disturb the entropic path. Clinamen – to reference Lucretius, touchstone for Michel Serres.

Real Time

Theopraxesis facilitates digital immersion, as an interface for living in Real Time – speed of light continuous feedback loops, surveillance and smart environments, ubiquitous computing, mobile mixed and hybrid realities, AI database, web ontologies. In short, enhanced Surrealist utopia. The point is that in electracy entertainment is the interface engaging the lifeworld, not an escape or pastime, but the hegemonic apparatus. Iterations of loops, continuous recursive self-reference, augment auto-affection and generate the conditions of chaos complexity, including emergent (unpredictable) conditions. Chorography in the interface supplements GPS with EPS (existential positioning system), recognizing that netizens (egents of konsult) are positioned in time as well as space, generating causality through their own faculties. The temporality of Real Time is Moment (Augenblick), Kairos, that Ulmer refers to as flash reason (reasoneon). Modernist letters modelled the formal designs of Moment in the theory and practice of correspondences (Baudelaire), illumination (Rimbaud), epiphany (Joyce), Weltinnenraum (Rilke), objective correlative (Eliot), vortex (Pound), gest (Brecht), dreamwork (Freud), dialectical image (Benjamin), bricolage (Lévi-Strauss) – creating an archive and textshop for Heidegger's Aletheia. The genre for designing EPS is mystory: individual image of wide scope as distributed collective cognitive map (chorography).

Topology tracks invariant positions, which remain relationally consistent even as space-time-cause are stretched, distorted, twisted, crumpled (folded), passing through many transformations (the wide image is invariant through all experience, whether personal, social, disciplinary). The parallax power of predator vision, emblematized in first-person shooter games, put into a machine, enables tracking any relational dynamic. Destiny is topological. Gestalt, plasma or shape turn out to be relative to mobile position, separated from subject, delegated to cinematic imaging. A donut and a coffee mug are the same shape, topologically, which may have been the poetics of David Lynch's Twin Peaks. The insight of edge is that in topology Idea (Form) is subordinated to position (not what but where matters), a crucial factor in sublime conditions of generalized sprawl erasing literate categorization. The issue is important for chorography, since Plato introduced chora to account metaphysically for how eternal universal Form and ephemeral physical Matter interact. Order emerges from chaos within the receptacle of chora, with four elements sorted out of the inchoate in the manner of threshing grain. Such figures of thought multiply around chora, from Plato to the present. Plato's figurative sieve returns in Deleuze's plane of immanence. This sieve resonates with the trellis pattern emergent natively within sprawl.

Meteorology is a better Modulor for choral ratio than human anatomy (open flow rather than closed organic shape), relative to the operation of chora as chaosmos. The electrate lifeworld is folded, Idea is emergent out of formless. The implication of folded metaphysics is that Real Time innervation makes it necessary to live and tell (deemed impossible until now). Kierkegaard complained that life in novels is narrated backwards, but unfortunately must be lived forwards. As a life works through the narrative tests of experience it passes from indeterminate to necessary (fate is retrospective). In Real Time, these two paths of past and future become simultaneous. Writers have documented that immediate experience never was possible, that retrospection is an essential feature of event. Studies of PTSD confirm the aporia: traumatic incidents return and repeat because they remain outside experience. This reference to trauma reminds us that our chora is a scene of predation. Living requires cultural doubling (or more). Lacan explained that when you hear a shoe drop, it is already the second shoe (the first one didn't register). Such also is the ancient rationale of "cosmetics" and contemporary fashion: to appear the body must be adorned.

In Real Time interface it will be possible/necessary in principle to live retrospectively (at light speed), to control through postproduction tools the differance (delay and displacement) of event, to eliminate the gap between occasion and wit (l'esprit de l'escalier), to write phenomenology (previously considered a contradiction in terms). Thus is resolved Derrida's deconstruction of Husserl, or rather, deconstruction becomes a guide to how to design perception, not to mention the misunderstanding between Prometheus and Epimetheus adopted by Bernard Stiegler to remind us of Pandora's grammatological box. Theopraxesis requires an authored spontaneity, method acting, improv, similar to what Zen calls No-Mind. No-Mind is nothing mystical (although it has been ornamented with mystery, including by the Beat poets). Phenomenologists noted it as the performance of expertise, having embodied or internalized in oneself a skillset such as chess, sailing, or any other expertise, such that it may be produced without thought. Massively Multiplayer Online Games (MMO) innervate players for this lifeworld interface. They suggest a model by which mixed reality might function, including augmenting differance to counter dromosphere dimension collapse. Players toggle between living and telling, as in the game they toggle between immersion in the narrative situation and managing the control panel of game devices (story and discourse). The three modal axes of Greimas's actantial system (savoir, vouloir, pouvoir) become articulated in this interface, for which Dungeons and Dragons is an apprenticeship.

Otaku, fans of Japanese pop culture, have developed identity behaviors that further blur the border between live/tell, replacing socially functional ideology with the diegeses of their preferred fictional world, to the point of a fan legally marrying his favorite character, as happened in one case (unfortunately, the relationship recently ended in divorce). This possibility is anticipated in the tradition of Pygmalion's love bringing Galatea to life. In one version, the artist marries his creation. The point to emphasize is that this entertainment order is becoming hegemonic, holding the power in the world exercised previously by nation states and world religions. However skeptical one might be, we have to wonder if entertainment might have better success than its predecessors in accomplishing human essence (happiness)? The implication for education is that native designers of flesh will have internalized the formulas of entertainment (just as literacy internalized the logics of science), such that egents are their own auteurs. The identity formation of electracy, it should be recalled, is not soul or self, but brand. A self in choral interface is an image (modelled by stars, celebrities, influencers). Apps will have automated the post-production tools, screenplay composition, AI special effects and the like, leaving living human capability free and available for the play of creativity.

Tradition warns that techne (craft) may be taught, but not ingenium (daimon). This truism will be tested in konsult. The further caveat is that schooling as presently constituted is doubling down on science, not recognizing that the digital apparatus is entertainment. Apparatus history suggests that it is inappropriate to expect school as the institution of literacy to convert to apprenticeships for electracy, just as church did not swap liturgy for empirical experiments. The appropriate role for school within electrate hegemony is not to become entertainment corporations, but to invent theopraxesis for attuning all four apparatus. A table work textshop exercise to outline this program might begin with reading Hannah Arendt's rewriting of Kant. She found her answer in Kant's Critique of Judgment, the aesthetics of taste, to be applied to organization of popular culture as a public sphere. Meanwhile, Arts and Letters educators might save their disciplines by alerting students to the skillset actually needed to undertake world-building for Franchise brands.

Values

One further point may be noted. Entertainment design foregrounds the fundamental element requiring political and ethical invention. The primary force of resistance, the obstacle to reimagining the lifeworld, the greatest opportunity for education, is the order of values grounding culture and even civilization (axiology). In principle, electracy creates the possibility of designing values themselves, beginning with promoting qualities of aesthetic experience (intensity, participation, extimacy) as phenomena of value in their own right. There is an ascending intensity of bisociation from getting a joke, to anagnorisis in tragedy through epiphany in poetry, eureka in science, to satori in mysticism, that has illuminated civilization from the beginning. Bataille claimed that the moment of laughter captures this entire spectrum, however briefly, as when the forbidden behaviors of unofficial existence are made public in the aspect shift of a joke (diagrammed in Thom's cusp catastrophe). Perhaps the gift of digital technology will have been to put jouissance into a machine, as literacy put true/false into the computer, or orality put speech into writing. When literacy finally separated itself from religion, beginning in the sixteenth century, it introduced a split between fact and value, essentially determining that facts alone could be addressed, while values were not of this world (as Wittgenstein declared in his early positivism, whereof one cannot speak, thereof one must keep silent). Values are constructed in ideology. Every drama or narrative is organized by a premise, a particular value fundamental to the society, with the goal of the plot being to prove the validity of the value. Barthes referred to these constructs as mythologies. Self-help guides to authoring a screenplay describe this feature (see Lagos Egri, The Art of Dramatic Writing).

Values in literacy tend to be formulated at a level of generality associated with transcendental categories such as the Good, which have no properties, meaning that they are floating signifiers, abstractions open to dialogical struggles among powers contending for hegemony ("Freedom" is a recent example of a rift value). The MEmorial exercise of Electronic Monuments advised egents to identify the value structuring a disaster, that accounted for why it was irreducible and resistant to instrumental fixes. The debate surrounding the Second Amendment is paradigmatic, anchored in the Frontier Myth of a good man with a gun (creed of the NRA). The FRE analyzed this history in Miami Virtue, including a poetics of myth formation, applied to the proposal for an alternative American Myth of Jazz, based on the blues. Jam session collaboration displaces showdown confrontation. Guitars, not guns. The proposal of electracy is that "values" become accessible to rift design in the public sphere, as they always have been in the arts, exploring the limits of Edge (measure). As Baudrillard noted, there is a fatal strategy operating in life (necessity and fortune). The point is to recognize how this universal motor manifests itself (take lessons from Chinese Taoism). Heidegger declared his post-Dasein move to Ereignis (Event, Appropriation, Enownment) had only one precedent: the Tao (Dao), whose Modulor is flow. The challenge in the visceral apparatus of electracy, the formless conditions of chaosmos described by Bataille's General Economy, is that values function not only as ideals but also as base materialism, untransposed behaviors, such that the gun itself (for example), the guitar itself, or the border wall as such, is the value requiring expenditure, sacrifice, death. Designers of flesh (digital phenomenology) must understand that perception is a composite of external sensation and internal drives. Choragraphy as pedagogy of extimacy places Heidegger's Ereignis and the Dao in the context of Mihaly Csikszentmihalyi's study of creativity, Flow: The Psychology of Optimal Experience.

Electracy leaves to literacy the problematic of true/false in science or to orality right/wrong in religion, to family life/death in the home, to take up the dimension of human thriving in attraction/repulsion, emblematized in the polarity of electricity, or the activity mapped in quantum diagrams, as well as in human relationships. This polarity for Anthropos is experienced in the dialectic of fantasy/anxiety. Lacan's Seminar X makes explicit the interdependence of these states of mind. Anxiety is a direct line to the Real; it doesn't lie, an insight that gives evidence of Lacan's debt to Heidegger (Angst as the mood of Dasein). Each apparatus realized one human faculty, each one as fundamental as the others, but arising separately historically and seemingly antipathetic in their institutional projections, manifesting in their incommensurability what Lyotard called the differend. Identified by Csikszentmihalyi as the feeling of flow, happiness is the essence of humanity, Aristotle said: to become what we are we must be happy, achieved only collectively in a polis through choice and productivity, for which there is no science. Electracy augments fantasy against anxiety, sited in imagination. The modality of literacy was neutral, focused in the declarative proposition that could be proven true/false. The modality of electracy is "to be un/able," experienced as attraction/repulsion. The faculties of knowing, willing, and feeling do not automatically align. "Where there is a will, there is a way," is a utopian thought. Such is the dilemma to be addressed (again) in electracy. What base quotidian behaviors have we failed to recognize as values constituting the impossible? Fantasy has always imagined the impossible. Electracy brings fantasy into policy formation.

Works Cited

Freeman J. C., feat. Ulmer, G. L., *The Ulmer Tapes*, last modified June 23, 2022, https://johncraigfreeman.wordpress.com/ulmer-tapes/

Ulmer, G. L., Revelle, B. J. Tilson, W., & Freeman J. C. (2012). *Miami Virtue and the Ulmer Tapes*, Kamloops, BC, Canada: Small Cities Imprint.

Works Cited

Aarseth, Espen J. (1997), *Cybertext: Perspectives on Ergodic Literature* (Baltimore: Johns Hopkins University).

Adorno, Theodor W. (1994), *"The Stars down to Earth" and Other Essays on the Irrational in Culture*, Ed. Stephen Crook (New York: Routledge).

Agamben, Giorgio (1993), *The Coming Community*, Trans. Michael Hardt (Minneapolis: Minnesota).

Andreotti, Libero, and Xavier Costa, Eds. (1996), *Theory of the Derive and Other Situationist Writings on the City* (Barcelona: Museu d'Art Contemporain de Barcelona)

Appel, Alfred, Jr. (2002), *Jazz Modernism: From Ellington and Armstrong to Matisse and Joyce* (New York: Knopf).

Arieti, Silvano (1976), *Creativity: The Magic Synthesis* (New York: Basic Books).

Armer, Alan A. (1988), *Writing the Screenplay: TV and Film* (Belmont, CA: Wadsworth).

Aubert, Jacques, Ed. (1987), *Joyce avec Lacan* (Paris: Navarin).

Aumont, Jacques et al (1992), *Aesthetics of Film*, Trans. Richard Neupert (Austin: Texas).

Baker, Houston A., Jr. (1984), *Blues, Ideology, and Afro-American Literature: A Vernacular Theory* (Chicago: University of Chicago).

Ballard, Bruce (1991), *The Role of Mood in Heidegger's Ontology* (Lanham, MD: University Press of America).

Barabasi, Albert-Laszlo (2003), *Linked: How Everything Is Connected to Everything Else and What It Means for Business, Science, and Everyday Life* (New York: Plume Books).

Baraka, Amiri (1998), "Jazz and the White Critic," in O'Meally.

Baron, Fran, Ernst S. Dick, and Warren R. Maurer, Eds. (1980), *Rilke: The Alchemy of Alienation* (Lawrence, KS: University of Kansas).

Barthes, Roland (1977), *Image, Music, Text*, Trans. Stephen Heath (New York: Hill and Wang).

_____ (1981), *Camera Lucida: Reflections on Photography*, Trans. Richard Howard (New York: Hill and Wang).

Basho (1996), *Back Roads to Far Towns*, Trans. Cid Corman and Kamaike Susumu (Hopewell, NJ: Ecco Press).

Bataille, Georges (1985), *Visions of Excess: Selected Wrirings, 1927–1939*, Ed. Trans., Allan Stoekl (Minneapolis: Minnesota).

_____ (1995), *Encyclopaedia Acephalica* (London: Atlas).

Baudelaire, Charles (1964), *Baudelaire: Selected Verse*, Trans. Frances Scarfe (Baltimore: Penguin).

_____ (1970), *Paris Spleen*, Trans. Louise Varese (New York: New Directions).

Baxter, Charles (1997), *Burning down the House: Essays in Fiction* (Saint Paul, MN: Graywolf Press)

Beaujour, Michel (1991), *Poetics of the Literary Self-Portrait*, Trans. Yara Milos (New York: New York University).

Bell, Ian F. A. (1981), *Critic as Scientist: The Modernist Poetics of Ezra Pound* (New York: Methuen).

Benedikt, Michael, Ed. (1991), *Cyberspace: First Steps* (Cambridge, MA: MIT Press).

Benitez-Rojo, Antonio (1996), *The Repeating Island: The Caribbean and the Postmodern Perspective*, 2nd Ed., Trans. James E. Maraniss (Duke).

Benjamin, Walter (1969), *Illuminations*, Trans. Harry Zohn (New York: Schocken).

Berthoff, Warner (1986), *Literature and the Continuances of Virtue* (Princeton: Princeton University).

Biel, Steven (1996), *Down with the Old Canoe: A Cultural History of the Titanic Disaster* (New York: Norton).

Bloom, Harold, et al (1979), *Deconstruction and Criticism* (New York: Seabury Press).

Blumenberg, Hans (1983), *The Legitimacy of the Modern Age*, Trans. Robert M. Wallace (Cambridge, MA: MIT).

_____ (1997), *Shipwreck with Spectator: Paradigm of a Metaphor for Existence*, Trans. Steven Rendall (Cambridge, MA: MIT Press).

Borde, Raymond, and Etienne Chaumeton (1996), "Towards a Definition of Film Noir," in Silver and Ursini.

Bordowitz, Gregg (1994), "Dense Moments," in *Uncontrollable Bodies: Testimonies of Identity and Culture*, Eds. Rodney Sappington and Tyler Stallings (Seattle, WA: Bay Press).

Bosso, Christopher J., "The Contextual Bases of Problem Definition," in Rochefort, pp. 182–203.

Bowker, Geoffrey C. and Susan Leigh Star (2000), *Sorting Things out: Classification and Its Consequences* (Cambridge, MA: MIT).

Bracher, Mark, et al. Eds. (1994), *Lacanian Theory of Discourse: Subject, Structure, and Society* (New York: New York University).

Brecht, Bertolt (1964), Brecht on Theater: The Development of an Aesthetic, Trans. John Willett (New York: Hill and Wang).

Buci-Glucksmann, Christine (1994), Baroque Reason: The Aesthetics of Modernity, Trans. Patrick Camiller (Thousand Oaks, CA: Sage).

Buck-Morss, Susan (1989), *The Dialectics of Seeing: Walter Benjamin and the Arcades Project* (Cambridge, MA: MIT Press).

Butler, Jeremy G. (1996), "Miami Vice, The Legacy of Film Noir," in *Film Noir Reader*, Eds. Alain Silver and James Ursini (New York: Limelight).

Caputo, John D. (1993), *Against Ethics: Contributions to a Poetics of Obligation with Constant Reference to Deconstruction* (Bloomington: Indiana).

_____, Ed. (1997), *Deconstruction in a Nutshell: A Conversation with Jacques Derrida* (New York: Fordham University).

Carruthers, Mary (1998), *The Craft of Thought: Meditation, Rhetoric, and the Making of Images, 400–1200* (New York: Cambridge).
Caruth, Cathy (1995), "Introduction," in *Trauma: Exploration in Memory*, Ed. Cathy Caruth (Baltimore: Johns Hopkins).
Casey, Edward S. (1998), *The Fate of Place: A Philosophical History* (Berkeley: California).
Cheng, Francois (1994), *Empty and Full: The Language of Chinese Painting*, Trans. Michael H. Kohn (Boston: Shambhala).
Cheroux, Clement (2008), *Henri Cartier-Bresson*, Trans. David H. Wilson (New York: Abrams).
Chtcheglov, Ivan (1981), "Formulary for a New Urbanism," in Knabb, pp. 1–4.
Cohen, Margaret (1993), *Profane Illumination: Walter Benjamin and the Paris of Surrealist Revolution* (Berkeley: University of California).
Cooke, Mervyn, and David Horn, Eds. (2002), *The Cambridge Companion to Jazz* (New York: Cambridge University Press).
Cooper, Lane (1960), *The Rhetoric of Aristotle* (New York: Appleton-Century-Crofts).
Copjec, Joan (1993), *Shades of Noir: A Reader* (New York: Verso).
Corman, Cid (1996), "Introduction," in Basho.
Cornell, Drucilla (1992), *The Philosophy of the Limit* (New York: Routledge).
Cosgrove, Denis, Ed. (1999), *Mappings* (London: Reaktion Books).
Cowie, Elizabeth (1993), "*Film Noir* and Women," in Copjec (1993).
Crary, Jonathan (1990), *Techniques of the Observer: On Vision and Modernity in the Nineteenth Century* (Cambridge, MA: MIT Press).
Crease, Robert P. (2002), "Jazz and Dance," in Cooke and Horn.
Culler, Jonathan (1982), *On Deconstruction: Theory and Criticism after Structuralism* (Ithaca, NY: Cornell University).
Cutrofello, Andrew (1994), *Discipline and Critique: Kant, Poststructuralism, and the Problem of Resistance* (Albany: SUNY).
Davenport, Guy (1983), *Cities on Hills: A Study of I–XXX of Ezra Pound's Cantos* (Ann Arbor, MI: UMI Research Press).
Davis, Mike (1992), *City of Quartz: Excavating the Future in Los Angeles* (New York: Vintage).
Debord, Guy-Ernest (1989), "Two Accounts of the Derive," in Sussman, Ed. pp. 134–139.
de Certeau, Michel (1984), *The Practice of Everyday Life*, Trans. Steven Rendall (Berkeley: University of California).
_____ (1992), *The Mystic Fable*, Trans. Michael B. Smith (Chicago: University of Chicago).
Deleuze, Gilles (1988), *Foucault*, Trans. Sean Hand (Minneapolis: University of Minnesota).
_____ (1989), Cinema *2: The Time-Image*, Trans. Hugh Tomlinson and Robert Galeta (Minneapolis: University of Minnesota).

Deleuze, Gilles, and Felix Guattari (1977), *Anti-Oedipus: Capitalism and Schizophrenia*, Trans. Robert Hurley, Mark Seem, and Helen R. Lane (New York: Viking Press).

_____ (1986), *Kafka: Toward a Minor Literature*, Trans. Dana Polan (Minneapolis: University of Minnesota).

Derrida, Jacques (1993), *Aporias*, Trans. Thomas Dutoit (Stanford).

_____ (1994), *Spectres of Marxs: The State of the Debt, The Work of Mourning, and the New International*, Trans. Peggy Kamuf (Routledge).

_____ (1996), "Faith and Knowledge: the Two Sources of 'Religion' at the Limits of Reason Alone," in *Religion*, Eds. Jacques Derrida and Gianni Vattimo (Stanford: Stanford University).

_____ (1997), *Politics of Friendship*, Trans. George Collins (Verso).

_____ (1998), *Demeure: Maurice Blanchot* (Paris: Galilee).

Dery, David (1984), *Problem Definition in Policy Analysis* (Lawrence, KS: University Press of Kansas).

Devlin, Keith (1997), *Mathematics: The Science of Patterns* (New York: Scientific American Library).

Diggs, Stephen (1997), "Alchemy of the Blues," *Spring: A Journal of Archetypes and Culture*, 61.

Diller, Elizabeth and Ricardo Scofidio (1994), *Back to the Front: Tourisms of War* (Basse-Normandie: F.R.A.C.).

Doane, Mary Ann (1991), *Femmes Fatales: Feminism, Film Theory, Psychoanalysis* (New York: Routledge).

Dobyns, Stephen (1997), *Best Word, Best Order: Essays on Poetry* (New York: St. Martin's Griffin).

Donald, James (1999), *Imagining the Modern City* (Minneapolis: Minnesota).

_____ (2000), "Detectives," in Pile and Thrift.

Doyle, Richard E. (1984), *ATH: Its Use and Meaning. A Study in the Greek Poetic Tradition from Homer to Euripides* (New York: Fordham University).

Dreyfus, Hubert L. (2002), *On the Internet* (New York: Routledge).

Duttmann, Alexander Garcia (1996), *At Odds with Aids: Thinking and Talking about a Virus* (Stanford: Stanford University).

Early, Gerald (1998), "Pulp and Circumstances: The Story of Jazz in High Places," in O'Meally.

Eco, Umberto (1982), *The Aesthetics of Chaosmos: The Middle Ages of James Joyce*, Trans. Ellen Esrock (Cambridge, MA: Harvard).

_____ (1984), *Semiotics and the Philosophy of Language* (Bloomington: Indiana University).

_____ (1995), *The Search for the Perfect Language*, Trans. James Fentress (Oxford, UK: Blackwell Publishers).

_____ (1997), *Kant and the Platypus: Essays on Language and Cognition*, Trans. Alastair McEwan (San Diego: Harcourt).

Edelman, Murray (1988), *Constructing the Political Spectacle* (Chicago: University of Chicago).

Entrikin, J. Nicholas (1991), *The Betweenness of Place: Towards a Geography of Modernity* (Baltimore: Johns Hopkins).

Fink, Bruce (1995), *The Lacanian Subject: Between Language and Jouissance* (Princeton: Princeton).

Finn, Julio (1986), *The Bluesman* (New York: Quartet Books).

Flaxman, Gregory, Ed. (2000), *The Brain Is the Screen: Deleuze and the Philosophy of Cinema* (Minneapolis: University of Minnesota).

Floch, Jean-Marie (2000), *Visual Identities*, Trans. Pierre Van Osselaer and Alec McHoul (New York: Continuum).

Foster, Hal (1995), *Compulsive Beauty* (Cambridge, MA: MIT Press).

Foucault, Michel (1973), *The Order of Things: An Archaeology of the Human Sciences* (New York: Vintage).

Fynsk, Christopher (1986), Heidegger: Thought and Historicity (Ithaca: Cornell).

Gabbard, Krin, Ed. (1995), Representing Jazz (Durham, NC: Duke University).

_____ (2002), "Images of Jazz," in Cooke and Horn.

Garber, Frederick (1995), "Fabulating Jazz," in Gabbard.

Gasche, Rodolphe (1986), *The Tain of the Mirror: Derrida and the Philosophy of Reflection* (Cambridge, MA: Harvard).

_____ (1999), *Of Minimal Things: Studies on the Notion of Relation* (Stanford: Stanford University).

Gebhardt, Nicholas (2001), *Going for Jazz: Musical Practices and American Ideology* (Chicago: University of Chicago).

Gefin, Laszlo (1982), *Ideogram: History of a Poetic Method* (Austin: University of Texas).

Gelernter, David (1992), *Mirror Worlds: Or the Day Software Puts the Universe in a Shoebox* (New York: Oxford University Press).

Gibson, William (1987), *Count Zero* (New York: Ace).

Giles, Cynthia (1992), *The Tarot: History, Mystery and Lore* (New York: Simon & Schuster).

Gilloch, Graeme (1996), *Myth and Metropolis: Walter Benjamin and the City* (Cambridge, England: Polity Press).

Godzich, Wlad (1986), "Foreword: The Tiger on the Paper Mat," in *Paul de Man: The Resistance to Theory* (Minneapolis: University of Minnesota Press).

Goldman, Robert, and Stephen Papson (1996), *Sign Wars: The Cluttered Landscape of Advertising* (New York: Guilford Press).

Greimas, Algirdas Julien, and Jacques Fontanille (1993), *The Semiotics of Passions: From States of Affairs to States of Feeling*, Trans. Paul Perron and Frank Collins (Minneapolis: University of Minnesota).

Grossmann, Reinhardt (1983), *The Categorial Structure of the World* (Bloomington: Indiana University).

Guattari, Felix (1984), *Molecular Revolution: Psychiatry and Politics*, Trans. Rosemary Sheed (New York: Penguin Books).

_____ (1995), *Chaosmosis: An Ethico-Aesthetic Paradigm*, Trans. Paul Bains and Julian Pefanis (Sydney: Power Publications).

Guenther, Irene (1995), "Magic Realism, New Objectivity, and the Arts during the Weimar Republic," in *Magical Realism: Theory, History, Community*, Eds. Loia Parkinson Zamora and Weendy B. Faris (Durham, NC: Duke University).

Gusfield, Joseph R. (1981), *The Culture of Public Problems: Drinking-Driving and the Symbolic Order* (Chicago: University of Chicago).

Hacking, Ian (1999), *The Social Construction of What?* (Cambridge, MA: Harvard University Press).

Haineault, Doris-Louise, and Jean-Yves Roy (1993), *Unconscious for Sale: Advertising, Psychoanalysis, and the Public*, Trans. Kimball Lockhart with Barbara Kerslake (Minneapolis: Minnesota).

Hamilton, Cynthia S. (1987), *Western and Hard-Boiled Detective Fiction in America* (Iowa City: University of Iowa).

Hass, Robert, Ed. (1994), *The Essential Haiku: Versions of Basho, Buson, and Issa* (Hopewell, NJ: Ecco).

Havelock, Eric A. (1967), *Preface to Plato* (New York: Grosset and Dunlap).

_____ (1986), *The Muse Learns to Write: Reflections on Orality and Literacy from Antiquity to the Present* (New Haven: Yale).

Hawkes, David (1996), *Ideology* (New York: Routledge).

Heidegger, Martin (1959), *An Introduction to Metaphysics*, Trans. Ralph Manheim (New Haven: Yale).

_____ (1988) "The Basic Problems of Phenomenology," trans. Albert Hofstadter, in *Pratt Journal of Architecture*, 2.

_____ (1995), *The Fundamental Concepts of Metaphysics: World, Finitude, Solitude*, Trans. WilliamMcNeill and Nicholas Walker (Bloomington: Indiana).

Hermassi, Karen (1977), *Poetry and Theater in Historical Perspective* (Berkeley: University of California).

Hersh, Seymour M. (2003), "Selective Intelligence," in *The New Yorker*, May 12: 44–51.

Hine, Robert V. and John Mack Faragher (2000), *The American West: A New Interpretive History* (New Haven: Yale University).

Hirshfield, Jane (1998), *Nine Gates: Entering the Mind of Poetry* (New York: HarperCollins).

Hutton, Patrick H. (1993), *History as an Art of Memory* (Hanover, NH: New England).

Hyde, Lewis (1998), *Trickster Makes This World: Mischief, Myth, and Art* (New York: North Point)

Janicaud, Dominique, and Jean-Francois Mattei (1995), *Heidegger from Metaphysics to Thought*, Trans. Michael Gendre (Albany: State University of New York).

Jameson, Fredric (1981), *The Political Unconscious: Narrative as a Socially Symbolic Act* (Ithaca: Cornell).

_____ (1988), "Cognitive Mapping," in *Marxism and the Interpretation of Culture*, Eds. C. Nelson and L. Grossberg (Urbana: U of Illinois).

Joyce, James (1964), *A Portrait of the Artist as a Young Man* (New York: Viking Press).

Jullien, Francois (1995), *The Propensity of Things: Toward a History of Efficacy in China*, Trans. Janet Lloyd (New York: Zone).

_____ (2000), *Detour and Access: Strategies of Meaning in China and Greece*, Trans. Sophie Hawkes (New York: Zone Books).

Karcher, Stephen (1997), *The Illustrated Encyclopedia of Divination* (Rockport, MA: Element Books).

Kekes, John (1995), *Moral Wisdom and Good Lives* (Ithaca: Cornell).

Kenner, Hugh (1971), *The Pound Era* (Berkeley: University of California).

Kerenyi, Carl (1967), *Eleusis: Archetypal Image of Mother and Daughter*, Trans. Ralph Manheim (Princeton: Princeton University).

Khatib, Abdelhafid (1996), "Attempt at a psychogeographical description of Les Halles," in Andreotti and Costa, Eds. pp. 72–76.

Kleinberg, Howard (1985), *Miami: The Way We Were* (Miami: Miami Daily News).

Knabb, Ken, Ed. (1981), *Situationist International Anthology* (Berkeley, CA: Bureau of Public Secrets).

Knight, Arthur (1995), "Jammin' the Blues, or the Sight of Jazz, 1944," in Gabbard.

Kouwenhoven, John A. (1998), "What's 'American' About America," in Robert G. O'Meally.

Kracauer, Sigfried (1960), *Theory of Film: The Redemption of Physical Reality* (New York: Oxford University Press).

Krauss, Rosalind E. (1994), *The Optical Unconscious* (Cambridge, MA: MIT).

_____ (1999), *Bachelors* (Cambridge, MA: MIT Press).

Kroker, Arthur (1992), *The Possessed Individual: Technology and the French Postmodern* (New York: St. Martin's).

Lacan, Jacques (1978), *The Four Fundamental Concepts of Psycho-Analysis*, Ed. Jacques-Alain Miller, Trans. Alan Sheridan (New York: Norton).

Lakoff, George and Mark Johnson (1999), *Philosophy in the Flesh: The Embodied Mind and Its Challenge to Western Thought* (New York: Basic Books).

Latour, Bruno (1993), *We Have Never Been Modern*, Trans. Catherine Porter (Cambridge, MA: Harvard University).

Lecercle, Jean-Jacques (1990), *The Violence of Language* (New York: Routledge).

Lee, Jung Young (1994), *Embracing Change: Postmodern Interpretations of the* I Ching *from a Christian Perspective* (Scranton: University of Scranton).

Lefebvre, Henri (1991), *The Production of Space*, Trans. Donald Nicholson-Smith (Cambridge, MA: Blackwell).

Lipsitz, George (1994), *Dangerous Crossroads: Popular Music, Postmodernism and the Poetics of Place* (New York: Verso).

Loftus, Elizabeth F. (1979), *Eyewitness Testimony* (Cambridge, MA: Harvard).

Lorca, Federico Garcia (1955), *The Selected Poems of Federica Garcia Lorca*, Eds. Francisco Garcia Lorca and Donald M. Allen (New York: New Directions)

Lyotard, Jean-Francois (1990), *Heidegger and "the Jews,"* Trans. Andreas Michel and Mark S. Roberts (Minneapolis: Minnesota).

_____ (1994), *Lessons on the Analytic of the Sublime*, Trans. Elizabeth Rottenberg (Stanford: Stanford University).

MacIntyre, Alasdair (1981), *After Virtue: A Study in Moral Theory* (Notre Dame: University of Notre Dame)

McQuire, Scott (1998), *Visions of Modernity: Representation, Memory, Time and Space in the Age of the Camera* (London: Sage).

Mann, Wolfgang-Rainer (2000), *The Discovery of Things: Aristotle's* Categories *and Their Context* (Princeton, NJ: Princeton).

Marx, Werner (1987), *Is There a Measure on Earth? Foundations for a Nonmetaphysical Ethics*, Trans. Thomas J. Nenon and Reginald Lilly (Chicago: Chicago).

Matibag, Eugenio (1997), "Ifa and Interpretation: An Afro-Caribbean Literary Practice," in Olmos.

Meltzer, David (1999), *Writing Jazz* (San Francisco: Mercury House).

Mirzoeff, Nicholas (1999), *An Introduction to Visual Culture* (New York: Routledge).

Mitchell, W. J. T. (1983), "Metamorphoses of the Vortex: Hogarth, Turner, and Blake," in *Articulate Images: The Sister Arts from Hogarth to Tennyson*, Ed. Richard Wendorf (Minneapolis, MN: University of Minnesota).

Moakley, Gertrude (1966), *The Tarot Cards Painted by Bonifacio Bembo for the Visconti-Sforza Family* (New York: New York Public Library).

Naremore, James (1995), "Uptown Folk: Blackness and Entertainment in *Cabin in the Sky*," in Gabbard.

Naremore, James (1998), *More Than Night: Film Noir in Its Contents* (Berkeley: University of California).

Nichols, Ashton (1987), *The Poetics of Epiphany: Nineteenth-Century Origins of the Modern Literary Moment* (Tuscaloosa: Alabama).

Noriega, Chon A. (1996), "Talking Heads, Body Politic: The Plural Self or Chicano Experimental Video," in *Resolutions: Contemporary Video Practices*, Eds. Michael Renov and Erika Suderburg (Minneapolis: Minnesota).

Nuti, Lucia, "Mapping Places: Chorography and Vision in the Renaissance," in Cosgrove, pp. 90–108.

Olmos, Margariet Fernandez and Liazbeth Paravisini-Gebert, Eds. (1997), *Sacred Possessions: Vodou, Santeria, Obeah, and the Caribbean* (New Brunswick: Rutgers).

Olson, David R. (1994), *The World on Paper: The Conceptual and Cognitive Implications of Writing and Reading* (New York: Cambridge University Press).

O'Meally, Robert G., Ed. (1998), *The Jazz Cadence of American Culture* (New York: Columbia University).

Paravisini-Gebert, Lizabeth (1997), "Women Possessed: Eroticism and Exoticism in the Representation of Woman as Zombie," in Olmos.

Pasolini, Pier Paolo (1983), *Lutheran Letters*, Trans Stuart Hood (Manchester: Carcanet New Press).

Perez, Evan (1998), Associated Press story, *Gainesville Sun*, May 13.

Peter, Rozsa (1961), *Playing with Infinity: Mathematical Explorations and Excursions*, Trans. Z. P. Dienes (New York: Dover).

Peters, F. E. (1967), *Greek Philosophical Terms: A Historical Lexicon* (New York: New York University).

Pile, Steve, and Nigel Thrift, Eds. (2000), *City A–Z* (New York: Routledge).

Pound, Ezra (1960), *A B C of Reading* (New York: New Directions).

_____ (1965), "Vorticism," in *The Modern Tradition: Backgrounds of Modern LIterature*, Eds. Richard Ellmann and Charles Feidelson, Jr. (New York: Oxford University Press).

Poundstone, William (1992), *Prisoner's Dilemma: John von Neumann, Game Theory, and the Puzzle of the Bomb* (Doubleday).

Proust, Marcel (1970), *Swann's Way*, Trans. C. K. Scott Moncrieff (New York: Vintage Books).

Rasula, Jed, "The Jazz Audience," in Cooke and Horn.

Reale, Giovanni (1985), *A History of Ancient Philosophy, III: The Systems of the Hellenistic Age*, Trans. John R. Catan (Albany: SUNY).

Reid, David, and Jayne L. Walker (1993), "Strange Pursuit: Cornell Woolrich and the Abandoned City of the Forties," in Copjec.

Ricoeur, Paul (1985), *Time and Narrative*, Volume 2, Trans. Kathleen McLaughlin and David Pellauer (Chicago: University of Chicago).

Rilke, Rainer Maria (1958), *The Notebooks of Malte Laurids Brigge*, Trans. M. D. Herter Norton (New York: Capricorn).

Roberts, Allen F. (1993), "Insight, or Not: Seeing is Believing," in *Secrecy: African Art That Conceals and REveals*, Ed. Mary H. Nooter, et al (New York: Museum of African Art).

Robins, Kevin (1996), *Into the Image: Culture and Politics in the Field of Vision* (New York: Routledge).

Rochefort, David A., and Roger W. Cobb, Eds. (1994), *The Politics of Problem Definition: Shaping the Policy Agenda* (Lawrence: Kansas University).

Rossi, Paolo (2000), *Logic and the Art of Memory: The Quest for a Universal Language*, Trans. Stephen Clucas (Chicago: University of Chicago).

Rugoff, Ralph, Ed. (1997), *Scene of the Crime* (Cambridge, MA: MIT).

Sadler, Simon (1998), *The Situationist City* (Cambridge, MA: MIT Press).

Santillana, Giorgio de, and Hertha von Dechend (1977), *Hamlet's Mill: An Essay on Myth and the Frame of Time* (Boston: David R. Godine).

Sartre, Jean-Paul (1965), *The Philosophy of Jean-Paul Sartre*, Ed. Robert Denoon Cumming (New York: Vintage).

Schwartz, Sanford (1985), *The Matrix of Modernism: Pound, Eliot, and Early Twentieth-Century Thought* (Princeton: Princeton University).

Scott, Clive (1999), *The Spoken Image: Photography and Language* (London: Reaktion Books).

Serres, Michel (1982), *The Parasite*, Trans. Lawrence R. Schehr (Baltimore: Johns Hopkins University).

Silver, Alain, and James Ursini, Eds. (1996), *Film Noir Reader* (New York: Limelight).

Simon, Josef (1995), *Philosophy of the Sign*, Trans. George Heffernan (Albany, NY: State University of New York Press).

Slomkowski, Paul (1997), *Aristotle's* Topics (New York: Brill).

Slotkin, Richard (1992), *Gunfighter Nation: The Myth of the Frontier in Twentieth-Century America* (New York: HarperCollins).

Smithson, Robert (1979), "A Tour of the Monuments of Passaic, New Jersey," in *The Writings of Robert Smithson*, Ed. Nancy Holt (New York: New York University).

Sokel, Walter H. (1980), "The Devolution of the Self in *The Notesbooks of Malte Laurids Brigge*," in Baron, Ed.

Spitzer, Leo (1963), *Classical and Christian Ideas of World Harmony: Prolegomena to an Interpretation of the Word "Stimmung"* (Baltimore: Johns Hopkins).

Strauss, Walter A. (1980), "Rilke and Ponge: 'L'Objet c'est la Poetique," in *Rilke: The Alchemy of Alienation*, Ed. Frank Baron, Ernst S. Dick, and Warren R. Maurer (Lawrence: Kansas).

Stuckenberg, J.H.W. (1986), *The Life of Immanuel Kant* (Lanham, MD: University Press of America).

Sudnow, David (1981), *Ways of the Hand: The Orbanization of Improvised Conduct* (New York: Harper Colophon Books).

Surette, Leon (1979), *A Light from Eleusis: A Study of Ezra Pound's* CANTOS (Oxford: Clarendon Press).

Sussman, Elisabeth, Ed. (1989), *On the Passage of a Few People through a Rather Brief Moment in Time: The Situationist International, 1957–72* (Cambridge, MA: MIT).

Telesco, Patricia (1998), *Future Telling: A Complete Guide to Divination* (Freedom, CA: Crossing).

Tierney-Tello, Mary Beth (1999), "Testimony, Ethics, and the Aesthetic in Diamela Eltit," *PMLA* 114: 1 (January).

Tindale, Ian, Paul Macdonald, James Rowley (2001), *Flash XML Studiolab* (Birmingham, UK: Friends of ED).

Tisdall, Caroline (1971), "Historical Foreword," in *Metaphysical Art*, Massimo Carra, et al., Trans. Caroline Tisdall, (New York: Praeger), pp. 7–16.

Titon, Jeff Todd (1994), *Early Downhome Blues: A Musical and Cultural Analysis*, Second Edition (Chapel Hill: University of North Carolina)

Trilling, Lionel (1971), *Sincerity and Authenticity* (Cambridge, MA: Harvard University).

Ueda, Makoto (1995), "Basho on the Art of the Haiku: Impersonality in Poetry," in *Japanese Aesthetics and Culture: A Reader*, Ed. Nancy G. Hume (Albany: SUNY).

Ulmer, Gregory L., (1989), *Teletheory: Grammatology in the Age of Video* (New York: Routledge).

_____, (1992), "Metaphoric Rocks: A Psychogeography of Tourism and Monumentality," in *The Florida Landscape: Revisited*, Christoph Gerozissis, Ed. (Lakeland, FL: Polk Museum).

_____, (1994), *Heuretics: The Logic of Invention* (Baltimore: Johns Hopkins).

Vance, Eugene (1987), *From Topic to Tale: Logic and Narrativity in the Middle Ages* (Minneapolis: Minnesota).

Ventura, Michael, "Hear That Long Snake Moan," in *Shadow Dancing in the USA* (Tarcher, 1985).

Vernet, Marc (1993), "*Film Noir* on the Edge of Doom," in Copjec.

Vidler, Anthony (2000), *Warped Space: Art, Architecture, and Anxiety in Modern Culture* (Cambridge, MA: MIT Press).

Virilio, Paul (1997), *Open Sky*, Trans. Julie Rose (New York: Verso).

_____ (1999), *Politics of the Very Worst; An Interview by Philippe Petit*, Trans. Michael Cavaliere (New York: Semiotexte).

_____ (2003), *Unknown Quantity* (New York: Thames & Hudson).

Walter, Eugene Victor (1988), *Placeways: A Theory of the Human Environment* (Chapel Hill: North Carolina).

Weir, David (1996), *James Joyce and the Art of Mediation* (Ann Arbor: Michigan).

Wilhelm, Richard (1967), "Introduction," in *The I Ching, or Book of Changes*, Trans. Richard Wilhelm, 3rd Edition (Princeton: Princeton University).

Williams, William Carlos (1995), *Paterson*, Revised Edition (New York: New Directions).

Williams, John (1993), *Into the Badlands: Travels through Urban America* (London: Flamingo).

Wilson, Robin (1999), "A Challenge to the Veracity of a Multicultural Icon," *The Chronicle of Higher Education* (January 15).

Wollen, Peter (1997), "Vectors of Melancholy," in Rugoff.

Yates, Frances A., (1966), *The Art of Memory* (Chicago: University of Chicago).

Zizek, Slavoj (1989), *The Sublime Ojbect of Ideology* (New York: Verso).
_____ (1991), *Looking Awry: An Introduction to Jacques Lacan through Popular Culture* (Cambridge, MA: MIT Press).
_____ (1992), "Hitchcockian 'Sinthoms,'" in *Everything You Always Wanted to Know about Lacan (But Were Afraid to Ask Hitchcock)*, Ed. Slavoj Zizek (New York: Verso).
_____ (1994), "A Hair of the Dog that Bit You," in Mark Bracher, et al.
_____ (1997), *The Plague of Fantasies* (New York: Verso).
_____ (2001), *The Fright of Real Tears: Krzysztof Kieslowski between Theory and Post-Theory* (London: British Film Institute).
_____ (2001a), *On Belief* (New York: Routledge).
_____ (2002), *Welcome to the Desert of the Real! Five Essays on September 11 and Related Dates* (New York: Verso).

www.ingramcontent.com/pod-product-compliance
Lightning Source LLC
Chambersburg PA
CBHW071411230926
51846CB00001B/60
9789004534629